T0179175

# Sterilization of Drugs and Devices

## Technologies for the 2000s

*Edited by*

# Fred M. Nordhauser
### and
# Wayne P. Olson

CRC Press
Taylor & Francis Group
Boca Raton  London  New York

CRC Press is an imprint of the
Taylor & Francis Group, an **informa** business

CRC Press
Taylor & Francis Group
6000 Broken Sound Parkway NW, Suite 300
Boca Raton, FL 33487-2742

First issued in paperback 2019

ISBN-13: 978-1-57491-060-5 (hbk)
ISBN-13: 978-0-367-40047-7 (pbk)

Invitation to Authors

Interpharm Press publishes books focused upon applied technology and regulatory affairs impacting Healthcare Manufacturers worldwide. If you are considering writing or contributing to a book applicable to the pharmaceutical, biotechnology, medical device, diagnostic, cosmetic, or veterinary medicine manufacturing industries, please contact our Director of Publications.

Social Responsibility Programs

Reforestation

Interpharm Press is concerned about the impact of the worldwide loss of trees upon both the environment and the availability of new drug sources. Therefore, Interpharm supports global reforestation and commits to replant trees sufficient to replace those used to meet the paper needs to print its books.

Pharmakos-2000

Through its Pharmakos-2000 program, Interpharm Press fosters the teaching of pharmaceutical technology. Under this program, complimentary copies of selected Interpharm titles are regularly sent to every College and School of Pharmacy worldwide. It is hoped that these books will be useful references to faculty and students in advancing the practice of pharmaceutical technology.

Library of Congress Cataloging-in-Publication Data

Visit the Taylor & Francis Web site at
http://www.taylorandfrancis.com

and the CRC Press Web site at
http://www.crcpress.com

# Contents

# Preface

The dedication to the development, production, and distribution of pharmaceutical products is directed to the safety of the patient or user of the pharmaceutical preparation. Within the confines of the safety of the product is the need to provide the product in a clean, safe formulation that is ultimately free of unwanted or adventitious microbiological contamination. These unwanted contaminating agents could result in undefined and unnecessary complications that are not documented nor defined with respect to the performance of the pharmaceutical product, nor associated with the specific performance of the pharmaceutical formulation.

The efforts to generate the safety factors with respect to biological contamination has led to the growth of the technology and improvement of the tools needed to produce "sterile" pharmaceuticals over the last two decades. Associated with the growth of technology has been the validation of the specific process. In support of this approach, much effort has been made to understand and clarify the effect and nature of the method of sterilization.

The inclusion of final sterile processing as an integral part of the overall production methodology benefits from the efforts of harmonization. This global approach to the production of pharmaceuticals also opens the way for the incorporation of techniques and procedures that have been utilized in many and varied locations.

Previous volumes encompassing the topic of aseptic manufacturing have been published in the 1980s. These publications foresaw the development of the importance of closure integrity, sterile powders, bulk pharmaceutical compounds, processing strategy, the inclusion of robotics to reduce human contamination, and the improvements in cleaning and sterilizing in place. All of these procedures and venues have been improved, and the result is a defined control of the processing environment.

To quote from the preface of an earlier work, "the technology associated with aseptic processing is an expanding and developing area" (Olson and Groves 1987). This expansion is demonstrated in this work where we document the increased variation but effective methodology in use today and with a look toward the next decade.

Fred M. Nordhauser
Wayne P. Olson
October 1997

## ACKNOWLEDGMENT

We the editors would like to thank the contributors for their dedication in providing you the reader with a view of current and future technology. We believe that you will be able to add to your knowledge and made aware of the varying approaches that can be applied to the handling of sterile pharmaceuticals.

| | |
|---|---|
| Ross A. Caputo | AbTox Inc., Mundelein, IL |
| V. Robert Carlson | VRC Co., Inc., Cedar Rapids, IA |
| Vance Caudill | O'Neal Engineering Inc., Raleigh, NC |

| | |
|---|---|
| Joseph Dunn | PurePulse Technologies, Inc., San Diego, CA |
| Barry P. Fairand | SteriGenics International, Inc., Hayward, CA |
| Victoria Galliani | AbTox Inc., Mundelein, IL |
| Gary Graham | AbTox Inc., Mundelein, IL |
| Michael J. Groves | Institute of Tuberculosis Research, University of Illinois at Chicago, Chicago, IL |
| Paul J. Haas | Despatch Industries, Minneapolis, MN |
| Michael Jordan | Lockwood Greene, Spartanburg, SC |
| Mark Kaufmann | O'Neal Engineering Inc., Raleigh, NC |
| Phillip A. Martens | AbTox Inc., Mundelein, IL |
| Hans L. Melgaard | Despatch Industries, Minneapolis, MN |
| Wayne P. Olson | Oldveco, Beecher, IL |
| Brian D. Reid | GamRay CONSULTING, INC., Merrickville, Ontario |
| James Wilson | Pharmaceutical Systems Inc., Mundelein, IL |
| Jack H. Young | The Pennsylvania State University, Erie, PA |

# Foreword

The goal of sterile pharmaceutical manufacturing is to make a drug product that lacks viable microorganisms capable of reproduction in the drug product itself or after injection into a patient. The manufacture of parenteral solutions is typically accomplished by either terminal sterilization or aseptic processing. Terminal sterilization of the drug product is performed in its final container, and is a U.S. Food and Drug Administration (FDA)-accepted manufacturing procedure. Aseptic manufacturing processes consist of a series of highly controlled steps.

Both processes have limitations. Terminal sterilization cannot be used for products that are thermally liable, thus limiting the scope of drug products that can be sterilized. Aseptic processing is the commonly used alternative for products that are thermally liable, but it has historically been associated with far more sterility failures and product recalls. Mandatory terminal sterilization of all aqueous-based parenteral drugs was proposed by the FDA (*Federal Register*, Friday, October 11, 1991). Two years later, a public meeting known as the Aseptic Fill Conference was held in Bethesda, Maryland, to discuss the proposed rule. Although no final rule has been published to date, the conference was successful in providing the FDA with an abundance of information about various types of sterile parenterals that cannot be terminally sterilized (e.g., liposome preparations and ophthalmic solutions packaged in plastic bottles). Subsequent discussions and actions taken by pharmaceutical manufacturers are probably more meaningful in terms of manufacturing a variety of sterile products than in the publishing of a rigid, final rule.

Thus, it seems that the prospect of required terminal sterilization has prompted researchers to define alternative, aseptic processing technologies. Subsequent to the Aseptic Fill Conference, much effort was put into trying to increase the sterility assurance levels (SALs) of aseptic manufacturing processes from the accepted standard of $10^{-3}$ to those approaching terminal sterilization ($10^{-6}$). That effort was unrewarded. But the fact remains that the majority of parenteral drug products currently manufactured, and those that will be manufactured in the future, cannot be terminally sterilized by moist heat, irradiation, or any other means currently available. Thus, aseptic processing of some sort will be required to manufacture newer products, such as biopharmaceuticals. A fairly unique combination of sterilization procedures may be required to yield a suitable parenteral product. Each manufacturing step will require a different sterilization procedure (e.g., sterilization-in-place of the filling machine, gamma irradiation of the active pharmaceutical ingredient, etc.). For example, the manufacture of sterile suspensions may require sterilization of the active ingredient by aseptic recrystallization in a Class 100 hood because the active cannot be filtered or gamma irradiated. All subsequent steps in the manufacturing procedure must then be carefully controlled to avoid microbial contamination. A carefully planned, validated, and controlled aseptic manufacturing process should be capable of yielding a sterile pharmaceutical product.

The *Sterilization of Drugs and Devices* is a "must read" for anyone involved in sterile pharmaceutical process development and/or manufacture. The following topics of continuing concern to the pharmaceutical industry and the FDA are discussed in-depth: steam sterilization-in-place of deadlegs (Chapter 3), filter sterilization of drug solutions (Chapter 7), chemical decontamination/sterilization of surfaces (Chapter 8), viral inactivation and removal (Chapter 12), and parametric release of drug products (Chapter 14). Isolated aseptic filling lines (Chapter 9) have recently received a great deal of interest. Isolators prevent the biggest source of microbial contamination (human contact with the filling line). Pulsed white light (Chapter 11) is capable of killing large numbers of a wide spectrum of microorganisms and may be used for product–contact surface sterilization or in-line terminal sterilization of drug solutions. This method holds great promise but is not fully proven for its proposed use. Recent developments in sterility testing, including the isolation of injured organisms and new compendial test requirements, are authoritatively discussed in Chapter 13. The broad range of in-depth information provided in *Sterilization of Drugs and Devices* contains numerous technologies and ideas that will be helpful in designing sterile manufacturing processes.

Kenneth Muhvich
November 1997

# 1

## Introduction

*Fred M. Nordhauser*

Nordco Consulting
Brighton, MI

Pharmaceutical parenteral preparations are, by standard requirements, sterile and very often devoid of particulate contaminants. Included (by nature and requirement of use) are ophthalmic preparations, surgical lavages, total parenteral nutrition [TPN] formulations, APD formulations, dialysis formulations, cell culture formulations, and various fluids used for patients with autoimmune or carcinogenic diseases.

The production of these compounds and formulations to produce a sterile product requires environmental and personnel control, as well as direct manipulation of the solution. To accomplish these techniques, established regimens for the use of containment of the manufacturing area (including the use of high efficiency particulate air [HEPA] filtration); incoming raw material monitoring; gowning; container cleaning and sterilization; the use of ultraviolet (UV), gamma, or filtration for microbial control; and explicit standard operating procedures (SOPs) are usually advised.

1

These processes generally involve working with cold sterilization, and work has proceeded in this area to examine the impact of proper gowning to prevent accidental microbial contamination of the working environment. For example, work performed in Scotland (by William White) has demonstrated that sterile gowns are capable of allowing the shedding of particles up to 17 μm in size that are usually contaminated with *Staphylococcus aureus* and *Staphylococcus epidermis* organisms. There was also the historical question of defective or damaged microorganisms (those exposed to partially effective UV or, as suggested by my co-editor [Wayne Olson], disinfectant-damaged organisms).

To address this issue, various regulatory actions have been initially established for further penetration into the production of sterile pharmaceuticals. These address barrier isolation methodologies, terminal sterilization processes, container and closure procedures, clean-in-place (CIP) and sterilize-in-place (SIP) controls, filtrative definition and documentation, and so forth.

## HISTORICAL DEVELOPMENTS

Processes and testing in the pharmaceutical industry have changed dramatically during the period when the author was employed at American Cyanamid, Pall Corporation, and Gelman Sciences. During these "early" years, filtration at 0.45 and 0.2 μm was the method of cold sterilization, and it primarily consisted of pumping solutions through 293 mm diameter membrane filter discs. When volumes were larger than 10–50 L that could be processed through individual steam-sterilized (or autoclaved) disc filters, they were assembled in a manifold nature. This was accomplished as aseptically as possible, and then the total volume processed in a serial or parallel manner. The testing of these "process filters" is not yet completely understood. Regardless of the filter used, there were sterility test failures of some product lots that were due either to the filter itself or the method of handling and manipulation. As a result, the failures were attributable to both vendors and filter users or pharmaceutical manufacturers.

Improvements were in both testing and filter products. Filter sterility testing became more consistent with procedures developed by Frances Bowman of the U.S. Food and Drug Administration (FDA) and Charles Schaufus of Millipore Corp. Improvements in nondestructive filter testing were developed by Retty and Leahy of Millipore Corp., together with Dr. Pall of the Pall Corporation. The last of this critical triad of developments in filters was the introduction of the first 10-inch, pleated, *membrane* filter cartridge of 3.5 ft² in a single device by Gelman Sciences. These accomplishments allowed for the processing of larger production lots without the complicated manipulations needed for disc filter assemblies.

## CURRENT DEVELOPMENTS

The improvements in sterilizing filtration have been significant and have played a major role in the cold sterilization of pharmaceutical preparations. Sterilization in the final container is limited to heat stable compounds and to radiation (currently and historically gamma irradiation). However, about 80 percent of solutions for injections now are filter sterilized.

Additional techniques and procedures in the handling of preparations, stock cultures, and master cell lines are being used with *the emphasis on controlling the bioburden* (whether viral or bacterial), with the result of a final sterilization that is more effective, economical, and reliable. The added emphasis on process and raw material control is crucial because retesting as a measure of assuring sterility is becoming obsolete. Retesting for viral components in biologics will also become passé.

Questions and unresolved topics regarding sterility must continue to be addressed. Some of these apply to level limits of product contamination, the efficacy of barrier techniques on control of bioburden, the ability of current testing to address slow-growing or damaged organisms, and so forth. Are there further issues to examine with "classical" sterilizing procedures? These issues are addressed in the following chapters of this book. If the topic is treated with a cavalier attitude, then there will be a limitation to improvement, change, and approach.

Answers with ultimate resolutions are not the purpose of this text, but it is intended to address new and current issues and supplement the discussion found in various texts, for example, Meltzer (1987), Morrissey and Phillips (1993), Olson and Groves (1987), and Wagner and Akers (1995). Therefore, the present text address issues that over the next decade hopefully will have been solved to everyone's satisfaction. Additional issues will probably present themselves for review and consideration.

## REFERENCES

Meltzer, T. 1987. *Filtration in the pharmaceutical industry.* New York: Marcel Dekker, Inc.

Morrissey, R. F., and G. B. Phillips, eds. 1993. *Sterilization technology: A practical guide for manufacturers and users of healthcare products.* New York: Van Nostrand Reinhold.

Olson, W .P., and M. J. Groves, eds. 1987. *Aseptic pharmaceutical manufacturing: Technology for the 1990s.* Buffalo Grove, IL: Interpharm Press, Inc.

Wagner, C., and J. Akers, eds. 1995. *Isolator technology: Applications in the pharmaceutical and biotechnology industries.* Buffalo Grove, IL: Interpharm Press, Inc.

# 2

# The Sterilization of Pharmaceutical Solutions, with Emphasis on the Kinetics of the Bacterial Death Process

*M.J. Groves*

Institute for Tuberculosis Research
University of Illinois at Chicago
Chicago, IL

## SOME REMINDERS: THE STRUCTURE AND FUNCTIONS OF MICROORGANISMS

Before discussing the mechanisms of death, it will be beneficial to summarize briefly the form and functions of living microorganisms. (For more detail, the interested reader is referred to modern textbooks, such as that by Nester et al. 1995).

All microorganisms consist of single cells covered with a cellular membrane that separates the cell from the surrounding environment. However, cells can be divided into two broad groups: those in which the nuclear material is separated from the rest of the cellular interior by another membrane, eukaryotic cells, or those in which the cellular contents are not separated, prokaryotic cells. The latter tend to be smaller in size (0.3–2.0 μm in diameter) and this classification encompasses all bacteria. Eukaryotes, on the other hand, have true nuclei, are larger (2–20 μm in diameter) and include algae, fungi, and protozoa. Both categories face much the same problem when attempting to stay alive because they all must reproduce copies of themselves. In order to duplicate itself, the cell must contain genetic material that will ensure that all of the required parts of the cell are synthesized. The basic function of living must be to acquire simple foodstuffs from the surrounding environment and generate enough energy to carry out all of the necessary synthetic and degradative processes associated with maintaining an existence (i.e., living). Genetic information is stored in deoxyribonucleic acid (DNA) sequences and the functional enzymes, which are usually proteins, are synthesized by ribonucleic acid (RNA). Enzymes are essentially catalysts, accelerating the processing of one material to another, and usually require smaller molecules, such as coenzymes, synthesized from vitamins, for efficient activity. Energy for the process is essentially stored by adenosine triphosphate (ATP) so that, for example, ATP measurement of a bacterial suspension is generally a good indicator of how many bacteria are actually "alive" or functioning in the system.

For the most part, the following discussion will be confined to bacteria. Eukaryotic cells are more readily destroyed by violent changes in their external environment, such as the application of heat. Viruses, strictly speaking, are not prokaryotes because they lack the cell membrane associated with unicellular organisms. Most viruses contain a piece of genetic material (DNA or RNA, but not both) that is surrounded by a relatively simple protein coat. A viroid, also an infectious agent, is even simpler, with just a small RNA sequence without a coat. Because they lack protection, these agents, in isolation, are readily "inactivated" or destroyed by heat or some chemical agents. On the other hand, prions, about which very little is known, are considered to be without DNA or RNA, and while they are associated with disease and appear to be capable of resisting extreme heating conditions, their mode of survival and replication is poorly understood. Diseases such as scrapie in sheep and bovine spongiform encephalopathy (BSE) in cattle have been associated with prions and they may also be the causative agent of human Creutzfeldt–Jacob disease (CJD), a degenerative disease of the brain that has an incubation period of 5–40 years. The connection between these three diseases is unknown, although infected sheep tissues fed to cattle appear to have been the prime cause of BSE. Scrapie has been known for at least 250 years, and there does not appear to be a connection with the exceptionally rare CJD in humans. However, recently, there have been some indications that prions of bovine origin may be able to infect humans, with potentially alarming implications. This digression serves to draw attention to the fact that human or veterinary disease is often caused by a microorganism even though not all microorganisms cause disease. Disease-causing organisms or pathogens are of principal concern when considering pharmaceutical products, especially those preparations administered through the otherwise protective skin, parenteral products. The other point, of course, is the fact that drugs are given to patients who, by definition, are not well and, in many cases, are susceptible to infection. The emphasis of this chapter will be to demonstrate how products can be considered to be free from living microorganisms. This must be considered to be all microorganisms, pathogens or not. If a nonpathogen is found in a product, the chances are that pathogens may also be present. Even nonpathogens, in the wrong place at the wrong time under optimal growth conditions, can cause disease. Hence, the technology of sterile products must emphasize the total removal or destruction of all microorganisms, not just a selected few.

It should be noticed that this last statement is a summary of conventional wisdom. Recently, Gilbert and Allison (1996) have questioned the whole concept since, as they point out, the issue is one of public health in which an association has been made between the presence of microorganisms in parenteral products and induced disease. An aseptically produced product is assumed to be "aseptic," that is, it does not produce disease. This aseptic requirement has been extended for the most part by regulatory agencies into a requirement for sterility in which, by definition, all microorganisms have been eradicated or eliminated. The result, according to these authors, is a situation in which a limit is being imposed for which there are no meaningful or appropriate test methodologies. Attempts over the past 50 or more years to develop sterility tests have been doomed to failure. All of the the current tests being carried out by manufacturers under pressure from regulatory agencies in order to "prove" sterility provide a somewhat imprecise evaluation of asepsis rather than sterility for the limited number of products tested in a batch and, thereby, destroyed. The manufacturer is, therefore, assessing the asepsis of the entire batch of the product by extrapolation to the remaining, untested, containers. Moreover, the test results relate only to bacteria or fungal contamination for a particular growth medium that is incubated at a particular temperature for a minimum of 7 days (to be extended to 14 days). Measurements obtained under these conditions are effectively meaningless because of the statistical limitations imposed on the experimental design. The comment was made that no other public health limits require lower limits than those that can be measured. An additional point was made since there is a need for parenteral products to meet the most stringent definitions of sterility rather than just asepsis, yet the need for this requirement has not been demonstrated. The authors pointed out that, with very few exceptions, the vast majority of invading organisms are killed on introduction into the mammalian body and, from a clinical perspective, unless the numbers of invading bacteria were extremely high—> $10^{10}$–$10^{12}$ for instance—the response produced in the recipient animal is likely to be very low. This minimum infective number (MIN) can be measured as the number of organisms that, by a particular route, have a 50 percent chance of causing an infection. For some organisms, such as *Salmonella typhi,* the MIN is 1,000–10,000. Of more general concern is *Pseudomonas aeruginosa,* which has a MIN of 2,000–3,000 organisms in healthy eye tissue but considerably less if the tissue is damaged. One might add that anthrax may have a MIN of 1–10 organisms to cause a fatal disease but, for the most part, the general argument is that most organisms do not cause disease. Gilbert and Allison cite *Bacillus stearothermophilus* as an organism with an MIN so low that it cannot be measured at body temperature and, therefore, is effectively nonpathogenic because it simply will not grow under these conditions. However, this organism is used as a worst case or "gold standard" for the application of a heat sterilization process because it is resistant to heat, not because it is a dangerous pathogen.

These arguments will be revisited by way of a coda at the end of this chapter. There are some obvious advantages in taking these suggestions seriously, although there may be some disadvantages too.

Returning to bacterial cells (prokaryotes), there is generally a clear relationship between form and function that is generally essential for "life." Here we must attempt to define "life" since it is the opposite of "death," another function that is essential to define. A cell must have an often rigid cell wall that encloses the contents, protecting and separating them from the surrounding environment.

The contents, on the other hand, must be able to generate, utilize, and store energy, generally with the help of ATP. The energy is used to synthesize cellular components and, eventually, allow replication of the complete genetic material. Although not as well separated and defined as they are in eukaryotic cells, these functions are clear enough; without these basics, the cell cannot survive. Beyond these basic elements, the cell may move around in its environment, often swimming with flagella or whiplike appendages to the cell wall, which also requires energy. Some cells may transfer genetic material to other cells, which may be seen in the invasive disease process. Many cells may need to store reserve energy sources, such as carbohydrates, and some cells, threatened with a hostile environment, may form an endospore that is a "living" but essentially nonfunctional form of a cell in the normal sense. Other cells will secrete extracellular materials, such as mucus or slime, as a capsular layer around the outside of the wall. Indeed, some *Mycobacteria* are capable of secreting a polysaccharide coat or integument around the cells to offer protection against excessive oxygen levels during growth. This integument may be part of the reason why these microorganisms have excèssively long growth times, with doubling times of 6,000 or more minutes compared to only 20 minutes for *Escherichia coli.*

One factor that is often discussed with reference to bacteria is their categorization according to whether or not the cell wall will stain with iodine, the so-called gram-positive or gram-negative organisms. Gram-positive organisms have external cell walls consisting mainly of multiple layers of peptidoglycans, which are polymeric backbone structures of sugars and amino sugars. Gram-negative cells, on the other hand, have only a single peptidoglycan layer, with an outer membrane consisting of phospholipids, proteins, and lipoproteins with lipopolysaccharides (LPSs). It is the presence of LPSs that triggers a febrile reaction on the administration of gram-negative organisms (dead or alive) to many mammals, especially humans. Purified LPS or endotoxin is now used as a standard pyrogen, although not all pyrogens are endotoxins. The cell wall must be rigid because the cell contents have a high osmotic pressure. The enzyme lysozyme and some antibiotics, such as penicillin, can cause destruction of the bacterial cell wall, which will either cause the cell to burst or, if it is surrounded by an iso-osmotic medium, to develop into a spherical or balloon-shaped spheroplast. Unfortunately, some organisms, such as members of the genus *Mycoplasma,* actually lack rigid cell walls and are both small (~ 0.20 μm) and flexible enough to pass through porous membranes used to "sterilize" solutions.

The living cell will need nutrients and water to pass through the cell wall into the cytoplasm and for the waste products to pass out. Very often, this is achieved by gating—or transport—proteins that structurally pass through the cytoplasmic wall and the outer wall, opening or gating as required to allow nutrients to pass through. Some materials can also cause the development of pores in the cell wall, allowing transport of, for example, nucleic acids from the cell. Water and other small molecules tend to be able to pass through cell walls without much difficulty by passive diffusion. Since this is a two-way process, many vegetative bacteria are readily dried out if removed from their normal aqueous environment.

The living organism is, therefore, a very busy place, likened by some authors to a functioning city inside a protective wall. The analogy is appropriate for the metabolizing cell, busy producing products to be used inside the walls or, in some cases, exported outside. The analogy loses some of its validity when considering growth, since cities grow bigger by expanding whereas bacteria grow and divide. The genetic information that must be carried from one cell to another is contained in the DNA, which is tightly coiled in the fibrils known as chromosomes. Most bacteria have plasmids that also contain DNA and are smaller than chromosomes. Plasmids replicate independently and, unlike chromosomes, many copies of a plasmid may be present in a cell. Not all of the genetic information required for survival is bound up in plasmids, but factors connected with bacterial resistance to, for example, antibiotics, may be associated with the DNA in plasmids. Plasmids are found to varying degrees in a large number of bacteria and vary in size as well as providing or controlling different enzymic functions. Moreover, they are readily transferred between different bacteria, accounting in part for transfer of bacterial resistance to antibiotics. In nature, outside the hospital environment, plasmids probably provide some flexibility, allowing a more rapid response to environmental factors without, at the same time, putting the essential DNA in the chromosomes in jeopardy.

Because DNA molecules are so essential to "life," it is not surprising that bacteria have developed mechanisms to protect and repair them. For example, modification enzymes are present that methylate certain bases in fragments of DNA, preventing restriction enzymes, which otherwise degrade foreign or unfamiliar DNA, from recognizing the DNA as a substrate. DNA damage is often caused by light (ultraviolet radiation) or chemical agents that produce mutations. Some bacteria have excision enzymes that can remove damaged DNA fragments. In addition, light causes activation of an enzyme that repairs the thymine-thymine bonding produced by light, so this represents a feedback mechanism. However, damaged DNA that is still able to replicate can cause a mutation of the organism to develop. It can go either way: A mutant cell may die out very rapidly or may be able to survive and replicate in the new environment that caused the mutation in the first place. The mutation rate may be low, perhaps one cell in a million, but if that cell is able to replicate, it would not take very long for other cells in a culture susceptible to a new environmental stress to be replaced by a new generation of cells that can resist that stress. Not all mutations are bad in the real world; however, from the narrow perspective of this present chapter, if insufficient antibacterial chemical is added to a culture to completely destroy all bacteria present, some mutants will survive and proliferate, and the purpose of adding the antibacterial in the first place is frustrated.

The basic picture of a dynamic and busy cell, developing metabolism and growing, ultimately dividing to give birth to new cells to continue the process is satisfying as a mental image of "life." A more relevant and serious question is determining what "death" is—a much more difficult question to answer. In principle, a single cell never grows old; it simply divides. In practice, however, a collection of cells in a growing culture behaves in a very different fashion.

## THE BEHAVIOR OF BACTERIAL POPULATIONS

If a single cell of a genus that divided into two equally sized daughter cells was placed in a nutrient medium, it would seem likely that, after a period of incubation for growth and division, all of the cells in the culture would be at the same stage of development. If only a bud formed from the original cell, then it is possible that the parent will grow old and die. Whatever the reason, cultures of single species of organisms behave as if they developed and then grew old. A growing culture will generally show four clearly defined stages of growth, Figure 2.1. The first phase shows no net growth in viable cell numbers, after which, there is an exponential growth phase, followed by yet another stationary phase where the overall numbers of cells do not change very much. Finally, there is a decline in numbers, characterized by an

**Figure 2.1.** Bacterial growth rate curve.

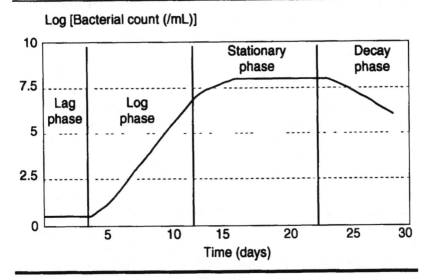

Log [Bacterial count (/mL)]

exponential decrease in viability. Here cells are considered to die off because they can no longer divide and grow. Some of these characteristics may be due to the bacteria themselves; they simply become old and "die" of exhaustion. Nevertheless, it is perhaps ironic that, experimentally, the system in which the organisms are growing can be considered to be a closed system in which nutrients are provided initially but are not renewed. One could therefore argue that the system itself becomes atrophied—insufficient nutrients are available and toxic waste products increase to the point where the organisms are unable to sustain themselves. The stationary phase, reached when the cell population is around $5 \times 10^9$/mL for many species, is the point at which either the organisms stop growing and are just existing, or the growth and death rates become equalized.

Another point that needs to be taken into account is the fact that the number of viable (i.e., living) cells are measured by taking a sample from the culture and adding it to a fresh culture medium, usually in or on an agar plate. Single cells are then required to divide enough to become visible as single colonies after several days incubation. A cell in stasis may still be alive in the sense that it is metabolizing nutrients, but it may have damaged DNA and be unable to divide or grow enough for colonies to become visible to the naked eye during the relatively short period required for the incubation of an undamaged cell. Given enough time, the damaged DNA might be repaired and the cell recover enough to grow again into a "normal" (i.e., dividing) cell. The experimental behavior of a growing culture, containing a myriad of cells, cannot be said to mirror the behavior of a single microorganism; it represents a statistical aggregate of the behavior of many bacteria over a period of time. The "lag phase" is said to represent a culture shock experienced by the organisms as they are transferred from one culture medium to another. The length of the lag phase will depend on nutritional and environmental factors, supposedly because the organisms require enough time to produce enzymes to cope with a new nutritional or environmental requirement. Taking a sample in the exponential growth phase and placing it into an identical but fresh medium at the same temperature usually results in a very short lag phase, suggesting that the cells were vigorously growing before and after sampling.

Bacteria growing in a closed system will eventually die out completely—or do they? What is the situation if a small number of microorganisms or endospores are accidentally introduced into a parenteral product? Do these cells pass through a prolonged lag phase, followed by a short exponential growth phase, and does the culture become sterile over a period of time as the contaminating cells die off? With the exception of nutritional amino acid and sugar solutions, most parenteral solutions would represent a hostile environment for most bacteria. For example, Moldenhauer et al. (1995) evaluated the heat resistance of endospores of *Bacillus coagulans* in 30 different parenteral solutions, using a survivor curve method to measure the time required to reduce the endospore population by 90 percent at a specified temperature. The endospores had no resistance in parenteral solutions already known to be bactericidal or inhibitory to growth, but noninhibitory solutions allowed the spores to survive a little longer. However, the authors considered that this organism was a suitable biological indicator for moist heat sterilization applications.

There are always exceptions; many species of the genus *Pseudomonas* are capable of growing in dilute saltwater solutions. In a hostile environment represented by the average aqueous solution of a drug, it is possible to envisage that a few microorganisms could grow or just survive, only to die off over a period of time as the product is stored prior to sale. It is possible, but extremely difficult to prove, that this has been the situation with aseptically prepared parenteral solutions that have not been exposed to an overkill sterilization process. One cannot rely on chance; it is well recognized today that processes must be designed to exclude all possibilities of chance contamination with any viable microorganisms.

## THE FORMATION OF ENDOSPORES

When considering spore formation by bacteria, one is reminded of the use of lifeboats in ships, or pods and capsules in spaceships, where essential personnel (or information, genetic or otherwise) are placed in small units that are designed to assist survival in unfriendly environments.

Endospores are produced inside certain gram-positive bacilli and are later found in culture medium, following lysis and destruction of the parent cells. They are a unique cell type that is only produced by vegetative cells exposed to hostile environments, such as a medium deficient in essential components. Generally, endospore formers are found only in the genera *Bacillus* and *Clostridium,* both serious pathogens. The formation of spores inside the cell goes through several characterizable processes, starting with the separation of the DNA into individual chromosomes. A cortex of lipoidal material, together with an unusual component, not usually found in vegetative cells, dipicolinic acid (DPA), is deposited around the DNA element. The DPA is in the form of the dicalcium salt, and this is apparently important for resistance to heat stress. The capsule, formed by invagination of the cytoplasmic membrane and containing DNA and RNA as well as ribosomes, continues to be coated with lipids and is eventually separated from the parent cell by lysis of the cell wall. The endospore has certain properties that differentiate it from the parent viable vegetative cell. For one, it is dormant and is unable to multiply without first reverting to another vegetative cell. Although "alive," they do not degrade compounds to generate energy, nor do they synthesize cell components, and residual water is unable to pass through the wall. The dormancy is quite remarkable, Sykes (1965) quoting an example of canned roast veal that was found to be quite edible after 113 years but nevertheless contained spores that could be recovered. There are other perhaps less well-authenticated examples where viable spores have been claimed to have been recovered from coal or from the contents of a jar of food found in the tomb of Tutankhamen, nearly 3,000 years old. However, it does seem likely that endospores are capable of surviving adverse conditions for many years. Moreover, it is well recognized that endospores will resist killing by heat, freezing, or drying as well as exposure to many toxic chemicals and radiation. As an example, boiling water will kill most vegetative cells in just a few minutes, whereas endospores can survive when being boiled at 100°C for as much as 15 hours or when exposed to dry heat at 160°C for 3 hours. The real challenge of applying any sterilization process is to determine if endospores originally present in the product are truly inactivated.

## THE DEATH PROCESS

We need to evaluate the process by which a cell dies and to realize that a cell can be damaged or become dysfunctional without necessarily being dead. If a cell is lysed or totally incapacitated from any functional process, it can be said to be dead; it is not metabolizing and it is not capable of reproducing itself, which is a true definition of dead. However, we need to consider a situation where a cell can be said to be damaged. It need not necessarily be capable of reproducing itself in that particular medium or at that particular time. Some metabolic processes are capable of repair; given enough time, the cell might recover its capacity to metabolize, grow, and divide. If the functionality of a spore is considered, it might be thought to be dead—the metabolic processes are certainly at a very low rate and a spore, as such, is incapable of dividing. Given enough time and the right environment, a spore will develop into a vegetative cell that does metabolize and grow. A spore is not considered to be nonfunctional or dead; it is simply dormant. We must also be aware that a vegetative cell, shocked by an applied environmental stress, may simply shut down its functions. By definition, it is "dead," but under the right conditions it might recover. Marginal sterilization processes, therefore, may be incapable of completely inactivating all of the organisms present in a closed system (inside an ampoule, for example). For the most part, sterilization processes are designed to kill all organisms in the system by "overkill," thereby removing the marginality of the process. Unfortunately, killing processes also damage the product in most cases, and it has been necessary to explore the limits of the sterilization procedure in order to maximize the killing but minimize the damage to product.

A nonfunctional, inactivated, or dead cell is what is needed following the application of a sterilization process; the danger is that a dormancy phase may have been triggered in the organism that will not necessarily show up in a short-term incubation test that is applied to a product.

## KILLING AGENTS

For the most part, killing or inactivation processes are limited to physical changes in the environment or the application of chemical stresses:

- Physical: dry heat, moist heat, ionizing radiation, intense light

- Chemicals: aldehydes, ethylene oxide, β-propiolactone, peroxides, chlorine dioxide, chlorine, chlorates, iodine, ozone, alcohols/phenols, quaternary ammonium compounds, chlorhexidine

Depending on the intensity of the killing process, for example, the temperature or length of time in the case of heat or the concentration, temperature, and length of time for a biocidal agent, microbes themselves have varying susceptibility as follows:

- Total inactivation: everything from bacterial spores to lipid encapsulated viruses are killed.

- Partial inactivation: most vegetative organisms (but not spores) are killed.

- Selective inactivation: only vegetative bacteria, most viruses, and fungi are affected.

Spore resistance to almost all inactivation mechanisms is the real challenge. If a process can be devised to kill spores, the likelihood is that all other microbial forms of life would be extinguished at the same time.

## Sites of Inactivation

Intuitively, it might be supposed that the application of heat would cause the precipitation and inactivation of proteins or that the addition of, for example, sufficient glutaraldehyde would covalently cross-link proteins on the bacterial surface. This intuition may be a little naïve in the sense that many biocidal agents and processes affect more than one site of interaction in the microbial cell; conversely, combinations of agents may be more effective than individual components.

If a hypothetical, nonspore-forming bacterium is considered, a biocidal agent may affect the external cell wall (or outer membrane in gram-negative organisms), the cytoplasmic membrane, or ribosomes and nucleic acids in the cytoplasmic contents. Some agents, such as those collectively identified as antibiotics, are often very selective in their inactivation mechanism. The β-lactam penicillins inhibit the formation of peptidoglycans, resulting in a loss of the rigid cell wall and the formation of pro-toplasts affecting organisms that are not necessarily inactivated, but merely dysfunctional. When subsequently placed in an appropriate medium, they may eventually recover from the applied stress. The cell wall (outer membrane) of gram-negative organisms is resistant to the transport of external agents, thereby either being resistant to most biocidal agents or requiring higher concentrations to be effective.

## External Cell Wall

Aldehydes, such as formaldehyde or glutaraldehyde, react with amino groups, especially at high pH, and cause subtle conformational changes in proteins by cross-linking. Enzymic proteins are, therefore, inactivated and functionality at the external surface of the organism is readily destroyed. Other agents, such as polycations or EDTA (ethylenediaminetetraacetic acid) are able to affect the permeability of the outer membrane of gram-negative organisms directly, allowing other biocidal agents to penetrate.

## Inner or Cytoplasmic Membrane and Membrane Enzymes

A number of biocidal agents, such as phenols, are able to penetrate to the cytoplasmic membrane and destroy or distort the essential phospholipids making up this membrane, resulting in a leakage of cytoplasmic contents or a complete loss of membrane protection. There are transport enzymes in this layer that are readily inactivated by phenols, quaternary ammonium compounds, chlorhexidines, and hexachlorophenes. The thiol (-SH) groups of proteins are affected by agents such as bronopol and organo-mercurials ("thiomersal"), and the net effect of these environmental additives is to produce a direct interference with several essential metabolic processes that may or may not be totally destroyed, depending, for the most part, on the concentration of the inactivating agent. Not all enzymic molecules are in exactly the same state of conformation, so a finite number of sites are in a protected or unavailable state. This consideration might lead to the supposition that functionality of the entire microorganism could be restored if there was an insufficient level of the biocide available for a sufficient length of time to destroy these enzymes permanently. The concept also helps to explain why some organisms in a closed system might be very susceptible to an environmental stress whereas some organisms of the same system might be more resistant.

## Cytoplasmic Contents

Cytoplasmic contents would consist of a number of proteinaceous enzymes, together with ribosomes that contain messenger-RNA components as well as proteins being synthesized. Finally, the DNA components of genes, actively controlling the RNA and the synthesis of proteins, are at the heart of the cell. Many chemicals act on cytoplasmic contents by a variety of mechanisms. Nevertheless, this assumes the agents have been able to diffuse through the many protective outer layers associated with a prokaryotic cell. The following inactivation processes may not always be realistic on a molecular level:

- Alkylation (ethylene oxide, propylene oxide, and β-propiolactone): these actively alkylate proteins, interfering with the spatial conformations necessary for enzymic activity.

- Oxidation (peroxides, hypochlorites): thiol groups are readily oxidized and, as noted, mercurials and other heavy metals react with thiol groups. Functionality is, therefore, affected by induced changes on the protein conformation.

- Cross-linking (aldehydes): proteins, DNA, and RNA are readily cross-linked, thereby losing their functionality.

- Disassociation (hydrogen peroxide and other agents): ribosomes are disassociated, resulting in a loss of synthetic capacity.

- Precipitation (phenols, quaternary ammonium compounds, and chlorhexidine): proteins are precipitated or, if the concentration is high enough, permanently denatured and become inactive. However, the denaturation process is sometimes reversible, suggesting one mechanism for recovery.

- Intercalation (acridines): DNA is intercalated by acridines and possibly other nucleic acid analogues or derivatives, thereby inhibiting some protein synthetic processes. However, these sensitive processes are designed to be repaired and self-regenerating after exposure to stress, requiring high intercytoplasmic concentrations of acridines for effectiveness as killing agents.

## THE EFFECT OF HEAT

The fact that moist heat is significantly more effective than dry heat as a sterilization process suggests that there may be a fundamental difference in the mechanism of loss of functionality. Undoubtably, as the environmental temperature is slowly raised, the activity of enzymic processes generally increases to a plateau that may be at or only slightly above the normal environment for the organism. After that, the rate process declines as the proteins (enzymes) become increasingly less functional. Initially, this functionality is likely to be due to slight conformational changes in the protein; at higher temperatures, hydrogen bonding is affected throughout the system, and proteins start to come out of solution as more and more hydrophobic groups are exposed to the aqueous environment around the molecule. Ultimately, the protein will become truly denatured, unable to function, and, more to the point, unable to recover its functionality. Similarly, the functionality of the DNA/RNA synthetic processes will be constrained or totally inhibited, with the net effect that there will be insufficient enzymes available to allow a completely effective metabolic process—death by any other criterion.

The essential difference between wet heat and dry heat may simply lie in the efficiency with which heat is applied to the product or the particular site where the microorganism is situated. Marcos-Martin et al. (1996) have recently pointed out the importance of dropwise condensation on surfaces for successful sterilization although, in terms of heat transfer, this factor was known much earlier. As a droplet of say, steam, condenses, latent heat of condensation is given up at that point on the surface and droplet condensation is more efficient than continuous film condensation for overall heat transfer across surfaces. Heat-induced inactivation of the microbe results in general precipitation of proteins across the board; some proteins (as enzymes) will be more essential than others, and the object of the exercise would be to apply sufficient heat to inactivate those enzymes that are essential for survival. This is not too difficult for many bacterial vegetative forms, exposure to temperatures of 50–60°C for a few minutes being sufficient to destroy functionality. On the other hand, in nature, there are some bacteria (thermophiles) that function optimally at temperatures in excess of 90°C or more, and, of course, endospores are considerably more resistant to heat than are most vegetative cells. In a functional vegetative cell, as the temperature is raised, some enzymic reactions stop. Both the external cell wall and the cytoplasmic membrane become ineffective at restraining the escape of cytoplasmic contents and the entry of water, bearing in mind the high osmotic pressure of the contents. RNA starts to break down, proteins come out of solution, and, finally, there is chromosomal injury. All structures and functions of the cell are damaged by heat, to varying degrees but repair and restoration of function is only possible if DNA functionality is not affected; once this occurs, the cell cannot recover. Dry heat, less efficient in terms of temperature and the time required for effectiveness, may initially simply dry out the organism, with sufficient moisture being retained around sensitive sites. Heat-damaged spores of *Bacillus subtilis* have been reported to mutate, perhaps as a result of depurination of DNA (Russel 1990). Dried organisms would, therefore, retain some comparison to bacterial spores that have exceptionally low levels of moisture. Other factors in the case of spores, such as a high lipid content of the surrounding walls and the association of dicalcium DPA with DNA, apparently stabilizing DNA, as well as local proteins, may also be involved. Interestingly, DNA itself becomes more tightly coiled, indicative of a lower water activity in the surrounding environment. The removal of calcium with EDTA increases the sensitivity of spores to heat. Beyond desiccation, the dry heat process also probably destroys functionality by the oxidation of proteins and by producing structural damage to DNA.

## THE EFFECT OF RADIATION ON BACTERIAL CELLS AND SPORES

Ionizing radiation, by definition, is radiation that causes the loss of an electron when an atom is struck. Four types have been identified—α-, β-, γ- and X rays. α-rays are large charged particles that do not penetrate into materials. β-rays are high-speed electrons and also have little penetration capability, but they are readily generated from either radioactive sources or machines. For sterilization, γ-rays are used since they penetrate readily and emanate from radioactive isotope sources, such as $^{60}$Co, that can be conveniently handled on a large scale. X rays, on the other hand, are readily produced but are not used for sterilization purposes. Ultraviolet radiation, at wavelengths of around 260 nm, is also used for sterilization, although it is less effective than ionizating radiation and demonstrates virtually no penetration of surfaces.

Ultraviolet and ionizing radiation have direct effects on DNA. With ionizing radiation, the ionization of water, to produce short-lived but highly reactive protons ($H^+$) and hydroxyl ($OH^-$) ions, occurs because water is the main component in and around the cell. Internally, the reactive ions produce structural damage to DNA, resulting in synthesis of "errors" that produce mutations or, in extreme cases, leads to cellular death. If the level of radiation is low enough, the organism may be able to reconstruct and repair itself after the radiation source is removed. As noted earlier, for spores, DNA is in a form that is not associated with significant amounts of water. This suggests that DNA molecules associated with spores need to be struck directly by the radiation particle for damage to be manifested later. Irradiated spores are often capable of germinating and swelling, but are unable to grow any further. In addition, they become more readily affected by any additional applied environmental stresses.

There is at least one organism, *Deinococcus radiodurans*, that is able to repair damaged DNA following ionizing irradiation or ultraviolet treatment, using a specific enzyme for the process. However, most bacterial cultures contain cells that, for one reason or another, are more resistant than others, and, under some conditions, paradoxically, growth is actually briefly stimulated. As might be anticipated, bacterial spores are more resistant to radiation. There is also an oxygen effect; cells grown in an oxygen-enriched atmosphere are often more sensitive to irradiation.

## PULSED LIGHT STERILIZATION

Another form of radiation that is currently of interest is the use of pulsed bright light (PureBright®, PurePulse Technologies, San Diego, CA). This process uses very intense, short duration pulses of broad spectrum white light to kill microorganisms, including bacterial spores, viruses, and even *Cryptosporidium* oocysts. Based on technologies originating in the Reagan era Star Wars initiative, these high-intensity flashes are produced using pulsed power processing techniques that can result in highly bactericidal effects not seen with conventional light sources or even low-energy continuous ultraviolet sources. Initially explored for the sterilization of food products, the technology has been extended to the sterilization of form/fill/seal products for which the pulsed characteristic of the process is ideally suited for application to a production line. As described by Dunn (see Chapter 11), an electrical storage capacitor is discharged through a xenon flash lamp to provide a 300 μs flash of bright light (20,000 × normal sunlight) every second. The light can be focused by using surrounding reflective chambers or mirrors, with each container being exposed to 20 or 10 successive flashes, each of 1.5–4.0 Joules/cm$^2$. A wide range of microorganisms was shown to be sterilized up to 7 logs, and filled, sealed, polyethylene containers of Water for Injection (WFI) were effectively terminally sterilized.

Although an investigation of this novel process is ongoing, the potential for further development is considerable. The spread of the radiation is approximately 25 percent in ultraviolet radiation, 45 percent in visible radiation, and 30 percent in near-infrared radiation. The radiation passes through clear packaging and light transmissive fluids. It is nonionizing, there is no applied heat, and it does not penetrate opaque materials, although it has been demonstrated to be extremely effective on surfaces coated with bacterial spores. The application to drug solutions is only now beginning to be explored, and it is possible that materials strongly absorbing in the ultraviolet range might also deteriorate when treated by this process (F. Leo, private communication, 1997). The limitations of this effect remain to be evaluated. The process is otherwise very attractive since it does not require anything other than regular manpower, is ideally suited for a production line process, can be retrofitted into an existing facility, and is estimated to cost 0.1 cent/sq. foot of treated area. Although used to sterilize surfaces, intravenous solution bags have been sterilized by pulsed light after high levels of spore suspensions (up to 8 logs/mL) had been added, indicating that sealed containers of prefiltered solutions should be readily processed by this method. One can be optimistically enthusiastic about this novel method of sterilization. Obviously at an early stage of development, the limitations are only just beginning to be explored, but already sterilization of some products appears to be feasible, including large-volume parenteral solutions in both glass and plastic containers. Savings in sterilization energy alone would appear to be considerable by comparison with conventional steam autoclaves, and the possibility of complete automation of the process must be very attractive.

## STERILITY TESTING

The current sterility tests need to be considered since there is a reasonable chance that the pharmacopeial requirement of seven days incubation after the sample is taken is not long enough.

There is general agreement that the growth cycle of an organism being placed in a fresh growth medium consists of a lag phase, followed by an exponential growth phase to a plateau, after which the culture starts to decay and die, as illustrated in Figure 2.1.

If we consider an organism inadvertently present in a pharmaceutical solution that is unlikely to be congenial for growth (parenteral nutritional solutions being an exception), the lag phase will be variable, but is likely to be prolonged. We can calculate the time it takes for, let us say, 10 *Escherichia coli* organisms/mL to grow to a concentration of $10^5$/mL, the point at which the solution becomes visibly cloudy.

The exponential growth rate, $\mu$, is defined by:

$$\mu N = dN/dt$$

where $N$ is the number of organisms/mL.

$$dN/N = \mu dt$$

$$\ln N = \mu t$$

$$\ln 10 = \mu t_o$$

$$\ln 10^5 = \mu tp$$

where $t_o$ is the time at which the exponential growth phase starts and $tp$ is the time at which growth becomes visible (not necessarily the plateau point). Thus,

$$\ln 10 - \ln 10^5 = \mu(t_o - tp)$$

$$2.303 - 11.513 = -\mu tp \quad (t_o = 0)$$

$$\therefore \mu tp = 9.21$$

Under optimal conditions, $\mu$ is 2.1 hours for *Escherichia coli* (Lugo 1995), but we can assume that, under unfavorable conditions, growth would be much slower, say, for the sake of discussion, $\mu = 0.1$ hours. Thus, $tp = 9.21/0.1 = 92.1$ hours = 3.83 days.

In order to demonstrate visible growth, the culture will need to be incubated for nearly 4 days, **plus** any time needed for the lag phase. The time becomes longer if there is only 1 organism present per mL, 4.8 days, and nearly 6 days if only

1 organism/10 mL; these counts are more likely to be typical of those to be found if the solutions have been thoroughly filtered during preparation. Thus, the success or failure of the incubation phase of the sterility test depends on the length of the lag phase and the number of organisms initially present in the product. In an inhospitable environment, it is likely that the lag phase would be so prolonged that the resultant net growth rate becomes very small.

These considerations become important if we are trying to determine if a particular sterilization process has been successful, or even to validate the process at a later stage. Organisms originally present in the system are likely to have been shocked or stressed, with a likely extension of their lag phase and a slowing of the growth rate. The issue becomes even more critical if only a few organisms are present in the first place. The lower the bioburden, the more effective a sterilization process is likely to be but, as can be seen from the above discussion, so also is the chance of determining if the process has been successful.

## FILTRATION "STERILIZATION"

In any discussion of sterilization methods, filtration is the one method that does not provide an environmental stress on organisms. Any contaminating microorganisms are removed by passing the drug solution through a thin porous membrane or screen, generally made of cellulose or some other high molecular weight polymer, but occasionally of sintered metal or ceramic. Like all porous assemblies, there is a range of pore sizes present in the separation system, and conventional wisdom holds that the nominal "sterilizing" cut-off size should be 0.2 μm. However, because there will be a range of sizes in any filter membrane or surface, this must be regarded as a *maximum* size as determined by some externally applied physical test, such as measurement by a mercury intrusion method. More relevantly, in practical terms, it should be the minimum "size" of a particle that is totally retained by the membrane. Because of surface adsorptive effects that are, parenthetically, critically affected by environmental factors, such as electrolyte concentration and pH, it is possible that particles much smaller than the nominal pore size of the membrane are retained; however, for all practical purposes, no reliance should be given to this factor. In practical terms, what is needed is the assurance that the membrane will remove and retain all bacteria or fungi present in the parenteral solution being passed through it. Not only must the user be aware of the fact that the membrane in use may contain pores that are larger than the "nominal" or "maximum" size quoted by the filter manufacturer, but also that there may be leaks or pathways around the filter and its supporting system that allow liquid to pass around, not through, it. This means that assemblies of filters and the retaining and supporting equipment must also be tested. This is usually by some form of bubble test in which gas is forced back through the filter wetted with product and the pressure at which the first bubbles appear measured—the so-called "bubble point." Naïvely, this is assumed to be a measure of the largest pore (or hole) in the assembled system, although this is not always correct.

When sterilization using filters was attempted for a number of years, it was assumed most bacteria were removed with membranes with pore sizes around 1.0 $\mu$m (i.e., 1,000 nm). With this perfectly valid assumption in mind, it therefore seemed reasonable to use 0.45 $\mu$m (450 nm) pores as a guarantee of success. Later, it was discovered that some organisms, under unfavorable environmental conditions, tended to become smaller or even, in some extreme cases of starvation, developed into a form known as somnicells, which are structurally different from endospores. The organism studied in depth from a pharmaceutical perspective was *Pseudomonas diminuta*, which is currently believed to provide a useful challenge or test organism for validating a 200 nm membrane filter. The environment in which the organism is grown needs to be rigorously controlled to ensure that the "size" is more or less reproducible. Nevertheless, somnicells may be significantly smaller than 200 nm in diameter; viruses are most certainly smaller on average, down to 25 nm, and prions are probably smaller still. In addition, organisms without cell walls, protoplasts, have been found to pass or insinuate themselves through porous filters. This leaves us to answer the question, "Does a porous membrane filtration really sterilize the

solution passed through it?" In absolute terms, the answer must be "no." although pragmatically, the solution may well have been rendered "aseptic," following the argument of Gilbert and Allison (1996).

## KINETIC CONSIDERATIONS IN THE DEATH PROCESS

Irrespective of the nature of the applied environmental stress, the bacterial cell or, exceptionally, the endospore will be exposed to a new situation in which the desired endpoint is death. This will be brought about by effects on enzymatic proteins or the cellular DNA produced by heat, some chemicals, and radiation. If the dose of the chemical agent is insufficient or the heat stress is inadequate, the organism may be able to repair itself; although it may require time before it becomes completely functional again, the purpose of the application of stress is defeated. This type of situation has been best studied for heat sterilization. Most sterilization processes are designed to kill the most-resistant organism present in the system several times over, an overkill situation. The cookbook approach to steam sterilization found in most early pharmacopeias and pharmaceutical textbooks— 15 minutes at 121°C (10 lb psig)—is an example of this overkill situation.

The overkill approach in pharmacy was, until quite recently, operated in an absolute fashion, just like a cookbook. There was a tendency to regard product heated for 14 minutes at 121°C as being "nonsterile" and 15 minutes or longer as "sterile." As noted earlier, the mechanism of death does not work like this; there is no one point in the process where organisms are alive and all of a sudden they are dead—the square wave approach. The process is a probabilistic one; as the stress is applied to a given unit in the sterilization process over a longer and longer period of time, organisms are dying off (whatever that means) at an increasing frequency or rate.

One other aspect about the older literature that also needs to be illuminated concerns the use of pressure when, in fact, it is temperature that is the applied stress. If the autoclave or vessel is truly evacuated so that the steam contains no air, then the reading of internal pressure will allow the temperature to be read as a reasonable approximation from saturated steam pressure/temperature tables. This necessity became redundant when reliable methods of measuring temperature inside the pressure vessel or, indeed, inside the product itself became available. The true effect of pressure on microorganisms has not been measured adequately, and the applied stress within an autoclave is temperature, not pressure.

Unfortunately, although the contaminating microorganisms will be rendered nonfunctional, by the same token, effects will be produced on the active ingredient(s) of the product undergoing sterilization. As a result, investigators have explored less heroic but more marginal sterilization conditions in order to minimize product damage. As evident from the previous discussion, there are some obvious dangers associated with this approach, since microorganisms do not all suddenly cease to function after the application of so many minutes of a particular stress. The process is a random or kinetic situation; some cease relatively rapidly, some relatively slowly, and everything in between the two extremes. An example of this situation would be a bacterial suspension containing both vegetative cells and endospores; both subpopulations would also have inherent variability. The problem is to define the combination of time and stress conditions that results in the death of *every* cell in the system—sterilization.

It should be evident from the previous discussion that the bacterial death process is complex and variable. Given a simple culture of one single organism, at any one time during the growth of that culture, some organisms will be at a very sensitive or susceptible phase of their own individual growth cycle. Taking heat stress as an example, the DNA in chromosomes could be affected, or a proteinaceous enzyme at some stage down the metabolic line. All enzymes are equal but some are more equal than others—to paraphrase George Orwell—but not all are necessarily exactly susceptible to the same applied stress as experienced by any one organism or by one of its neighbors. The situation becomes more complex when dealing with a culture containing more than one species of organism, especially so for cultures of vegetative cells and their associated endospores. The result is that the sterilization process must be regarded as a probabilistic situation; there is a **chance** that an individual organism may die (or cease to function), but not all the organisms will die at the same time. As the stress is applied, the response is probabilistic, or exponential, and not the sudden change associated with a square-wave response. In a mixed culture of different species, some species are more resistant than others; the situation, in other words, is even more complex.

Experimentally, the death rates (actually measured as "survival rates") are exponential, which suggests that the net process involved is most likely to be first order in which two components react to give a product.

$$A + B \rightarrow \text{product (or "death")}$$

This may only be a resultant or apparent first-order response to the applied stress. The experimentally measured rate of the reaction will be determined by the rate of the slowest or controlling step in the overall degradation, deterioration, or death process. If $N_o$ is the number of organisms per unit volume (analogous to a concentration term in a chemical reaction) at zero time, $t_o$ (i.e., at the time the killing stress starts to be applied), we can define a killing rate as

$$-dN/dt = kN \qquad \text{[eq. 1]}$$

where $k$ is the rate of killing or the velocity constant according to a first-order reaction. Integrating [eq. 1] between limits $N_o$ at $t_o$ and $N_t$ at a later time $t$, we get

$$\int_{N_o}^{N_t} dN / N = -k \int dt$$

$$\text{or } \ln N_t - \ln N_o = -k(t - 0)$$

$$\therefore \ \ln N_t = \ln N_o - kt \qquad \text{[eq. 2]}$$

Note that for this and subsequent arguments, all logarithms will be left as natural (to the base $e$) logarithms, ln, and not to the base of 10, common logarithm, log. The relationship between the two is a simple constant, 2.303 (ln 10), but the manipulation as natural logarithms is much easier. Eq. 2 can be written in exponential form:

$$N = N_o e^{-kt} \qquad \text{[eq. 3]}$$

Bear in mind that each enzymatic inactivation reaction or DNA degradation reaction will have its own unique value of $k$, the rate constant for that particular reaction. In terms of the death process, we can also write characteristic rate constants for the chance of each organism in a system dying using the following subsets:

- Chance of an organism dying: $k_1$

- Chance of an organism being damaged but not recovering: $k_2$

- Chance of an organism being damaged but recovering: $k_3$

- Chance of an organism recovering: $k_4$

We need to apply a sterilizing stress to the system such that

$$k_1 \gg k_2 \gg k_3 \gg k_4$$

The overall process is controlled by the limiting stage. By designing a process in such a way as to guarantee "death," $k_1$ (and $k_2$) must be many times greater than either $k_3$ or $k_4$. In a marginal situation, this may not always be the case, and it may be necessary to adjust the conditions so that at least $k_2$ is the rate-limiting or controlling step. In practice, it is not possible to determine each individual rate for each organism or each species of organism present in a closed system. All we can hope to do is to measure the resultant rate, determined as the total number of organisms originally present (per unit volume) and the number of organisms present after a given time following an applied stress. Nevertheless, although it may be crude, we can measure a definable rate at which the organisms are becoming nonfunctional—or dying—and design the way in which the stress is applied so that, eventually, we can say with reasonable certainty that all the microorganisms originally present are dead.

From a microbiological point of view, we can measure the time that it takes for 90 percent of the organisms present to be damaged irrevocably, or for the population to drop to one-tenth of the original number, a decimal reduction time under isothermal conditions. From eq. 3,

$$N = N_o t^{-kt}$$

$$\therefore k = 1/t \, (\ln N_o - \ln 0.1 \, N_o)$$

$$= 1/t \, (\ln (1/0.1))$$

$$t = 2.303/k$$

This $t$, the time to reduce the population (irrespective of the size of $N_o$ itself) to one-tenth of its original value, is termed the D value (i.e., D = 2.303/k). It is readily measured by exposing a culture of the organisms to a stress, usually heat, and counting the surviving cells over periods of time. A plot of the logarithm of the numbers of surviving organisms against the time of exposure should be linear, with a slope of $k = 2.303/D_{av}$. This has been experimentally observed since the turn of the present century and is true for a number of different sterilization processes, including heat, radiation, or chemicals. The heat sterilization process should be designed such that, at a given temperature, the applied stress here will reduce the number of organisms present per unit volume by at least 12 log units, to at least one-billionth of the original concentration. The necessary time of exposure will therefore be to 12 D.

This might be taken as an overkill situation that is not difficult to determine for vegetative cells, since these only require a few minutes at elevated temperature to die, but it is much more difficult to measure when endospores are present. Indeed, should a microorganism be obtained from a product that has been exposed to a marginal sterilization process, it needs to be tested in order to determine what its D value actually is, and the process adjusted accordingly to ensure that this value is exceeded. Usually, however, a resistant, endospore-forming model organism, such as *Bacillus stearothermophilus* ($D_{121} \approx$ 4–5 min) for moist heat or *Bacillus xerothermodurans* ($D_{170} \approx$ 6 min) for dry heat conditions would be selected as a biological indicator and the D value measured under the exact conditions proposed for use. This is effectively a worst-case scenario that is generally acceptable, assuming that a more resistant, hitherto unknown, species is not present. However, it might be noted in passing that a vegetative organism would be likely to have D values < 1 minute at temperatures above 100°C.

An exception to the linear semilog plot might be mentioned here, since, under dry heat conditions, the plot is not linear for the first five or so minutes as the organisms appear to dry out; after that point linearity is observed. Calculations for effective D values under these conditions need to be made *after* the first five minutes of the process in order to allow for the semilog linear phase of the killing process to be established.

An important point needs to be made here about bacterial numbers. Working on the survival of the botulism organism *Clostridium botulinum* in canned foods, it was discovered that the highest concentration of endospores that could be harvested, cleaned, and suspended in phosphate buffer was approximately $10^{11}$/mL. If this suspension was subjected to heat stress for at least 12 D, there would, therefore, be a reasonable chance that the heated suspension would contain no survivors. As it happens, the D value at 250°F (121°C) was calculated to be 0.21 minutes, a total time of 12 × 0.21 = 2.52 minutes being required.

Since, in this situation, it would mean that the numbers of endospores present in the food must be somewhat smaller than $10^{11}$/mL, any time in excess of 2.52 minutes would result in overkill and, in the case of food, probably, with a loss of quality or flavor. In the case of a pharmaceutical product, it could lead to degradation of the active component and one must, therefore, look for less-damaging criteria when designing a sterilization process.

Two further points need to be made. The first is that the minimum heat (or stress) exposure of 12 D has been firmly rooted historically for the past 70 years based on work involving the sterilization of foods. It could be argued that most pharmaceutical solutions are less hospitable to microorganisms and it might be feasible to minimize heat trauma to the product by requiring less stringent criteria, perhaps 6 D. This, however, needs to be tightly bound to the *number, $N_o$,* of organisms originally present since, if it is low initially, the chance of success (i.e., reducing $N_t$ to zero and below) would be proportionately higher. The concept of sterility assurance level (SAL) arises from this thought. If the appropriate semilog plot of survivors/mL versus time of exposure is made for a particular microorganism and vehicle, a point will be reached where the number of microorganisms is zero (i.e., in principle, the overall system is sterile). However, if the plot is extended below the zero point, we start to get into the area of the graph that estimates the chance or probability of an organism surviving, 1 in 10, 1 in 100, and so on. The SAL (or, in some literature, the probability of survivors per item, SPI) is considered to be the probability of an organism surviving a given sterilization process. It is generally agreed that an SAL of at least $10^{-6}$ should be the aim. Since it is not feasible that one-tenth or one-millionth of a part of an organism will survive the treatment, the concept has been expanded to mean that there is a chance of 1 microorganism surviving in 10, or 1 in 1,000,000 processed items *after the treatment,* in other words 1 in 10 or 1 in 1,000,000 containers in the lot could be contaminated. This concept introduces us again to the fact that, when dealing with microorganisms, one cannot be absolute, but can only estimate the chance or probability that a product is actually sterile, that is, containing no viable or surviving microorganisms in an absolute or real sense.

## CONCEPTS VERSUS REALITY, AND SOME SPECULATION

The concept of SAL has been welcomed by regulatory authorities as a standard by which the effectiveness of pharmaceutical sterilization processes may be judged. However, a great deal of confusion seems to have resulted in this field, and the net effect has been that bad scientific judgments have been made in both industrial and regulatory circles. If measurements cannot be made, it seems fair to argue that scientific judgments cannot be made either. Although it is true that we can now measure with reasonable precision the temperature and time parameters involved in a heat sterilization process, the one question that cannot be ignored is, What happens to the microorganisms in the product and how can this be *measured*? As noted, the SAL is, strictly speaking, a concept based on an earlier, irrelevant set of conditions, namely measurements made on the spoilage of food and not on pharmaceutical solutions containing few, if any, microorganisms in the first place. Moreover, even if we can suggest that an SAL of $10^{-6}$ is sufficient for heat-sterilized processes, how can we justify an SAL of $10^{-3}$ for aseptically processed products since we cannot measure it? The regulatory authorities are clearly implying that this means that it is acceptable to suggest that 1 container in each 1,000 produced may be contaminated. This becomes very personal if you happen to receive that one container and, equally, is not acceptable. What it means is that each container has an SAL of $10^{-3}$, not the entire batch. This raises the question of just how useful or valid is a media-fill test of an aseptic process because *any* container that happened to be contaminated during a media fill run is evidence of the fact that the process is inadequately designed for the purpose. It is time to seriously question the application of a concept derived from the heat treatment of foodstuffs to the cold production of thoroughly filtered pharmaceutical solutions into sterile container components.

One aspect seems to be obscured in the argument, the application of a *probability* function. To illustrate this, let us assume that we have made a small batch of a solution that we know will support microbial life, say 10 ampoules, and we suspect that at least one ampoule in the batch could be contaminated (i.e., it has an SAL of $10^{-1}$). If we test the first ampoule and find no growth, the odds of finding the contaminated ampoule in the remaining 9 are obviously going to be higher. If we now sequentially test all of the remaining ampoules, we can see how the odds change, Table 2.1. It will be evident that the probability of finding the contaminated ampoule really does not change all that much initially, only doubling when over half the batch has been sampled.

**Table 2.1. Determining the Odds of Finding a Single Contaminated Ampoule in a Batch of 10 with an SAL of $10^{-1}$**

| Ampoule Number | If No Growth, New Odds | Probability |
|:---:|:---:|:---:|
| 1 | 1 in 9 | 0.111 |
| 2 | 1 in 8 | 0.125 |
| 3 | 1 in 7 | 0.143 |
| 4 | 1 in 6 | 0.167 |
| 5 | 1 in 5 | 0.200 |
| 6 | 1 in 4 | 0.250 |
| 7 | 1 in 3 | 0.333 |
| 8 | 1 in 2 | 0.500 |
| 9 | 1 in 1 | 1.000 |

If we repeat the same exercise for a batch size of 1,000 with an SAL of $10^{-3}$, Table 2.2, we see that the odds change even more slowly. Effectively, it is not difficult to appreciate that if we take 20 ampoules from this batch and test them for sterility (with all the associated problems that this test has), the chance of finding even 1 contaminated ampoule remains at 0.001, and these odds do not change very much until we have tested at least one-tenth of the lot size. The real effect of this SAL concept is that we can regard each individual container within the lot as having a probability of, say, 1 in 1,000 of being contaminated; it does not mean that, in every 1,000 containers, there is 1 that is contaminated. As evident from the data shown in Tables 2.1 and 2.2, the chance of finding a contaminated unit is effectively an exponential process. Statistically, the chance of finding one nonsterile unit during, for example, a media fill of $N$ units in a batch is given by

$$\text{probability} = P = 1 - (1 - x)^N$$

where $x$ is the expected contamination rate and $N$ is the number of units produced.

**Table 2.2. Determining the Odds of Finding Contaminated Ampoules in a Batch of 1,000 with an SAL of $10^{-3}$ as a Function of Sample Size**

| Ampoule Number | If No Growth, New Odds | Probability (to three decimal places) |
|---|---|---|
| 1 | 1 in 999 | 0.001 |
| 2 | 1 in 998 | 0.001 |
| 3 | 1 in 997 | 0.001 |
| 4 | 1 in 996 | 0.001 |
| 5 | 1 in 995 | 0.001 |
| 6 | 1 in 994 | 0.001 |
| 7 | 1 in 993 | 0.001 |
| 8 | 1 in 992 | 0.001 |
| 9 | 1 in 991 | 0.001 |
| 10 | 1 in 990 | 0.001 (01) |
| 20 | 1 in 979 | 0.001 (02) |
| 100 | 1 in 899 | 0.001 (12) |
| 900 | 1 in 99 | 0.010 |
| 1,000 | 1 in 1 | 1.0 |

To amplify this point, if a lot size is 3,000 and no growth is found during a media fill, the probability that the SAL is at least $10^{-3}$ is

$$P = \leq 1 - (1 - 0.001)^{3000} \leq 0.95 \ (029)$$

In this situation, we can be reasonably certain (with 95 percent confidence) that the entire batch is sterile (or, more accurately, insufficient microorganisms were recovered or could grow to provide *visible* evidence of growth). On the other hand, if 3 containers from a batch of 3,000 showed evidence of growth, the situation changes somewhat.

In a large population, an estimate can be made of the distribution around that mean from a binomial standpoint with a standard deviation of $(pq/N)^{1/2}$, where $pq$ is the proportion of the population that is contaminated, leaving the uncontaminated as $(1 - p) = q$, then

$$\text{estimated upper limit} + 1.96\,(pq/N)^{1/2}$$

$$\text{estimated lower limit} - 1.96\,(pq/N)^{1/2}$$

$$\therefore \text{ upper limit (to } P = 0.95)$$

$$\frac{3}{3000} + 1.96\frac{[3/3000\times(1-3/3000)]^{1/2}}{3000} = 0.00213\ (0.21\%)$$

lower limit

$$\frac{3}{3000} - 1.96\frac{[3/3000\times(1-3/3000)]^{1/2}}{3000} = -0.000131\ (-0.013\%)$$

Since the lower limit is below zero, we can say that the chance of a container in this lot being contaminated is 0.21 percent with a confidence limit of $P = 0.95$. This is well above an SAL of $10^{-3}$ and, for this reason, the process is unlikely to be satisfactory. More recently, regulatory authorities have suggested that a process be validated by running at least 3 media fills of at least 3,000 containers. Why 3, and not 30? All this does is marginally increase confidence and significantly increase costs. Ongoing media fills are also required, again without any real factual basis.

Nevertheless, the bottom line remains that SAL is a concept that is not subject to measurement with the precision and accuracy that is required in any scientific investigation. A process is "validated" by running a media fill; the product, for the most part, cannot be run in the same way, and we have to depend on sterility tests carried out after the process has been finished, with all the attendant difficulties of interpretation. In some senses, it is not a valid procedure to separate the production process from the product; we should only be concerned with the safety of the final product.

This leads to the position that the application of an aseptic process, involving aseptic assembly of a solution and container, needs to be considered much more carefully. The container components can be considered to be "sterile"; they have been separately terminally heat sterilized. The solution, on the other hand, will have been repeatedly filtered through porous membranes of decreasing size so that there is a probability of very few particles with diameters in excess of 200 nm being present in the system. Notice that one can be deliberately nonspecific. "Particles" can include but are not limited to viable particles, such as viruses or nanoforms of bacteria. If, instead of viability or growth *measurements* being made on the system, we started to examine a solution about to be filled into a sterilized container by a nonspecific particle-sizing procedure, such as photon correlation spectroscopy (PCS) (which measures average size but not numbers/unit volume), a number of advantages might suddenly become evident. For example, sizing data could be made available instantly or "in real time." To explain in a little more detail, coherent light from a laser is scattered to a detector by the particles encountered in the beam. The signal is accumulated over a brief period of time and the correlation curve calculated as a mean "size" of the particle, sometimes with an indirect measure of the size distribution. The instrument, therefore, provides a measurement of the average particles present in the system, which in the case of a "clean" solution would be very small. This would enable the operator to appreciate the fact that the bulk solution contains few, if any, "particles" (viable or otherwise) above the threshold of the smallest pore size of the last filtration membrane. Moreover, the software could be modified to warn if the porous membrane was failing during the procedure or extraneous particulate matter was appearing by detection of superthreshold particulate. This might be a valuable quality control tool that would provide records of the progress of the batch. The point is that a pharmaceutical solution is by any criterion "clean" and is ideally suited to evaluation by any existing on-line light-scattering equipment. The equipment is certainly available off the shelf and some modest changes in software would enable a great deal of valuable information to be obtained and stored.

There needs to be a distinction between this suggestion and what is already required for measuring particulate matter by pharmacopeial methods, such as the U.S. Pharmacopeia use of light blockage and microscopy techniques. These are criteria established as measures of "quality" for the complete product and its package. There the particles are counted per unit volume and generally originate from residuals or by the degradation of seals and container walls as well as the product itself. This situation is now well understood and in some senses is irrelevant to the present suggestion. What is proposed here is a continuous monitoring of submicrometer particulate as a function of size and time. Modern laser light-scattering methods are extremely sensitive, but cannot count or quantitate particulate in quite the same way as is needed for a compendial assay. PCS would effectively provide a size distribution of material in the liquid down to, say, 10 nm and could certainly be used to detect significant quantities of material with diameters of 200 nm. Since it is reasonably certain that most of this < 200 nm material will not be viable, we could use this procedure to suggest that the solution being filled is "aseptic" (Gilbert and Allison 1996).

It seems that by concentrating on irrelevant and arbitrary "numbers"—SAL—that cannot be measured, we may be concentrating on the wrong things. However, we can measure particles, and equipment is already available for the suggested purpose. Caution may be needed initially when attempting to interpret the data, but with time and experience, a greater degree of comfort would develop. For example, some biotechnology products, such as genetically engineered proteins, would have characteristic molecular sizes large enough to be measured by the analytical equipment and, therefore, provide "size" signatures that would be valuable quality assurance information. Viruses, on the other hand, would also be likely to have characteristic "signals" or signatures that it might be possible to deconvolute from the general background noise or information that is likely to be measured.

The object of this approach would be to take attention away from the process involved to the product itself, using realistic measurements, readily obtained with accuracy and precision in real time. Perhaps we should stop attempting to rationalize our processes in terms of parameters, such as SAL, that cannot be measured. As Gilbert and Allison (1996) point out, if we concentrate on the "aseptic" side of aseptic processing and stop demanding "sterility" that cannot be measured or even guaranteed with absolute certainty, then we might make some progress in understanding and improving what we are trying to achieve.

## Coda: Some Thoughts on Gilbert and Allison (1996)

The contribution made by these authors to the overall topic of sterilization and, especially, the production of aseptic pharmaceutical products needs to be discussed in greater depth. As Groves and Murty (1995) noted in the introduction to their book, up to 87 percent of small-volume parenterals in the United States are currently being filled aseptically, and this percentage will grow higher as more biotechnology-derived therapeutic drugs enter the marketplace. Demands by regulatory authorities on both sides of the Atlantic for terminal heat sterilization of all parenteral products seems rather poorly timed, although the net effect of these demands has been to focus attention on the issues involved.

Gilbert and Allison start by drawing attention to the fact that it is virtually impossible to have an absolute state of "sterility," defined biologically as "a total absence of all viable life." Useful as a general starting point, this definition has a number of failings from a pharmaceutical perspective and is inappropriate when compared to the absolute requirement of the space program, which was trying to avoid contamination of a somewhat larger environment. Tests for pharmaceutical sterility have failed miserably since there is a considerable sampling problem (e.g., Table 2.2) that requires statistical extrapolation back to the unsampled material actually administered to the patient. At best, therefore, such tests indicate the "asepsis" of the product, not the "sterility." For the most part, a few (say $< 10^2$) microorganisms per dose, even if they were known pathogens, introduced into a patient's body would be unlikely to have any detectable clinical effect. However, these authors do not mention the exceptions to this probably otherwise realistic assessment. These must include patients who have impaired immune systems, such as HIV and AIDS sufferers, and many cancer victims, especially if in their terminal phases. Moreover, anthrax is an exception in that single organisms may find an opportunity to develop under optimal conditions, and these would obviously be highly undesirable contaminants in any parenteral product. If we exclude all organisms from a product by definition, we have also excluded all pathogens. Nevertheless, Gilbert and Allison are correct when they note that sterile filtration is, at best, an aseptic process and is certainly not sterile in the absolute sense, since, in principle, viruses and prions could theoretically pass through the 100 or 200 nm porous membranes currently in use.

Another issue only lightly touched on by these authors is that of validation, since regulatory authorities now rightly demand that new or improved processes or process changes be validated to demonstrate their effectiveness. In many situations, properly carried out validation is scientifically challenging because it requires a complete knowledge of the underlying science as well as an intimate knowledge of the process itself. They state that an SAL of $10^{-3}$ can be demonstrated for aseptically prepared products by a broth or growth medium challenge, but this is incorrect since only the process is validated, not the product.

The process is, of course, an essential part of product design, but it is not an absolute sterilization or overkill process, and it is suggested that "sterilization" as a description might be better avoided for aseptically prepared products. Gilbert and Allison then go on to note that the food industry, with much more to lose than the pharmaceutical arena in terms of hazardous products consumed on a much larger scale, has for many years pioneered the use of flash heating or ultraheating processes that certainly do not conform to the absolute pharmacopeial definitions of sterilization. The resulting products are not "sterile" by any definition, but are certainly safe for their designated use. In the pharmaceutical context, it was suggested that the term *sterility* be avoided and products simply described as "Aseptic: Safe for its designated use."

Perhaps it is time to consider these suggestions seriously and determine if we really need to continue to increase the associated costs of marketing parenteral products. More debate, research, and education are needed.

## REFERENCES

Gilbert, P., and D. G. Allison. 1996. Redefining the sterility of sterile products. *Europ. J. Parent. Sci.* 1:19–23.

Groves, M. J., and R. Murty, eds. 1995. *Aseptic pharmaceutical manufacturing II: Applications for the 1990s.* Buffalo Grove, IL: Interpharm Press, Inc.

Lugo, N. M. 1995. Aseptic processing of biopharmaceuticals. *In Aseptic Pharmaceutical Manufacturing II: Applications for the 1990s,* edited by M. J. Groves and R. Murty. Buffalo Grove, IL: Interpharm Press, Inc., pp. 245–289.

Marcos-Martin, M. A., A. Bardat, R. Schmitt-Laesler, and D. Beysens. 1996. Sterilization by vapor condensation. *Pharm. Tech. Europe* 8 (2): 24–32.

Moldenhauer, J. E., S. L. Rubis, and I. J. Pflug. 1995. Heat resistance of *Bacillus coagulans* spores suspended in various parenteral solutions. *PDA J. Pharm. Sci. Technol.* 49 (5):235–238.

Nester, E. W., C. E. Roberts, and M. T. Nester, eds. 1995. *Microbiology: A human perspective.* Dubuque, IA: Wm. C. Brown.

Russel, A. D. 1993. Theoretical aspects of microbial inactivation. In *Sterilization technology: A practical guide for manufacturers and users of healthcare products,* edited by R. F. Morrissey and G. B. Philips. New York: Van Nostrand Reinhold.

Sykes, G. 1965. *Disinfection and sterilization: Theory and practice,* 2nd ed. London: Spon.

# 3

# SIP of Microbes in Deadlegs

*Jack H. Young, Ph.D.*

The Pennsylvania State University—The Behrend College
Erie, PA

Sterilization-in-place (SIP) utilizing saturated steam has become a widely used procedure in the parenteral and biotechnology fields. It offers a method of sterilizing equipment that is either too large or inconvenient to place in an autoclave and can enhance sterility assurance by reducing the amount of aseptic assembly. All surfaces to be sterilized must be exposed to an adequate combination of moisture and temperature for an appropriate length of time. The presence of moisture is critical, as sterilization can occur within 15 minutes in a 121°C saturated steam environment. Over 6 hours are required at 121°C when no moisture is present (Pflug and Holcomb 1983). Consequently, the removal of air or adequate air/steam mixing is required to assure sterilization within a reasonable time.

Dead-ended geometries (deadlegs) present a severe challenge for air removal because steam and air do not readily mix. Since relatively few steam SIP systems employ a vacuum to assist in air removal, deadlegs often become the most difficult-to-sterilize locations. In many cases, air bleeders need to be added to assure reproducible sterilization.

Although steam sterilization has been studied extensively by microbiologists in order to document the effects of temperature and time on various microorganisms, little has been done to provide the quantitative information necessary for the design of equipment that lends itself to SIP. Many published studies have focused on general SIP principles (Myers and Chrai 1981; Seiberling 1986; Berman et al. 1986; Agalloco 1990) or recommendations for specific pieces of equipment, such as filter cartridges (Myers and Chrai 1982). Recently, systematic studies have been undertaken to develop quantitative relationships among the parameters affecting sterilization. Young and Ferko (1992) have shown that the tube orientation, with respect to the gravitational vector, is critical. Young (1993a) has shown that the tube diameter has significant effects on sterilization. Using simple model systems of pipes and tanks, Noble (1992) has studied the effects of such parameters as pipe insulation, steam trap spacing, steam velocity, and scale factors for piping on steam SIP. He used steady state models for the analysis of tanks and interconnecting lines, but was unable to develop analytical solutions for deadleg models.

Design engineers have to rely on information accumulated from experience and historical guidelines, such as the aspect ratio (i.e., length to diameter ratio commonly referred to as $L/D$). A maximum $L/D$ of 6 has been quoted as being appropriate for SIP. As best as can be determined, this initiated from the U.S. Food and Drug Administration (FDA) (1976) citation of deadleg maximum $L/D$s of 6 for equipment subjected to clean-in-place (CIP). Young et al. (1994) and Young and Lasher (1995) have recently correlated microbiological sterilization data with the critical dimensionless parameters governing buoyancy-driven convective flow within dead-ended tubes and showed that no single parameter, such as $L/D$, can be used as a guideline for the steam SIP of deadlegs.

Many SIP cycles are developed after equipment is operational through trial and error testing with biological indicators

BIs). Heat-resistant bacterial spores are placed at the most difficult-to-sterilize locations within the equipment, and cycles are
run for varying lengths of time with the number of surviving
spores or number of positive BIs determined. From these data,
the time to kill all spores is calculated, and an appropriate safety
factor is added to yield the recommended sterilization exposure
time. If many runs are required, microbiological testing can become extremely expensive and time consuming.

Steam SIP is governed by microbiological, heat transfer,
and mass transfer principles, as well as knowledge of equipment
geometry and size. Understanding the principles governing air
displacement by steam is an essential element of SIP and can aid
in equipment design, minimize equipment modifications, reduce
microbiological testing, and reduce the time required to get
equipment into operation.

If all the air is removed such that saturated steam conditions
exist, then standard sterilization principles can be applied. The
kinetics of saturated steam sterilization are well understood and
documented. The process can be modeled as a first-order reaction with D values and Z values used to describe the rate and
temperature dependence of the reaction (Pflug 1987). The partial
removal of air results in mixtures of steam and air with varying
temperatures and steam concentrations. Sterilization at fixed
temperatures but with varying water vapor concentrations has
been studied but is less well understood and documented than
saturated steam sterilization (Murrell and Scott 1957; Murrell
and Scott 1966). Quantitative models are not currently available.

If temperature and water vapor concentration could be
monitored at specific locations, the time required for sterilization
could be estimated instantaneously. Unfortunately, sensors are
not available for monitoring water vapor concentrations at the
temperatures, pressures, and humidity levels required for steam
SIP. Therefore, BIs must be used to assure that sterilization conditions exist.

Steam SIP of deadlegs is a complex physical process with
time-varying mixing of a binary mixture within a cavity (i.e., the
tube). The mixture consists of a condensable vapor (steam) and a
noncondensable gas (air). Air displacement and subsequent

sterilization result from fluid density changes caused by temperature and solutal variations in the air/steam mixture within the tubes. Most studies concerning thermosolutal convection have examined free convection from vertical flat plates (Ostrach 1980), while comparatively few studies have considered flow patterns and characteristics of heat and mass transfer within cavities (Lai and Ramsey 1987; Weaver and Viskanta 1991a; Weaver and Viskanta 1991b), which exhibit more complex features than nonconfined flow adjacent to vertical surfaces. Flow patterns depend on the orientation of density gradients with respect to the gravitational vector, as well as whether buoyant forces due to temperature and concentration gradients augment or oppose each other.

Understanding and modeling steam SIP of deadlegs is further complicated by condensation of one component (steam) of the binary mixture on the tube wall. Empirical expressions have been developed for the convective heat transfer coefficient, $h$, associated with the condensation of a pure vapor such as steam on a vertical plate (Chin 1961; Sparrow and Gregg 1959), but $h$ can vary from 24 to 2500 W/m²-°C as the mixture composition varies from all air to all steam. Heat transfer reductions of over 50 percent can be brought about by the presence of 2–5 percent mass fraction noncondensable air during free convection (Sparrow and Lin 1964; Mori and Hijikata 1973). This results from a buildup of noncondensable air near the condensation site, which reduces the partial pressure of steam and the temperature at which it condenses. A resistive layer of air and condensate prevents steam from moving to the wall. If forced flow is present and condensate minimal, heat transfer may be substantially higher than predicted by free convective theory (Al-Diwany and Rose 1973; Sparrow et al. 1967).

Complexities of the physical process help to explain the difficulties encountered when trying to develop simple guidelines for the sterilization of deadlegs. This chapter is intended to provide the necessary information to engineers, microbiologists, and sterilization scientists to design, troubleshoot, and validate steam SIP processes. An overview of steam sterilization principles is included as the basis for the discussion of deadleg sterilization.

## STEAM STERILIZATION PRINCIPLES

Steam sterilization is the best understood and most dependable process for the destruction of microbial life. It is the preferred method of SIP for equipment that can withstand the required high temperatures of 120–135°C. Although there may be some dispute as to the mechanism of kill on the molecular level, universal agreement exists as to the three critical parameters effecting sterilization (time, temperature, and moisture). All activities related to steam SIP should be directed toward assuring that an adequate amount of moisture at the proper temperature be delivered for the required time to all sites requiring sterilization. This is critical during the design and testing of vessels, piping, and SIP cycles.

One of the most difficult geometries in which to assure adequate conditions for steam SIP is deadlegs. Consequently, each of the three critical parameters are briefly discussed below. More detailed information has been given by Joslin (1983) and Young (1993b).

### Time

When a large population of bacterial cells, spores, or viruses is exposed to saturated steam at a fixed temperature, the number of survivors can be determined as a function of exposure time. All organisms do not die at the same time, even when great efforts are taken to assure homogeneity of the exposed population. A linear or near linear curve, survivor curve, results when the logarithm of the number of survivors is plotted against exposure time at a constant temperature (Figure 3.1). Deviations from this linear kill curve on semilogarithmic graphs generally occur at the initial portion of the kill curve. For spores, this initial "lag" or "shoulder" has been attributed to the transformation of viable spores from their dormant state to the initiation of germination (Busta and Ordal 1964; Keynan 1964). This activation is not observed in survivor curves for vegetative cells.

Understanding and quantifying the survivor curve is critical in equipment design and in the design and validation of the

**Figure 3.1.** D value model for kill curve extrapolated to show probability of sterility.

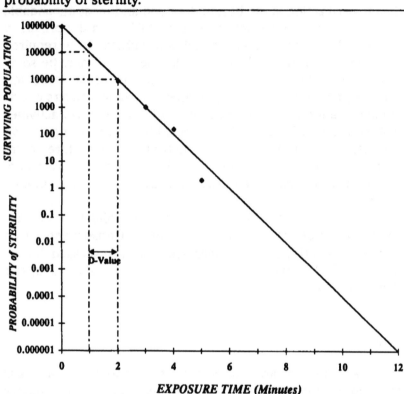

steam SIP process because it allows the determination of process efficacy and the establishment of quantifiable safety factors. When a linear survivor curve results from plotting the number of surviving microorganisms, $N(t)$, at any time $t$ on a semilogarithmic graph, $N(t)$ is given by:

$$N(t) = N_0 10^{-t/D} \qquad (1)$$

where $N_o$ = initial population and $D$ = D value.

The D value is proportional to the reciprocal of the semilogarithm survivor curve slope and is the time it takes to reduce the surviving population by 90 percent or 1 log when exposed to saturated steam at a constant temperature. For example, in Figure 3.1, the D value is 1.0 minute since the number of surviving spores is reduced from 100,000 ($10^5$) to 10,000 ($10^4$) in 1.0 minute. The surviving population can be estimated if $N_o$ and the D value are known. For example, with an initial population of $10^6$ and a 1.0 minute D value, one would expect 10 organisms to survive after 5.0 minutes and 1 organism would be expected after 6.0 minutes (Figure 3.1).

D values allow a direct comparison of the heat resistance of various organisms. Since a D value can be determined at any temperature, a subscript is used to designate the exposure temperature (i.e., $D_{121C}$). $D_{121C}$ values for *Bacillus stearothermophilus* spores, which are typically used to monitor steam sterilization processes, range from 1.0 to 3.0 minutes. These spores are one of the most resistant to steam sterilization. Most spores of obligate aerobic bacilli have $D_{121C}$ values less than 1.0 minute. Heat-resistant, nonspore-forming bacteria, yeasts, and molds have such small $D_{121C}$ values that they cannot be measured experimentally. Therefore, D values at lower temperatures are given. Typical $D_{65C}$ values for nonspore formers are in the range of 1.0 minute (Stumbo 1965).

Extrapolation of a D value survivor curve model below 1 surviving microorganism permits a quantitative determination of sterilization process efficacy. With an initial population of $10^6$ per location and 1.0 minute D value, one would expect 1 organism to survive after 6.0 minutes (Figure 3.1). Consequently, all locations with this initial population of $10^6$ would be unsterile after a 6.0 minute exposure. If the exposure time was increased to 7.0 minutes, 0.1 of an organism would be expected to survive at each location (Figure 3.1). Obviously, an organism either survives or is killed—a fraction of an organism cannot exist. Therefore, we need to interpret this value in a different fashion.

If 10 locations, each containing $10^6$ organisms, were exposed for 7.0 minutes, we would expect 1 organism to survive since the total number of organisms initially was $10^7$ (10 times $10^6$) and 7 logs would be killed in 7.0 minutes. The 1 surviving organism would have to be at 1 of the locations and the other 9 locations would contain no survivors. Therefore, 1 out of the 10 locations (0.1 or 10 percent) would be unsterile. The probability of a location being unsterile or the probability of at least 1 organism surviving is 0.1. Consequently, if the predicted number of surviving organisms per carrier is less than 1, this value can be interpreted as the probability of at least 1 organism surviving.

For terminal steam sterilization processes, $10^{-6}$ is the recommended probability of survival for bioburden on medical devices and in pharmaceuticals. This is also referred to as a sterility assurance level (SAL) of $10^{-6}$. A process should be so effective as to assure that less than one item in one million might be unsterile. This same concept can be used when validating a steam SIP.

D values are determined in specially designed test vessels, BIER vessels, that minimize the time to reach temperature and assure maintenance of constant, uniformly saturated steam conditions during exposure. The use of a D value to characterize the kill curve resulting from the steam SIP within process equipment may be inappropriate because the temperature may not be constant, and saturated steam conditions may not exist throughout the equipment during the exposure time. Figure 3.2 shows actual surviving population determined by serial dilutions and plate counts from a location 5.4 cm up an 1.0 cm inside diameter (ID), vertically orientated tube with a total length of 7.8 cm. No significant decrease in population occurs until approximately 50 minutes of exposure at 122.4°C. Linear regression analysis on the segment of the semilogarithmic plot showing decreasing populations resulted in the straight line shown. The slope of this line can be used to determine the time required to reduce the populations by 1 log. This is termed the cycle log reduction (CLR) time. The time at which the calculated line intersected the initial population value is termed the time to start of kill (TSK). The calculation of CLR time and TSK allows characterization of the kill kinetics at a specific location in process equipment where the temperature may not be constant.

## Temperature

Temperature dramatically effects the rate of kill, as evidenced by D value survivor curves for *Bacillus stearothermophilus* (Figure 3.3). At 121°C, 12.0 minutes would be required for a 6 log reduction, whereas this time could be reduced to 0.75 minute by using

**Figure 3.2.** Use of CLR time and TSK to characterize a typical kill curve in a deadleg.

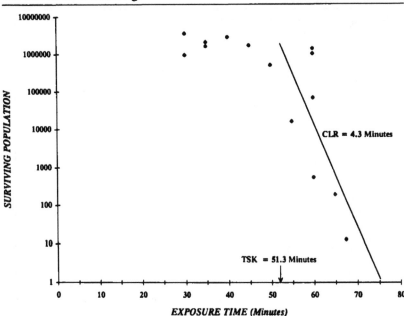

steam at 130°C. The logarithm of the D value linearly decreases with increasing temperature, and the Z value is used to describe this decrease. The Z value is the temperature change resulting in a 90 percent or 1 log change in D value. The D value at any temperature $T$ is given by

$$D_T = D_{ref} 10^{(T_{ref} - T)/Z} \qquad (2)$$

where $D_T$ = D value at temperature $T$, $D_{ref}$ = D value at some reference temperature $T_{ref}$, and $Z$ = Z value. For example, the D value at 110°C for a microorganism with a $D_{121C}$ of 1 minute and a Z value of 8°C is 23.7 minutes.

Typical Z values for bacterial spores range from 8.3 to 16.8°C (Perkins 1969). A Z value of 10.0°C is the standard value used for *Bacillus stearothermophilus* (PDA 1978). The criticalness

**Figure 3.3.** Effects of temperature on D value for *Bacillus stearothermophilus.*

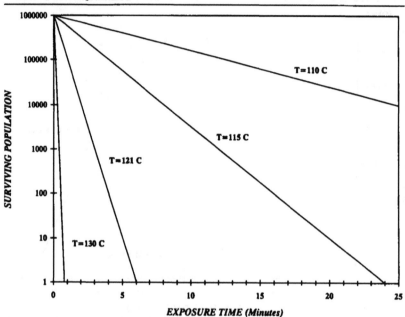

of temperature control in steam sterilization is illustrated by considering the effect of a 1.0°C temperature decrease on biological kill at 121.0°C. For an organism with a Z value of 10°C, this 1°C temperature decrease results in a 25 percent decrease in lethality.

D value and Z value models allow quantitative prediction of the effects of time and temperature, respectively, on microbial kill. The kill kinetic or lethality at a specific location can be predicted if these values and the temperature-time profile is known. This calculation requires that the temperature be that of saturated steam at all times.

The log decrease in population, $\Delta P_{log}$, or total predicted microbial kill when exposed to constant temperature $T$ for time $t$ can be calculated by

$$\Delta P_{\log} = \frac{t 10^{(T-T_{ref})/Z}}{D_{ref}} \tag{3}$$

For a time-varying temperature profile, the total $\Delta P_{\log}$ for the cycle can be calculated by integrating equation 3 over the cycle duration

$$\Delta P_{\log} = \frac{\int 10^{(T(t)-T_{ref})/Z} dt}{D_{ref}} \tag{4}$$

where $T(t)$ = actual time-varying temperature profile.

Usually, the continuous time-temperature profile is approximated by a series of discrete changes in temperature. The above integral is approximated by

$$\Delta P_{\log} = \frac{\sum_{i=1}^{n} t_i 10^{(T_i-T_{ref})/Z}}{D_{ref}} \tag{5}$$

where $t_i$ = duration in minutes of the $i$th step temperature change, $T_i$ = temperature of the $i$th discrete temperature step, and $n$ = total number of discrete temperature steps used in the approximation.

The predicted surviving population or SAL can then be determined from

$$SAL = N_o - 10^{\Delta P_{\log}} \tag{6}$$

## Moisture

The effect of moisture on heat sterilization is evidenced by the difference in temperature required for dry heat versus saturated steam processes. Typically, dry heat processes are run at 160°C or higher, whereas saturated steam process temperatures vary from 115 to 135°C. The enhanced kill in moist heat processes has been attributed to the effect of moisture on reducing the temperature required to denature or coagulate proteins. In the case of dry heat sterilization, microorganisms are dehydrated before the temperature rises sufficiently to cause protein denaturing. Consequently, dry heat is an oxidation process with different kinetics than moist heat or steam sterilization.

It has be demonstrated that spores are highly permeable. The free exchange of water occurs between a spore and its environment. Therefore, the water activity of a spore changes with the relative humidity of the environment. Although dry heat and steam represent the two limits of allowable moisture or water vapor concentration, varying conditions exist between these two states. Dry heat sterilization conditions can be characterized by a relative humidity or water activity, $A_w$, of 0, whereas steam corresponds to a relative humidity of 100 percent or $A_w$ of 1.0. Murrell and Scott (1966) and Angelotti et al. (1968) have studied the effect of relative humidity or $A_w$ on *Bacillus stearothermophilus* at 120°C. The D value was greatest at an $A_w$ of approximately 0.3. Further increases in $A_w$ result in decreases in heat resistance. These effects can result in over a 100-fold change in the D value.

During dry heat sterilization processes, the inital water activity of spores at room temperature can be high, but this temperature is too low for moist heat sterilization to occur. As the temperature is increased, the water activity decreases dramatically. The water activity is so low by the time moist heat sterilization temperatures are reached that no kill occurs. Due to the low water activity, the temperature must be further increased until dry heat sterilization temperatures are reached.

The greatest impediment to steam sterilization is inadequate air removal. Since air and steam do not readily mix, air significantly inhibits steam penetration and its associated moisture. All of the air does not have to be removed, but all surfaces requiring sterilization must be exposed to adequate moisture to allow the denaturing of proteins. Air can be removed by active means, such as pulling a vacuum or passive gravity displacement, whereby steam displaces the air. At sterilization temperatures, air is 1.6 times denser than steam and will be displaced downward. Therefore, one would expect the time required to sterilize deadlegs to depend on orientation. In addition, condensation will pool at the lowest site, and this liquid can inhibit heating and the achievement of desired temperature.

Unlike time and temperature that can be easily measured, moisture conditions surrounding a microorganism cannot be directly determined. The use of a BI is the only method currently available to assure that adequate moisture is present. A general indication of the quantity of air within the vessel can be obtained by comparing the vessel pressure to the saturated steam pressure calculated from the average vessel temperature. If the actual vessel pressure is greater than the calculated saturated pressure, the difference is due to the partial pressure of air remaining in the vessel. The presence of air also results in large temperature variations during empty vessel temperature distribution studies.

Good agreement between measured and saturated pressure does not assure uniform saturated steam conditions because small, localized air pockets can exist. For example, if $0.01 \text{ m}^3$ of air remained in a $5.0 \text{ m}^3$ vessel, the total pressure would be only 0.06 psi above the saturated steam pressure. This pressure difference would not be detected due to inaccuracies in temperature and pressure instrumentation. Sterility would be achieved if the air was thoroughly mixed with the steam. If the air existed unmixed in localized areas, sterility would not be achieved.

## FACTORS AFFECTING STEAM SIP

The factors most often resulting in steam SIP failures are condensate accumulation and air entrapment. Condensate accumulation can result in inadequate temperature, while air entrapment can result in inadequate moisture. Therefore, one would expect the orientation of deadlegs with respect to the gravitational vector, deadleg length, and deadleg ID to be critical parameters effecting steam SIP. It is desirable that the effect of each factor be understood when designing equipment or validating SIP processes. If quantitative guidelines are to be established for the steam SIP of deadlegs, this information is essential.

Quantitative thermal and microbiological studies have been conducted (Young and Ferko 1992; Young 1993a) using the test fixture shown in Figure 3.4 in which tubes of different IDs and lengths could be oriented at various angles with respect to the gravitational vector. Thermocouples and BIs were attached to a nylon string that ran from the top of the tube, down along the centerline, and exited the fixture. This allowed the determination of temperatures and biological kill rates as a function of position within the deadleg for different IDs, tube lengths, and orientations. Saturated steam with an average temperature of 122.4°C

**Figure 3.4.** Schematic representation of test fixture used to determine the effects of orientation, ID, and tube length on sterilization.

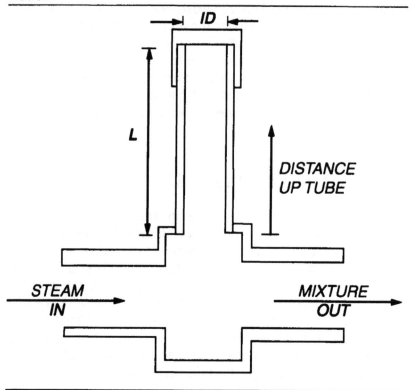

entered the fixture at a flow rate of 0.38 ± 0.03 gram per second. BIs consisted of 0.6 cm filter paper disks inoculated with $6 \times 10^5$ to $1.3 \times 10^6$ *Bacillus stearothermophilus* spores with a $D_{121C}$ of 3.4 minutes.

## Deadleg Orientation

Since air is denser than steam at sterilization temperatures, air will be displaced downward in the deadleg. Therefore, a vertically oriented deadleg with the closed end up (vertically up orientation) will be the most favorable orientation. A vertically oriented deadleg with the closed end at the bottom (vertically down orientation) will result in the collection of air and condensate within the deadleg. Figure 3.5 shows temperature profiles from a 1.0 cm ID tube with length equal to 7.8 cm. In the vertically up orientation, the first 1.5 cm of the tube reached the sterilization temperature of 122°C immediately upon pressurization with steam (Figure 3.5). This was also true for tubes orientated 5 and 45 degrees above horizontal (Young and Ferko 1992). Air within the tube is compressed on pressurization with steam, and it can be shown that a steam-air interface should be immediately established 1.5 cm up the tube for a tube of this length. Locations above this interface initially contain air, and time is required for buoyancy-driven convective flow and diffusion to result in displacement of the air. Thirty-nine minutes were required for the location 5.4 cm up the tube to reach 122°C. This same location was only at 60°C after 44 minutes when the tube was oriented vertically down (Figure 3.5). Due to the low temperatures within the vertically down tube, steam SIP will not take place due to the accumulation of condensate and air. The tube would fill with condensate in as little as 15 minutes. In this case, a steam bleeder is required to allow a pathway for the removal of condensate and air.

Although temperatures above 115°C are achieved, this does not mean that these temperatures are those of saturated steam and that sterilization will take place. The presence of saturated steam can only be demonstrated through the use of BIs. If the measured temperatures represent saturated steam conditions,

**Figure 3.5.** Temperature profiles from 1.0 cm ID tube with length of 7.8 cm when oriented vertically up (A) and vertically down (B).

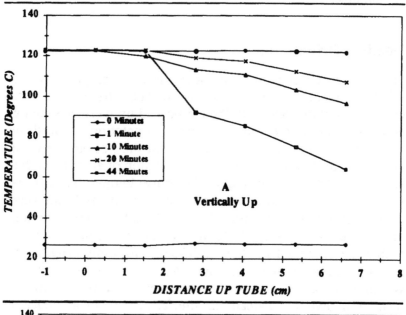

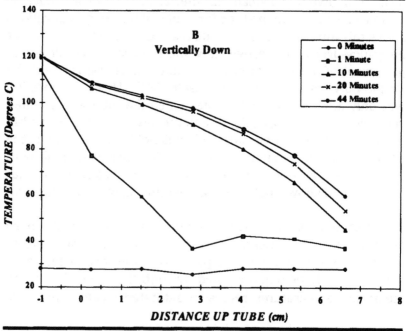

equation 6 allows the calculation of the surviving population as a function of time. Saturated steam conditions exist at that location if the calculated kill curve agrees with the microbiologically determined curve.

Figure 3.6 summarizes data from a 1.0 cm ID tube oriented vertically up. Excellent agreement exists between the microbiologically determined kill curve and that predicted by monitoring the temperature at a location 0.3 cm up the tube and assuming this temperature is that of saturated steam. This demonstrates near saturated steam conditions at this location during at least the first 20 minutes of exposure. Since this location is below the initial steam-air interface resulting from compression of air, these results are expected. Similar data were obtained at locations below the initial interface when the tube was oriented 5 and 45 degrees above horizontal.

**Figure 3.6.** Comparison of temperature predicted and biologically measured kill curves at two locations in 1.0 cm ID tube orientated vertically up.

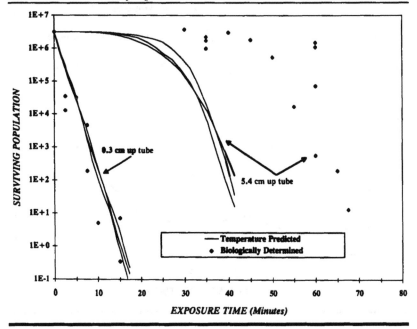

Temperature profiles predicted that kill should start approximately 15 minutes into the exposure time when BIs were located 5.4 cm up the tube (Figure 3.6). Actual kill did not start until
50 minutes into exposure. Equation 6 was not useful in predict-
ing microbiological kill. An air-steam mixture or all air existed
during the initial 50 minutes at this location. This mixture was
heated due to conduction up the tube and convection from the
tube wall to the mixture. The effect of the air was to lower the
moisture concentration and inhibit sterilization, even though the
temperature was above 121°C. After 50 minutes, enough mixing
had occurred such that saturated or near saturated steam conditions existed and kill was initiated.

Figure 3.7 shows the effect of orientation on kill within a
1.0 ID tube. CLR times were identical for all orientations at a location 0.3 cm up the tube. CLR times increased as distance up the

**Figure 3.7.** Effects of tube orientation on CLR times.

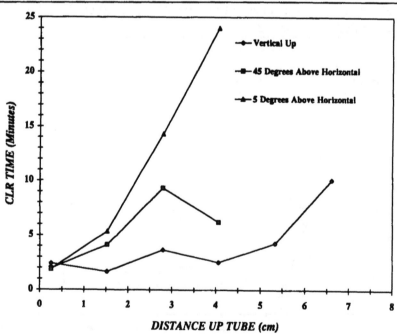

tube increased. This was most pronounced for the 5-degree orientation, where the CLR time increased by a factor of 13 over a distance of 3.8 cm. Corresponding increases for the 45-degree and vertical orientations were 3.1 and 1.04, respectively. For the vertical up position, CLR values were between 1.6 and 4.3 minutes for locations less than 5.4 cm up the tube. Given the D value of the spores used and an exposure temperature of 122.4°C, CLR times of less than 5.0 minutes indicate that saturated steam conditions of 119.4°C or higher were responsible for the kill. This implies that near saturation conditions were more easily achieved in the vertically up tube; these conditions existed further up the tube (5.4 cm up the vertical tube versus only 1.5 cm for the 45 and 5 degrees above horizontal orientations). These results are consistent with expected air displacement. In the vertical up position, buoyancy forces due to differences in density between the steam and air result in a downward displacement of the air. In the 5-degree up position, buoyancy forces will tend to keep the air in the tube concentrated along the lower half, with the only significant driving force for air removal being diffusion.

Bruch et al. (1963) have reported a $D_{120C}$ of 19 minutes and Z value of 24.4°C for *Bacillus stearothermophilus* spores when exposed to dry heat conditions. One would expect a CLR time of approximately 15 minutes when exposed to dry conditions at the average temperature used in these studies. This indicates that the kill being achieved in a portion of the 5-degree tube is due to dry heat and not to steam. Consequently, *Bacillus subtilis* would be the appropriate microorganism for monitoring sterilization at these locations.

Determination of a recommended sterilization time requires knowledge of the initial population, CLR time, and TSK of the most resistant microorganism, as well as the desired SAL. Sterilization times calculated in this chapter are based on an initial *Bacillus stearothermophilus* spore population of $1.0 \times 10^6$ and desired SAL of $10^{-6}$. Consequently, a 12-log reduction is required.

Figure 3.8 shows the total sterilization time required to achieve a 12-log reduction in *Bacillus stearothermophilus* spores using the experimentally determined CLR times and TSK values for a 1.0 ID tube with a total length of 7.8 cm. Only the vertically oriented tube could be sterilized throughout. The required

**Figure 3.8.** Effects of tube orientation on sterilization time.

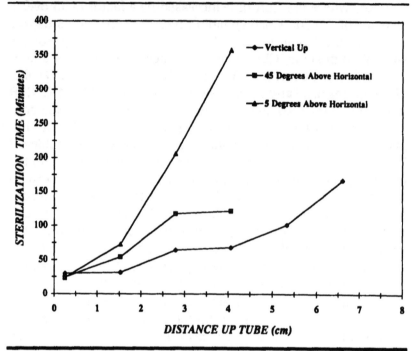

exposure time was 167 minutes. Locations greater than 4 cm up the tube, when oriented at 5 or 45 degrees above horizontal, were not sterilized in 6 hours. A steam bleeder or vacuum would be required to sterilize these tubes with 121°C steam.

## Deadleg Diameter and Length

Lacking published data on steam SIP, design engineers and validation scientists have relied on information accumulated from experience and data for the cleaning of deadlegs. The most commonly used parameter has been the ratio of length to inside diameter, which is commonly referred to as *L/D*. An *L/D* of 6 was proposed by the FDA (1976) for assuring the cleaning of deadlegs in parenteral manufacturing equipment. Seiberling (1986) proposed a value of 1.5.

Figure 3.9 shows the effects on CLR times of tube diameters from 0.4 to 1.7 cm with varying tube lengths. An 1.7 cm ID tube 18 cm in length showed CLR times less than 5 minutes throughout when oriented in the vertically up position (Figure 3.9). Only locations less than 5.4 cm up an 1.0 cm ID tube of length 7.8 cm showed CLR times less than 5 minutes. In the case of the 0.4 cm ID tube, only locations less than 1.5 cm up the 7.8 cm tube had CLR times of 5 minutes or less. The 23 cm long tube with 1.7 cm ID had CLR values throughout that were less than those observed at the ends of 0.4 and 1.0 cm ID tubes that were 7.8 cm long.

When tubes were oriented 5 degrees above horizontal (Figure 3.9), only the 1.7 cm ID tube with a length of 7.8 cm showed saturated steam kill throughout. For the 1.0 ID tube, steam kill occurred only 1.5 cm up the tube. Saturated steam conditions were not observed at any significant distance up the 0.4 cm ID tube.

Tube diameter had a significant effect on sterilization time (Table 3.1). The 0.4 cm ID tube with a length of 7.8 cm could not be sterilized in the vertical position, whereas 167 and 50 minutes were required for the 1.0 and 1.7 cm ID, respectively. 1.7 cm ID tubes with lengths up to 27 cm were sterilized in less than 245 minutes when oriented vertically up. Orienting tubes in the near horizontal position resulted in only the 1.7 cm tube with a length of 7.8 cm being sterilized throughout, with 63 minutes required for sterilization.

No trend was observed between $L/D$ and sterilization time. For example, a 1.0 cm tube with a $L/D$ value of 7.6 required 167 minutes to sterilize, whereas a 1.7 cm tube with $L/D$ of 7.8 required only 65 minutes (Table 3.1). A 7.8 cm long tube with 1.0 cm ID required the same time to sterilize as a 23 cm long tube with 1.7 cm ID. The $L/D$ values were 7.8 and 13.5, respectively. $L/D$ values alone do not provide a general guideline that can be used to predict sterilization.

## Flow Patterns Within Deadlegs

Air initially within the deadlegs is compressed on pressurization with saturated steam and can be displaced by buoyant-driven

**Figure 3.9.** Effects of ID and length on CLR times when tubes orientated vertically up (A) and 5 degrees above horizontal (B).

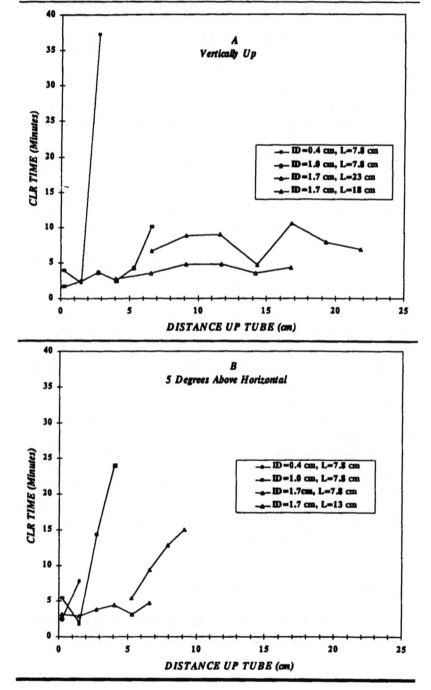

Table 3.1. Effects of ID, Length, and L/D on Sterilization
Times for Two Tube Orientations

| Tube Inside Diameter (cm) | Tube Length (cm) | L/D | Sterilization Time (Minutes) |
|---|---|---|---|
| Vertically Up | | | |
| 0.4 | 7.8 | 19.5 | NS |
| 1.0 | 6.8 | 6.8 | 85 |
| 1.0 | 7.8 | 7.8 | 167 |
| 1.7 | 7.8 | 4.6 | 50 |
| 1.7 | 13 | 7.6 | 65 |
| 1.7 | 18 | 10.6 | 71 |
| 1.7 | 23 | 13.5 | 161 |
| 1.7 | 27 | 15.9 | 245 |
| 5 Degrees Above Horizontal | | | |
| 0.4 | 7.8 | 19.5 | NS |
| 1.0 | 7.8 | 7.8 | NS |
| 1.7 | 7.8 | 4.6 | 63 |
| 1.7 | 13 | 7.6 | NS |

NS: Not Sterilized

convective flow, simple diffusion, or a combination of both. Air displacement resulting from diffusion is independent of tube orientation, whereas convective flow depends on orientation with respect to the gravitational vector.

Convective flow due to buoyant forces is caused by changes in density resulting from temperature and concentration (i.e., solutal) variations within the air-steam mixture. Temperature and solutal gradients both act to displace air in the direction of the gravitational vector, thereby resulting in more rapid air removal from vertical than horizontal tubes. Viscous forces oppose and tend to damp out the buoyant-driven flow. Since initial buoyant forces depend on temperature and concentration, not tube diameter, whereas viscous dissipation forces increase with decreasing tube diameter, buoyant-driven flow is more readily damped out in smaller diameter tubes.

In the case of 0.4 cm ID tubes, Young et al. (1994) have shown buoyant forces to be damped out; air displacement results primarily from diffusion. Temperature profiles for vertical and near horizontal orientations are nearly identical. Biological data in Figure 3.9 show little effect of orientation on CLR times.

For tubes greater than 0.4 cm ID, buoyant-driven convective flow becomes important. Temperatures below the initial interface show a larger temperature decrease followed by an increase, with final temperatures throughout the tube being higher in the larger diameter tubes. Decreases in temperature below the initial interface result from cooler air being displaced down the tube and mixing with steam. Enough air-steam mixing can occur near the interface such that saturated steam conditions may not be maintained, resulting in an ineffective mixture composition for sterilization.

Steam and air at the centerline of the tube are warmer than at the tube walls. Consequently, steam at the center of the tube near the initial interface rises; air along the tube walls flows downward (Young and Lasher 1995). This sets up counterrotating cells. Higher velocities are present in larger diameter tubes, causing more rapid transport of steam up the tube and subsequent displacement of air. Since tube walls have less of an effect on constraining buoyant-driven flow in vertical tubes, one would expect increasing tube length to have less of an effect on sterilization time for vertical than horizontal tubes, which was the case.

In the case of horizontal tubes, buoyant forces drive the more dense air perpendicular to the tube axis. This results in a series of counterrotating vortices whose diameters are constrained by the tube walls. Larger diameter tubes have larger vortices that result in more rapid displacement of air. A comparison of CLR times in Figure 3.9 for horizontally oriented tubes shows better air displacement in the 1.7 and 1.0 than 0.4 cm tubes, thereby indicating presence of buoyancy-induced flow that increases with increasing diameter.

The Grashof number is a ratio of buoyant to viscous forces acting on the air-steam mixture. It is customary to define both a thermal and solutal Grashof number, but an overall Grashof number, $Gr_{OV}$, equal to the sum of the thermal and solutal values can be used since the temperature and concentration induced buoyant effects augment each other. $Gr_{OV}$ is given by

$$Gr_{OV} = \frac{\rho_o^2 g D^3 (\beta_C \Delta C + \beta_T \Delta T)}{\mu^2} \tag{7}$$

where $\rho_o$ = mixture density at reference temperature and concentration, $g$ = gravitational constant, $D$ = tube diameter, $\mu$ = mixture dynamic viscosity, $\beta_C$ = mixture coefficient of concentration expansion, $\beta_T$ = mixture coefficient of temperature expansion, $\Delta T$ = change in temperature, $\Delta C$ = change in mass fraction of air.

$Gr_{OV}$ is proportional to the diameter cubed, thereby indicating a substantial increase in convective flow as tube diameter is increased. For saturated steam sterilization at 123°C, $Gr_{OV}$ for the 0.4, 1.0, and 1.7 cm ID tubes can be shown to be 460, 7926, and 38,943 respectively. This accounts for the increased ease of sterilizing the larger diameter tubes.

## SIP Guidelines

Steam SIP of deadlegs is dependent on tube diameter, length, and orientation with respect to the gravitational vector. All dead-ended tubes should be sloped for the drainage of condensate and positioned as upright as possible. Increasing tube diameter had the greatest effect on sterilization as it increased the ratio of buoyant forces to viscous dissipation forces. Tubes 0.4 cm ID and 7.8 cm long exhibited little buoyant-driven convective flow. The minimal air displacement observed was due primarily to diffusion. Saturated steam sterilization did not occur at any location above the initial steam-air interface. When these small diameter tubes are used, they should be equipped with an active aid for air removal, such as a steam bleed or vacuum cycle.

Saturated steam sterilization conditions are possible in 1.0 cm ID and larger deadlegs when positioned vertically. A sterilization time of 167 minutes, as defined by a 12-log reduction in *Bacillus stearothermophilus* spore population, could be achieved in 1.0 cm ID vertical tubes 7.8 cm long. Increasing tube diameter from 1.0 to 1.7 cm resulted in vertical tubes 3.5 times as long, 27 cm, being sterilized in 1.5 times longer, 245 minutes. Only the 7.8 cm long, 1.7 cm ID tube could be sterilized when positioned 5 degrees above horizontal. Consequently, tubes should be as large in diameter as practical. The initial cost of installing larger openings and tubing should offset the cost of installing and maintaining steam bleeds and vacuum systems. Shorter sterilization times, ease of validation, and greater sterility assurance favor the larger diameter deadleg geometry.

Increasing the sterilization temperature has many advantages for equipment that can withstand the higher pressure and temperature. Greater buoyancy forces result from a decrease in the density of the steam. The steam-air interface is further up the tube due to the increased sterilization pressure, and the rate of biological kill in saturated steam increases logarithmically with increasing temperature. The use of higher temperatures and pressures should be considered for complex equipment with small diameter deadlegs.

In many cases, CIP precedes SIP and requires that air be displaced with the liquid cleaning agent. This has resulted in the recommendation of a near horizontal orientation of deadlegs (Seiberling 1986). Increasing tube diameter should increase both CIP and SIP efficacy, but compromises on orientation will probably have to be made to accommodate both.

$L/D$ values do not provide a general guideline that can be used to predict sterilization. Tubes with similar $L/D$s, but different diameters, showed sterilization times varying up to 250 percent. Young and Lasher (1995) used results of microbiological spore testing and analysis of the governing conservation equations to identify critical parameters and their influence on sterilization of commonly used deadleg geometries. Nondimensionalizing the z-momentum equation resulted in identification of $L/D$ and $Gr_{OV}$ as governing variables. $Gr_{OV}$ was indicative of the magnitude of buoyancy-driven convective flow and was proportional to $D^3$. Convective flow was minimal in small diameter tubes with $Gr_{OV}$ less than $4 \times 10^3$ ($D$ less than 0.4 cm); air displacement resulted from binary diffusion. Increasing tube diameter to 1.7 cm resulted in a 77-fold increase in $Gr_{OV}$ and rapid air displacement.

Developing SIP guidelines requires the establishment of criteria to assure reproducible sterilization within a reasonable period of time. A maximum CLR time of 5 minutes for *Bacillus stearothermophilus* spores was established as the criteria when sterilizing with saturated steam between 121 and 123°C. This assured the air mass fraction to be less than 0.15 throughout the tubes and resulted in sterilization times less than 100 minutes for a 12-log reduction. For vertical tubes, acceptable *L/D* values increased from 7 to 12.2 as $Gr_{OV}$ increased from 0.7 to 2.5 × 10$^5$ (Young and Lasher 1995). This corresponds to a maximum acceptable tube length increasing from 7.0 to 20.7 cm as tube diameter increased from 1.0 to 1.7 cm.

## SUMMARY

The time required for SIP of deadlegs is dependent on tube diameter, length, and orientation with respect to the gravitational vector. Deadlegs must be oriented such that condensate and air is removed. Vertical up orientations are preferred. Deadlegs that are oriented downward require steam bleeders. For systems requiring both CIP and SIP, selection of orientation with respect to gravity will require compromises as CIP is best accomplished when deadlegs are oriented downward.

Increasing tube diameter significantly reduces sterilization time as it increases the ratio of buoyant to viscous dissipation forces. Tubes with IDs 0.4 cm and less exhibit little buoyant-driven convective flow and the minimal air displacement that does occur results from diffusion. Sterilization does not occur above the initial steam/air interface in 0.4 cm ID tubes 7.8 cm long. A sterilization time of 71 minutes, as defined by a 12-log reduction in *Bacillus stearothermophilus* spore population, can be achieved for 1.7 cm ID vertical tubes 18.0 cm long. 1.7 cm ID vertical tubes as long as 27 cm can be sterilized in less than 245 minutes. Only 1.7 cm ID tubes with lengths less than 7.8 cm can be sterilized when positioned 5 degrees above horizontal.

Increasing tube diameter from 1.0 to 1.7 cm results in vertical tubes twice as long being sterilized in 45 percent of the time.

$L/D$ values alone do not provide a general guideline that can be used to predict sterilization of deadlegs. Tubes with similar $L/D$ values but different diameters exhibit sterilization times varying up to 250 percent. Nondimensionalization of the z-momentum equation results in $L/D$ and $Gr_{OV}$ being identified as the parameters with the best potential for providing a general guideline. $Gr_{OV}$ is indicative of the magnitude of buoyance-driven convective flow and is proportional to the diameter cubed; thereby explaining the sensitivity of sterilization time to ID.

## REFERENCES

Agalloco, J. 1990. Steam sterilization-in-place. *J. Parenter. Sci. Technol.* 44 (5):253–256.

Al-Diwany, H. K., and J. W. Rose. 1973. Free convection film condensation of steam in the presence of non-condensing gas. *Int. J. Heat Mass Transfer* 16:1359–1369.

Angelotti, R., J. H. Maryanski, T. F. Butler, J. T. Peeler, and J. E. Campbell. 1968. Influence of spore moisture content on the dry heat resistance of *Bacillus subtilis* var. *niger. Appl. Microbiol.* 16:735–745.

Berman, D., T. Myers, and C. Suggy. 1986. Factors involved in cycle development of a steam-in-place system. *J. Parenter. Sci. Technol.* 40 (4): 119–121.

Bruch, C. W., M. G. Koesterer, and M. K. Bruch. 1963. Dry heat sterilization: Its development and application to components of exbiological space probes. *Dev. Ind. Microbiol.* 4:334–342.

Busta, F. F., and Z. J. Ordal. 1964. Heat-activation kinetics of endospores of *Bacillus subtilis. J. Food Sci.* 29:345–353.

Chin, M. M. 1961. An analytical study of laminar film condensation: Part 1-flat plate. *J. Heat Transfer* 83:48–54.

FDA. 1976. Current good manufacturing practice in manufacture, processing, packing, and holding of large volume parenterals. *Federal Register* 41 (106):22207.

Joslin, L. J. 1983. Sterilization by heat. In: Disinfection, Sterilization and Preservation, 3rd ed. edited by S. S. Block. Philadelphia: Lea and Febiger, pp. 3–46.

Keynan, J. H. 1964. Transformation of a dormant spore into a vegetative cell. In: *Spores III Symposium.* Ann Arbor, MI: American Society of Microbiology.

Lai, W. T., and J. W. Ramsey. 1987. Natural heat and mass transfer in a rectangular enclosure. In: *Natural Circulation,* vol. 92 edited by J. H. Kim and Y. H. Hassan. New York: ASME, pp. 361–372.

Mori, Y., and K. Hijikata. 1973. Free convective condensation heat transfer with noncondensable gas on a vertical surface. *Int. J. Heat Mass Transfer* 16:2229–2240.

Murrell, W. G., and W. V. Scott. 1957. The heat resistance of bacterial spores at various water activities. *Nature* 179:481–485.

Murrell, W. G., and W. V. Scott. 1966. The heat resistance of bacterial spores at various water activities. *J. Gen. Microbio.* 43:411–425.

Myers, T., and S. Chrai. 1981. Design considerations for development of steam-in-place sterilization processes. *J. Parenter. Sci. Technol.* 35:8–12.

Myers, T., and S. Chrai. 1982. Steam-in-place sterilization of cartridge filters in-line with a receiving tank. *J. Parenter. Sci. Technol.* 36 (3): 108–112.

Noble, P. T. 1992. Modeling transport processes in sterilization-in-place. *Biotechnol. Prog.* 8 (4):275–284.

Ostrach, S. 1980. Natural convection with combined driving forces. *PhysicoChem. Hydro.* 1:233–247.

PDA. 1978. Validation of steam sterilization cycles. Technical Monograph No. 1. Philadelphia: Parenteral Drug Association.

Perkins, J. J. 1969. *Principles and methods of sterilization in health sciences,* 2nd ed. Springfield, IL: Charles C. Thomas, p. 86.

Pflug, I. J., and R. G. Holcomb. 1983. Principles of thermal destruction of microorganisms. In: Disinfection, Sterilization and Preservation, 3rd ed. edited by S. S. Block. Philadelphia: Lea and Febiger, pp. 751–810.

Pflug, I. J. 1987. *Textbook for an introductory course in the microbiology and engineering of sterilization processes.* Minneapolis, MN: Environmental Sterilization Laboratory.

Seiberling, D. A. 1986. Clean-in-place and sterilize-in-place application in the parenteral solution process. *Pharm. Eng.* 6 (6):30–35.

Sparrow, E. M., and J. L. Gregg. 1959. A boundary layer treatment of laminar film condensation. *J. Heat Transfer* 81:13–18.

Sparrow, E. M., and S. H. Lin. 1964. Condensation heat transfer in the presence of a noncondensable gas. *J. Heat Transfer* (August): 430–436.

Sparrow, E. M., W. J. Mincowycz, and M. Saddy. 1967. Forced convection condensation in the presence of noncondensables and interfacial resistance. *Int. J. Heat Mass Transfer* 93:1829–1838.

Stumbo, C. R. 1965. *Thermobacteriology in food processing*. New York: Academic Press.

Weaver, J. A., and R. Viskanta. 1991a. Natural convection in binary gases due to horizontal thermal and solutal gradients. *J. Heat Transfer* 113:141–147.

Weaver, J. A., and R. Viskanta. 1991b. Natural convection due to horizontal temperature and concentration gradients-variable thermophysical property effect. *Int. J. Heat Mass Transfer* 34:3107–3120.

Young, J. H. 1993a. Sterilization of various diameter dead-ended tubes. *Bioengineering and Biotechnology* 42:125–132.

Young, J. H. 1993b. Sterilization with steam under pressure. In: *Sterilization technology—a practical guide for manufacturers and users of health care products,* edited by R. F. Morrissey and G. B. Phillips. New York: Van Nostrand Reinhold.

Young, J. H., and B. L. Ferko. 1992. Temperature profiles and sterilization within a dead-ended tube. *J. Parenter. Sci. Technol.* 46(4):117–123.

Young, J. H., and W. C. Lasher. 1995. Dimensionless parameters as design guidelines for sterilization of dead-ended tubes. *Biotechnol. Prog.* 11 (3):312–317.

Young, J. H., B. L. Ferko, and R. P. Gaber. 1994. Parameters Governing Steam Sterilization of Deadlegs. *J. Parenter. Sci. Technol.* 48 (3):140–147.

# 4

# Heat Processing for the Pharmaceutical Industry

*Vance Caudill, Ph.D.*

*Mark Kaufmann, PE*

O'Neal Engineering Inc.
Raleigh, NC

*Michael Jordan*

Lockwood Greene
Spartanburg, SC

The pharmaceutical industry has not sufficiently utilized heat processing methods to reduce microorganism loads or inactivation of reactive components. Several methods are available to heat process product at a minimum cost without affecting product quality. This is especially important within the realm of consumable and medicinal products. As federal laws become more regulatory and stringent, the technology for creating and maintaining product control via heat pasteurization or sterilization has evolved to comply with regulations. Effective heat process-ing requires the knowledge and understanding of fluid flow and heat transfer. These, with proper process definition, engineer-ing, equipment selection, and operational procedures are the critical basis for high-temperature short-time (HTST) or ultra-high temperature (UHT) processing of pharmaceuticals.

## BACKGROUND

### History of Heat Processing

The advantage of the HTST process was recognized in the early part of this century, as evidenced by early patents, which have been reviewed by Ball (1938). The first aseptic process for milk in cans was developed in Denmark (Burton et al. 1969), although the details were lost. The American Can Company developed a filling machine in 1933, called the Heat-Cool-Fill (HCF) system, which used saturated steam under pressure to sterilize cans and ends. Although technically successful, it did not survive com-mercially. In the 1940s, the Dole-Martin process was developed in which empty cans were sterilized by treatment with super-heated steam (Hersom 1985; Mek 1950). This process was later applied commercially to the aseptic packaging of split-pea soup (Havighorst 1951; Anon 1951).

The sterilization of packaged products is normally carried out by heating the product in a sealed, hermetic container for enough time to ensure that all portions of the product received a minimum heat treatment. This may be quantified by reference to a standard that takes account of time and temperature. The unit of measurement, designated Fo, is the heat treatment equivalent of one minute at 121.1°C (250°F). Since the product is heated in containers from the outside, different regions within the can re-ceive different heat treatments. To achieve an adequate Fo at the slowest heating point, other regions within the container are of-ten overprocessed. The advantage of using a heat exchanger to sterilize product uniformly is obvious. Additionally, the use of heat exchangers enables higher temperatures to be used for the sterilization process. Since there is a logarithmic relationship between time and temperature, drastic reductions in processing time are possible. For example, the rate of heat inactivation of *Bacillus stearothermophilus* spores increases 12.5 times for each 10°C rise in processing temperature (Burton 1975).

The quality advantage that accrues from using the so-called HTST or UHT process is explained by comparing the slope (z) of the thermal death time (TDT) curve for microorganisms with similar curves that depict the loss of desirable components. While the z value for bacterial spores is approximately 10°C, those for thiamine destruction (Jackson et al. 1945; Feliciotti and Esselen 1957; Mulley et al. 1975), the degradation of chlorophyll (Schanderl et al. 1962; Gupte et al. 1964), the oxidation of ascorbic acid (Hersom 1985), and the denaturation of serum proteins suggest a value averaging about 33°C. For each type of pharmaceutical product, thermal profile versus time must be developed to understand the heat process. Figure 4.1 represents two types of heat processing, aseptic and autoclave. As the figure illustrates, lethality and quality loss are a function of time and temperature and the UHT/aseptic probe is an improvement in processing conditions. Theoretically, the higher the processing temperature and the short the product exposure time the better the system will induce quality into the final product. In practice, this is limited by mechanical considerations and the danger of enzyme survival.

## Fluid Flow Characteristics

The primary requirement in pharmaceutical heat processing involves the characteristics of fluid flow within circular pipes. Fluids include both liquids and gases. The laminar flow of fluid moves in parallel elements, the direction of motion of each element being parallel to that of any other element. The velocity of any element is constant, but not necessarily the same as that of an adjacent element.

In turbulent flow, the fluid moves in elemental swirls or eddies, with the velocity (speed and direction) of each element changing with time. A violent mixing results, whereas there is no significant mixing in the case of laminar flow.

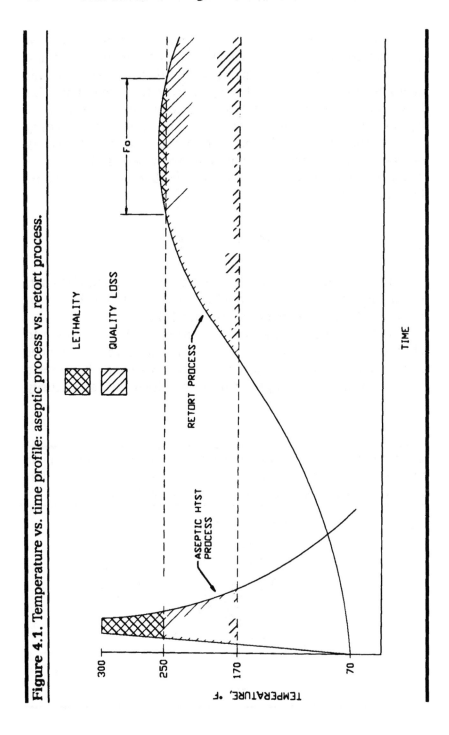

**Figure 4.1.** Temperature vs. time profile: aseptic process vs. retort process.

The distribution of velocity for laminar and turbulent flow is illustrated in Figure 4.2. Velocity is highest at the center and decreases as it approaches the surface until it reaches zero. This characteristic is true for either laminar or turbulent flow. This flow can be characterized by four variables: viscosity ($\mu$), velocity ($V$), density ($\rho$), and diameter ($D$) of the pipe. Fluid viscosity refers to the internal resistance of fluids to shear stress. Velocity relates to the flow rate or pump speed, whereas density is a product characteristic. The Reynolds number can be calculated from the above variable in this equation:

$$Re = \text{Reynolds number} = DV\frac{\rho}{\mu}$$

When Re is less than 2000, the flow is considered laminar; when it is greater than 4000, the flow is considered turbulent. In between, it is transitional flow (Harper 1979).

The Reynolds number defines the ratio of inertia force to viscous force on an element of fluid. A high Reynolds Number implies a thinner boundry layer at the pipe wall and, therefore, a lower resistance. The layer of resistance increases as viscosity

**Figure 4.2.** Velocity profiles of Newtonian flow in circular pipe where $v$ is the center velocity, $V$ is the average velocity, $E$ is the pipe roughness coefficient, and Re the Reynolds number.

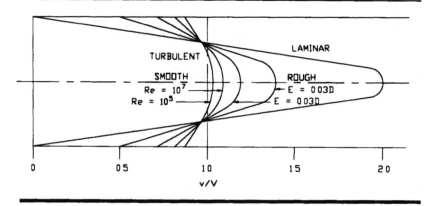

increases. A fluid is considered to be a Newtonian fluid if the relation of shearing stress to the rate of shearing strain is linear. A Newtonian fluid maintains a constant viscosity as the motion and agitation of the fluid are varied.

At the entrance of a pipe, the velocity of a fluid is uniform and constant. However, the velocity changes as it moves further into the pipe, until it is fully developed. The distance that is required for the fluid to become fully developed is the hydrodynamic entrance length ($Z_c$).

$$Z_c = 0.05 \, DRe$$

The velocity of a fully developed Newtonian fluid is represented by the follow equation:

$$V_z = V_{zmax} \frac{1 - r^2}{R^2}$$

Generally, a velocity of at least 1.52 m/s is desired for adequate cleaning in piping, regardless of pipe diameter.

As a fluid flows through a network of piping, it experiences head loss ($h_L$), given by this equation:

$$h_L = F \frac{lV^2}{D2g}$$

where $F$ is the friction coefficient, $l$ is the length of pipe section, $V$ is the velocity, $D$ is the pipe diameter, and $g$ is the force of gravity. Head loss is related to pressure change, $p$, by the equation

$$\Delta p = \gamma h_L$$

and $\gamma$ is the specific weight of the fluid. Head loss within a straight portion of pipe due to friction is called a major loss. Minor losses are those due to flow through components, such as valves, bends, and tees. In some cases, minor losses are greater than major losses.

## Heat Transfer

Within the processing industry, the aspect of heat transfer is one of great importance. This section will describe the fundamentals of heat transfer involved in pharmaceutical processing.

Heat ($Q$) is defined as the net transfer of energy across a system boundary due to a temperature difference across the boundary. Heat is positive in a system when the surroundings are at a higher temperature than the system. The SI unit for heat is the joule (J).

Heat transfer is, therefore, defined as the time rate of heat ($dQ/dt$) and has the SI unit of watts (W). Heat transfer occurs when convection of energy due to temperature difference is present, and the system boundary is at a solid-fluid interface (by the definition of heat). For convection of energy due to temperature difference to occur, the following criteria must be satisfied:

- There must be a fluid carrier in which conduction occurs.

- The fluid must be in contact with a solid carrier in which conduction is occurring.

- There must be relative motion between the fluid and the solid surface.

- A temperature difference must exist in the fluid.

Convection heat transfer is given by the equation

$$Q_n = h_c A (T_E - T_S)$$

where $h_c$ is the coefficient for convection, $A$ is the area perpendicular to the transfer, $T_E$ is the temperature of the surroundings (wall), and $T_S$ is the temperature of the system (fluid) (McMillan 1990).

There are three categories of industrial process heating applications: low temperature (below 290°C), medium temperature (290°C–590°C), and high temperature (above 590°C). There are many and varied applications of low-temperature heating processes, including pharmaceuticals.

A brief description will now be given of some variables, characteristics, and equations that may be considered when thermodynamically analyzing a process, such as pharmaceuticals. First, some of the flow characteristics such as the Reynolds number, velocity, and length required for the fluid to become hydrodynamically developed, may be studied (see section: Fluid Flow Characteristics). Next, the heat required for a certain section may be considered. This is given by the following equation:

$$q_w" = \frac{\rho VDC_p}{4L}\left(T_{bo} - T_{bi}\right)$$

where $q_w"$ is the heat transfer per time per area (wall-heat flux), $\rho$ is fluid density, $V$ is fluid velocity, $D$ is the duct diameter, $C_p$ is the constant pressure specific heat, $L$ is the length of the section, $T_{bo}$ is the bulk fluid temperature at the outlet, and $T_{bi}$ is the bulk fluid temperature at the inlet. The power required is then

$$q = q_w"A = q_w" \; \pi \; DL$$

The entrance length required for the flow to become fully thermally developed, $(Z_t)$, is found by the equation:

$$Z_t = 0.05DRePr$$

where $Re$ is the Reynolds number (See Section: UHT) and $Pr$ is the PRANDTL Number:

$$Pr = \frac{C_p \mu g_c}{k_f}$$

where $\mu$ is fluid viscosity, $g_c$ is a conversion factor, and $k_f$ is the thermal conductivity.

To determine the wall-temperature variation $(T_{wz})$, we must first consider the equation:

$$q_w" = h_z(T_{wz} - T_{bz})$$

where $h_z$ is the local coefficient, $T_{bz}$ is the bulk-fluid density variation, which is

$$T_{bz} = T_{bi} + \frac{2q''xz}{Vk_fR}$$

Then the wall-temperature variation (Janna 1986) can be found by

$$T_{wz} = T_{bz} + \frac{q''}{h_z}$$

## Engineering Data

Utilizing the knowledge of fluid flow and heat transfer, kinetic data of the process can be analyzed. Mathematical models have been developed to describe the reaction of the product under given conditions. Experimental data also may be obtained to aid in the process of kinetic study. The reaction to be studied must be properly identified, especially since multiple reactions may occur at the same time. With the known and calculated data, predictions can be made for the outcome of the process. For a heating process, product reaction (quality) and sterility can be expressed as a function of temperature and time, for a given flow rate and operating pressure.

## Process Drawings

A process is developed through many procedures. One of the first important steps is to develop process flow diagrams (PFD). PFDs are schematics that identify steps of the operation in the proper sequence. A conceptual diagram should identify the flow of elements from line staging to the operating areas, and then continue the flow through all associated processing equipment (Figure 4.3). Further development of the PFDs illustrates the fundamental size and shapes of various components, processing ranges and input-output requirements of each component. These requirements are based on the flow of materials, heat and material balances, unit operations, storage, and future expansion (Backhurst and Harker 1973). PFDs of the processing system should be completed before initiating vendor or equipment

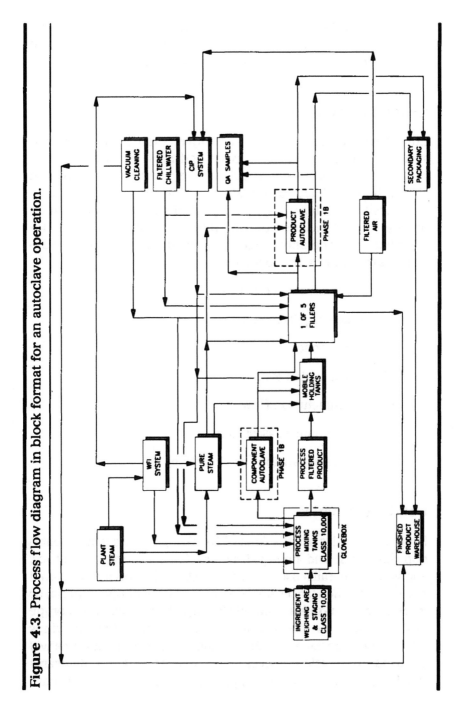

Figure 4.3. Process flow diagram in block format for an autoclave operation.

selection. The development of equipment specifications and detailed piping drawings, as well as equipment layouts, also should be postponed, pending completion of the PFDs.

## General Arrangements

Equipment layout or general arrangements (GAs) affect construction and manufacturing costs and must be engineered to prevent future problems in production. GAs of the operation should be completed to provide direction for the project team and possible vendors. A typical layout of an aseptic sterilization process should include both the filler and autoclave and autoclave handling operation (Figure 4.4). Preliminary layouts are developed to illustrate the following:

- System interface points

- Space requirements

- Ergonomics

- Construction issues

- Manufacturing flow

- Maintenance assessment

- Future expansions or alteration

## Detail Schematics

Detail schematics are the final PFDs, GAs, process instrumentation diagrams, and piping drawings (pipe routing, isometrics, and welding details) used for construction. The final PFD should present a detailed, accurate, and ordered flow of raw material or ingredient through each manufacturing phase, including storage of the finished products. Selected vendors and shop drawings must be incorporated into the product flow. These diagrams will be used to develop process and instrumentation diagrams (P&ID, or in some definitions, piping and instrument diagrams).

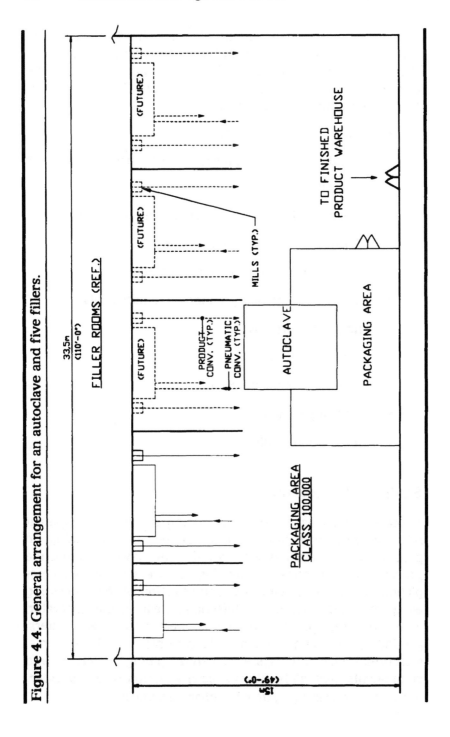

**Figure 4.4.** General arrangement for an autoclave and five fillers.

An instrument schematic should illustrate the detailed piping, electrical, and control requirements of the process. An example of a P&ID is provided in Figure 4.5. This diagram illustrates the heat and peroxide sterilization of a container (Caudill 1993). A P&ID provides the following details:

- Plans for drawing nomenclature
- Identification of equipment and instrumentation symbols
- Utilization of packaging PFDs
- Determination of relative location and type of sensors
- Illustration of single control loops
- Identification of critical control points
- Safety and emergency shutdown requirements
- Relation to process
- Equipment lists
- Materials of construction
- Design parameters
- Installation challenges

## Equipment Selection

Another critical aspect of a processing system is equipment selection. Process equipment must be selected depending on product requirements as well as regulatory laws. Defining the processing parameters and heating requirements allows completion of the preliminary GAs and selection of the processing equipment. WFI (Water for Injection), clean steam generator, water systems, processing pumps, heat exchangers, filters, product processing systems, clean-in-place (CIP) systems, sterilize-in-place (SIP) systems, valves, piping materials, and components must all be given special attention when considering a heat processing system for pharmaceuticals.

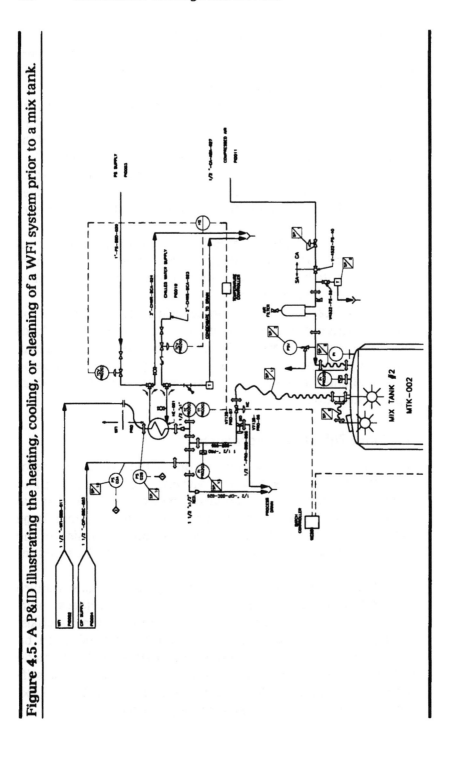

Figure 4.5. A P&ID illustrating the heating, cooling, or cleaning of a WFI system prior to a mix tank.

Equipment must also be specified to meet requirements for material, flow, production time, and size. Other aspects to consider when selecting process equipment are reliability, efficiency, and spare part's attainability.

Material must also be specified according to process requirements. For example, all equipment within the process must be rated for the pressure and temperature that is required of the production process, as well as the SIP, and CIP processes.

Equipment must be selected with present and future capacities in mind. As many components as possible should be sized to handle increased rates so to prevent their replacement as production requirements increase.

Since sterility of pharmaceutical products is of primary concern, so must it be for processing equipment as well. All components that are to be in contact with the product must be designed to prevent any contamination.

At the heart of any heat processing system is the heat exchanger. There are several types of equipment and methods available for the heat processing of liquids. The selection of heat processing equipment requires a knowledge of the fluid being processed and the characteristics and advantages of the various types of heat exchangers.

### Shell-and-Tube Heat Exchangers

Shell-and-tube heat exchangers are one of the oldest, most common designs. It is available in a variety of configurations and materials and provides a good heat transfer coefficient. The basic concept uses multiple small tubes sealed into a common large cylinder or shell. One fluid is passed through the inside of the tubes and the second flows around the outside of the tubes inside the shell.

Pharmaceutical and other sanitary applications require crevice-free construction, polished contact surfaces, self-draining design, and corrosion-resistant construction materials, see Figure 4.6. These applications require the product to be processed on the tube side of the exchange because the shell side has crevices and is not easily cleaned. To provide for a self-draining design, a single pass, straight tube or a U tube exchanger is usually used for sanitary applications.

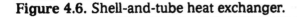
**Figure 4.6.** Shell-and-tube heat exchanger.

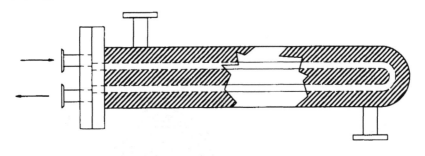

## Plate-and-Frame Heat Exchangers

Plate-and-frame heat exchangers are compact, low in cost, available in a variety of configurations, adaptable to many construction materials, and offers the highest heat exchange coefficients. The design utilizes multiple plates with a gasket between each, stacked and compressed between two heavy end plates. One end plate is usually fixed and is part of the frame that supports the plates and the movable end. The number of plates used on a given frame size and their arrangement can be varied for the specific application.

This design is widely used in the food and beverage industry for sanitary applications, such as processing milk, juices, and other liquids. In spite of their advantages and success in food and beverage processing, plate-and-frame heat exchangers have not gained wide acceptance in the pharmaceutical industry. Drainability, the many gasketed joints, and limited high temperature capabilities are among the concerns limiting pharmaceutical use. The design also has limitations for use with liquids containing solids.

## Tube-in-Tube Heat Exchangers

The tube-in-tube heat exchanger is conceptually a very simple design, consisting of one pipe concentrically placed inside another slightly larger pipe (Figure 4.7). The design has been

**Figure 4.7.** Tube-in-tube heat exchanger.

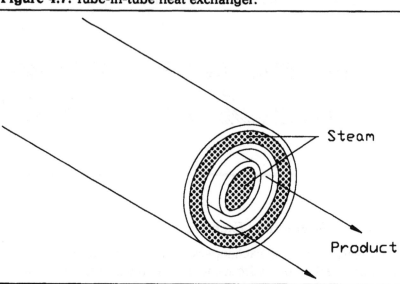

Steam

Product

embellished by adding a third concentric pipe, referred to as a triple tube, or the use of multiple tubes (usually limited to three or four) inside the larger tube. The latter design is similar to a long, thin, shell-and-tube heat exchanger.

Drainability and cleanability are key advantages of this design. When the product is confined to the inner tube of the unit, cleaning and draining are no different from the piping system used to transfer the liquid. This design is the logical choice for handling liquids containing solids.

These advantages are not without cost. The tube-in-tube heat exchanger is very large and expensive, and it generally has a lower heat transfer coefficient than the plate and frame or shell and tube. Several methods have been used to improve heat transfer in these units, including dents or dimples in the tube, corrugating the tube, spiral corrugations, triple concentric tubes, and the use of multiple tubes for the inner pass. Dimpling the surface of the tube increases the heat transfer by raising the Reynolds number, while the other methods increase both the

Reynolds number and the surface area. The triple concentric tube system increases the surface area available.

The tube-in-tube design is adaptable to systems where heat recovery from the hot product is desired. This requires that part of the system has product on both the inner and outer tube passes. Care in the design of the end connections, where the inner tube passes through the outer, is required to ensure a clean, drainable system.

## Steam Injection

Steam injection is a method of heating liquid by mixing it directly with steam. The steam must be injected into the liquid so that it is diffused as small steam bubbles in the liquid. If the steam is not diffused and large steam bubbles are injected into the liquid, implosions will occur, resulting in violent vibrations in the system. For sanitary systems, a source of clean steam is required. Figure 4.8 demonstrates that the inflow of steam becomes part of the product.

This method brings the liquid to the design temperature nearly instantaneously and uniformly. However, the process will add water to the liquid as a result of the condensed steam. The added water can be removed utilizing flash cooling of the product by passing it through a vacuum cooler where the water will flash to steam. The use of steam injection with flash cooling will provide very precise time-temperature process control for

**Figure 4.8.** Steam injection heat exchanger.

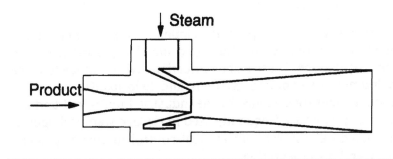

the product, but it is limited to applications where the addition of water or flashing of volatile liquids in the product is not a problem.

## Heated/Cooled Vessels

Vessels with jackets or coils for cooling or heating are commonly used for the heat processing of liquids. These vessels are used for both continuous and batch processing. In continuous processes, they are usually used to maintain temperature but can be used to heat and cool product. Because temperature cannot be changed as rapidly in a vessel of this type, it is not applicable to HTST systems.

## Controls

The automation of controls for an existing or new processing system is expensive. Many businesses find it difficult to justify automated systems based solely on process improvement. It is useful to establish other cost-saving applications of an automated monitoring and controlling system, including:

- Quality control of the product

- Improvements in product tracking and inventory

- Improvements in purchasing

- Preventative maintenance

- Rapid recall compliance

- Line efficiency

Automation has rapidly developed in cases where documentation is required by U.S. Food and Drug Adminstration (FDA) regulations. These same regulations will continue to force development of computer systems and help justify the use of automated control systems for the industry. Interlocked and automated systems are frequently used to replace standard operating procedures (SOPs) with manual control to ensure product quality and documentation.

Once a control system is decided on, a detailed control specification is required as a guideline to select possible control systems for the various equipment vendors. This functional specification is developed during preliminary engineering and is a "living document" intended to describe operational requirements for the systems to be integrated. When accompanied by logic diagrams or flowcharts, this document should contain all of the data required to facilitate final engineering and system programming. Once the processing systems, production monitoring or controlling points, and marketing requirements are determined, management must decide on the degree of automation desired and select an engineering firm or engineering staff to orchestrate the new, integrated system. Possible software and hardware systems, based on a developed business objective, should be studied. For example, if the project is financially justified by defined payback, cost would be a more dominant consideration in developing new market shares than a costly state-of-the-art facility (Caudill 1997).

## Plant Documents

The documentation support system that is to be used in conjunction with the heat processing procedures should include the following:

- Standard operating procedures (SOPs)

- Training program

- Preventive maintenance program

- Calibration program

- Change control program

### Standard Operating Procedures

The definition of standard operating procedures (SOPs) is the written, authorized set of instructions that describes (1) how to perform functions, not necessarily specific to a particular product, but of a more general nature, and (2) how to perform certain actions in a controlled way so that they are done the same each time. The applicability to heat processing is obvious. The repeatability of the performance of manual operations is just as important as the repeatability of automatic operations.

The types of SOPs for processing systems include training, cleaning, sampling, calibration, and user complaints. An SOP format would typically be outlined as seen in Table 4.1.

## Training

Personnel should be trained on the operation and/or maintenance of the equipment. Training must be documented. New employees must receive the same training as current employees. Refresher training should be included as part of the program.

## Preventive Maintenance Program

A documented preventive maintenance program must exist. Its goal is to prevent failure of equipment in a way that could affect quality of the water being produced. It should state what maintenance operations will be performed and their frequency. This could be time, operating hours, or triggered by an operating parameter. It should reference maintenance SOPs or work instructions for performing the maintenance tasks.

## Calibration Program

A documented calibration program is required to ensure that the process stays in control. It should cover acceptable accuracy and actions required whenever an instrument is found to be operating outside the required accuracy. Calibration equipment must be regularly calibrated and traceable to a known standard.

## Change Control Program

A change control program is required for any system to take advantage of advances in technology, improved equipment, engineering changes, or changes in incoming components quality

## Table 4.1. Standard Operating Procedure (SOP) Format

| | |
|---|---|
| Title of SOP | Simple and straight forward. |
| | Does it make sense to do? |
| | Examples: To describe . . . or How to . . . |
| Section 1: Purpose | Explains why procedure must be followed. |
| | Examples: To ensure . . . To prevent . . . To operate . . . |
| Section 2: Scope | Is it applicable to all similar equipment or is it specific? |
| | Is it for all products? |
| Section 3: References (Optional Section) | List supporting documents (i.e., equipment manufacturer's manual, work instructions, etc.); applicable regulations, other SOPs and MSDS. |
| Section 4: Required Forms | (Optional Section) |
| | List and forms/reports that will be generated by performing this procedure (i.e., batch records, lab analysis reports). |
| Section 5: Definitions | (Optional Section) |
| Section 6: Materials and Equipment | (Optional Section) |
| | List everything to carry out procedure (i.e., gloves, bags, scale, raw ingredients, etc.). |
| | List equipment being used (i.e., tank number 1, pump number 2, etc.). |
| Section 7: Responsibilities | List who performs the procedure. |
| | List who is responsible for ensuring that it is performed correctly. |
| Section 8: Procedures | List steps required to perform the task. |
| | Include safety note. |

such as water. Without a change control program, any change to the system would make it a new system that must be validated again. The change control system will set up a policy that allows changes with the proper reviews and approvals. Part of the review and approval process should include approval of a testing/validation program for the change equipment.

## DISCUSSION

### Sterilization Methods

There are two methods of heat processing: batch sterilization and continuous sterilization. In batch sterilization, the liquid is pumped into a storage/mixing vessel that is equipped with an agitator and a heating jacket or heating coils. The product is then heated to the sterilization temperature, held for the required time, and then cooled.

In continuous sterilization, the liquid is pumped at a constant flow rate through two heat exchangers. The first heat exchanger heats the product to the desired sterilization temperature. The liquid then passes through an intermediate area, usually a vessel or large area of piping, to allow the liquid to remain at the temperature for the recommended time requirement. The product then passes through the second heat exchanger for cooling (Jackson 1991). Temperature profiles for batch and continuous sterilization processes are shown in Figure 4.9.

### HTST

High-temperature short-time (HTST) pasteurization is the most important operation in milk processing. By applying the research developed in the dairy industry, we can apply some of the same principles to the pharmaceutical industry. The extent of microorganism inactivation by heat depends on a time temperature relationship for any liquid product. Minimum temperature and time relationships for pasteurization are based on thermal death

**Figure 4.9.** Process comparison of a batch temperature profile to a continuous temperature profile.

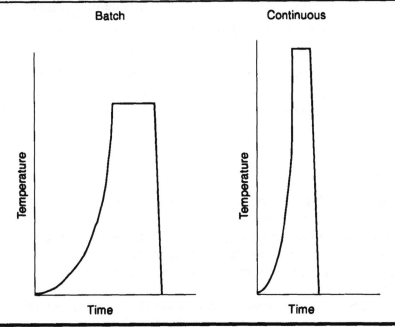

time studies with heat-resistant microorganisms. The specific death rate constant (*k*) should be determined for the sterilization temperature (*T*), using the following equation:

$$k = A\exp\left(-E/RT\right)$$

where *A* = constant, *E* = activation energy, *R* = universal gas constant, and *T* = absolute temperature.

Some typical values are given in Table 4.2 (Jackson 1991). These temperatures and corresponding times are the minimum requirements for pasteurization. Any temperature or any holding time greater than any of these standards is satisfactory. Higher processing temperatures can be utilized to increase shelf life and to address compatibility concerns. For the pasteurization of products with added dissolved solids, more severe heat treatments are required than for a homogenous liquid.

**Table 4.2. Typical Values for the Constants in the Specific Death Rate Equation**

| Organism | E (kcal/mol) | A (min$^{-1}$) |
|---|---|---|
| *Bacillus stearothermophilus* (FS1518) | 67.48 | 4.93 × 1037 |
| *Bacillus subtilis* (FS5230) | 68.7 | 9.50 × 1037 |
| *Clostridum sporogenes* (PA3679) | 68.7 | 1.66 × 1038 |
| Vegetative cells | 20 max | 1.20 × 1021 |

## UHT

The ultrahigh temperature (UHT) treatment is a process that has been developed during the past 30 years. In the United States, no official definition exists regarding UHT sterilization of pharmaceuticals. To produce sterile products, the heat treatment must be great enough to be stored at room temperature without losing its quality.

There are two basic methods for UHT processing—direct heating and indirect heating. The direct method involves heating the product by steam injection or infusion; this results in product dilution. This is followed by evaporative cooling under vacuum to remove the added water, which restores the product to its original composition. The second method, indirect heating, involves heat transfer across a heat exchange surface. In this method, steam does not contact product. Figure 4.10 illustrates typical temperature-time curves for direct and indirect heat processing.

Plate-and-frame, shell-and-tube, and tube-in-tube heat exhange units are possible indirect systems. Several types of commercial systems are available for aseptic processing. The sterile product side of the exchanger always must operate at a higher pressure than the regenerative or cooling portion of the exchanger. Often the raw product is used for cooling while reclaiming heat from the sterilized product. A secondary medium at lower pressures may be used during regeneration and cooling to reduce concerns of product contamination, but the thermal

**Figure 4.10.** Direct-autoclave and indirect-UHT processing.

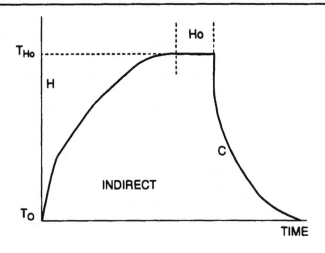

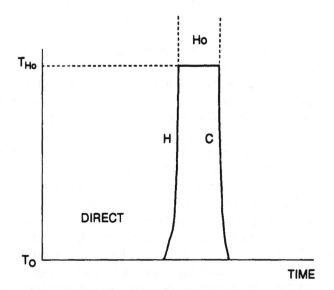

Key: H = heating, Ho = holding, C = cooling

efficiency of the system will be lower. A shell-and-tube unit consists of tubes mounted in a cylindrical heating chamber (Figure 4.6).

Product cooling can be accomplished in three steps. Primary cooling occurs in the regeneration section in which water is utilized as a intermediate carrier of heat (a product-to-water-to-product heat recovery network). The absorbed heat is transported to heat the incoming product. A throttling valve on the discharge side of the circulating pump can be used to regulate the recirculating water flow that controls the extent of thermal regeneration. From the regeneration shell-and-tube heat exchangers, the product has an option of passing through a one stage, remote homogenization valve. Further cooling takes place in three cooling tower water shell-and-tube coolers and one chill water shell-and-tube cooler. Product passes through a flow diversion valve assembly and on to a filler or storage tank. Product flow velocities of 2.4–6.7 meters per second are used to give good heat transfer and reduce deposit formation (Biziak et al. 1982). These units are common in ultrahigh processing of pharmaceuticals.

A tube-in-tube heat exchanger utilizes concentric stainless steel tubes, one inside the other. Two concentric tubes are employed to preheat during regeneration. Three concentric tubes are used for conducting system energy into increasing product temperature. Product flows through the second tube or the center tube of the triple tube system. The steam heating medium flows in a counter current direction in the outer and inner tubes. A differential pressure recorder must be installed to monitor and continuously record pressures. Since this system incorporates the use of product-to-product heat regeneration, triple tubes are used for cooling. The operating pressures and temperature are an indication of product flow rate, product properties, and back pressure resistance. A minimum back pressure of 69 kPa (10 psig) over the operating steam pressure is required in the holding tube. This would prevent any two-phase flow caused by boiling of the product in the tube. Noncondensable gases would displace product in the holding tubes, resulting in reduced resident time, as well as reducing heat transfer in the heat exchangers. This requirement to back pressure is similar in all UHT systems.

## Processing Data

Pharmaceutical processing system require the collection of data to verify the equipment operating parameters and processing settings. The first stage normally involves the testing or processing of water. The raw data, as well as calculated results, are to be included in the file, along with status definition relating to the data using terminology of "Pass," "Fail," or "No Test." In the event of impracticality of archiving raw data (e.g., material composition, etc.), summary recording documentation and signature will suffice. Three consecutive successful runs resulting with a "Pass" are required to compose a validated "state-of-control." Any failure in one trial will disqualify the other trials. Failed validation trials will be documented and include recommendations for corrective actions which will be completed before the validation trials resume.

Studies that are disqualified for technical reasons, as in lab accidents or equipment breakdowns, are considered "No Tests" and will have to be repeated. However, they will not require a documented investigation (for the purpose of the qualification) and do not disqualify any acceptable test runs performed prior to the "No Test." All studies performed, regardless of final disposition, should be retained as part of the permanent process validation file.

### Water Contaminants

Water entering processing equipment contains many different contaminants. It helps to categorize contaminants according to the technology that removes them. Below is a list of six categories into which all contaminants fit. Some contaminants may fit into more than one category, but one technology may be more appropriate than another. For example, you would use carbon adsorption to remove trihalomethanes, alcohols, solvents or other naturally occurring or synthetic organics. Bacteria also contain organic and inorganic contaminants, but carbon adsorption would not be used to remove bacteria from the water. Hence, bacteria would be classified as a microorganism and is primarily sterilized or removed from the water by ultraviolet sterilization or ultrafiltration. Catergories of contaminants are listed in Table 4.3.

## Water Standards

The challenge in setting and achieving water quality specifications comes in the microbiological quality specifications. The USP does not give specifications for bacteria levels for WFI and the PW (purified water) specification is very loose. PW must meet EPA drinking water bacteriological quality with a recommended action limit set at 100 CFU/mL. Microbiological quality for WFI is indirectly specified by the requirement that the level of pyrogenic substances be low (endotoxin ≤ 0.25 EU/mL). Most WFI systems are designed to produce water with very low levels of microorganisms, typically less than the 10 CFU/100 mL, which is the action limit recommended by the FDA.

The level of microbiological purity required for PW depends on how the water is to be used. Water that is used in a

**Table 4.3. Contaminants**

| Categories of Contaminants | Examples |
| --- | --- |
| Inorganics | Dissolved solids, such as sodium, cal-. cium, and magnesium cations or chloride, sulfate, and bicarbonate anions |
| Organics | Dissolved carbon containing compounds such as gasoline, isopropyl alcohol, insecticides, and trihalomethane |
| Particulates | Sediment or solids floating in water, such as silt, sand, and clay |
| Microorganisms | Bacteria and algae. *Psuedomonas* is the most common species of bacteria found in ultrapure water |
| Microorganism by-products | Endotoxins or pyrogens, enzymes, amino acids |
| Dissolved Gases | Oxygen, nitrogen, and carbon dioxide |

product that goes on to terminal sterilization is processed at high temperature, or results in a dry tablet will have a relatively loose quality specification, typically 100 CFU/mL. Water used in a formulation that may be susceptible to microbiological growth, such as liquid antacids or nasal solutions that are cold processed, should have very low microbiological levels or even be essentially sterile.

While the USP gives test methods for ensuring the sterility of water, test methods used to determine the total viable microorganisms present are not specified by the USP.

### Processing Documentation Failures

A review of processing inspections that gathered data over a three-year period gives some insight into potential problem areas that can be avoided by a good testing process (Caudill 1997a). Objections typically fall into four areas: construction, documentation, calibration, and operations. Typical problems in construction are deadlegs, nonsanitary fittings, insufficient sampling ports, unprotected atmosphere breaks, and inadequate temperature control and monitoring. Documentation-related issues include lack of written validation protocols, lack of SOPs, no established microbial limits and insufficient data or validation of test methods. Failures to calibrate conductivity meter, temperature recorders, and pressure gauges were frequently cited. Operation problems included failure to follow SOPs, inadequate sanitization and maintenance records, inadequate documentation of operations, and inappropriate microbiological tests.

## CONCLUSION

Regulation and validation in the pharmaceutical industry have slowed the development of heat processing technology. Once a product process has been developed, a great amount of inertia must be overcome before a heat processing change will occur. As energy prices increase, economical conditions will increasingly provide reasons to improve heat processing methods. With planned product research and proper engineering, future systems will provide higher product quality at lower energy costs.

## REFERENCES

Anon. 1951. *Food Engineering* 23:68.

Backhurst, J. R. and J. H. Harker. 1973. *Process plant design*. London: Heinemann Educational Books Ltd.

Ball, C. O. 1938. *Food Res.* 3:13–21.

Biziak, R. B., K. R. Swartzel, and J. A. Jones. 1982. Energy evaluation of a UHT shell and tube processing system. *Journal of Food Science* 47:1875–1878.

Burton, H. 1975. UHT and microbial considerations. *Proceedings of the 6th European Symposium—Food: Engineering and Food Quality*, Cambridge, London.

Burton, H. 1970. Ultra-High-Temperature processing of dairy products. *Proceedings of the Seminar Organized by the Agricultural Institute*, edited by T. R. Ashton, E. L. Crossley, F. O'Connor, and A. C. O'Sullivan, May 1969, in the Republic of Ireland.

Caudill, V. E. 1997a. Water system validation. *Proceeding of Pharmaceutical Waters*. Atlantic City, NJ.

Caudill, V. E. 1997b. Monitoring systems for an aseptic line. *Proceeding of Interphex Conference*. Philadephia, PA.

Caudill, V. E. 1993. *Engineering: plant design, processing, and packaging*, edited by Y.H. Hui. New York: VCH Publisher, Inc.

Feliciotti, E. and W. B. Esselen. 1957. *Food Technology* 11:77.

Gupte, S. M., H. M. El-Bisi, and F. J. Francis. 1964. *Journal of Food Science* 29:379.

Harper, J. L. 1979. *Elements of food engineering*. Westport, CT: The AVI Publishing Company, Inc.

Havighorst, C. R. 1951. *Food Ind.* 23:72.

Hersom, A. C. 1985. *Aseptic processing and packaging of food*. New York: Marcel Dekker, Inc.

Jackson, A. T. 1991. *Process engineering in biotechnology.* Englewood Cliffs, NJ: Prentice-Hall, Inc.

Jackson, J. M., J. F. Feaster, and R. W. Pilcher. 1945. *Proc. Inst. Food Technology* 81.

Janna, W. S. 1986. *Engineering heat transfer.* Boston MA: PWS Publishers.

McMillan, H. K. 1990. *Thermodynamics.* Columbia, SC: Kinko Publishing.

Mek, M. W. 1950. *Aseptic canning system embodying short high temperature sterilization.* Institute of Food Technologists Conference, in Chicago, IL.

Mulley, E. A., C. R. Stumbo, and W. M. Hunting. 1975. *Journal of Food Science* 40:985.

Schanderl, S. H., C. O. Chichester, and G. L. Marsh. 1962. *Journal of Organic Chemistry* 27:3865–3869.

# 5

## In-Line Sterilization of Liquids, from the Viewpoint of Food Processing

*V. Robert Carlson*

VRC Co. Inc.
Cedar Rapids, IA

## PASTEURIZATION AND STERILIZATION

### Pasteurization of Milk (and Similar Dairy Products)

In-line thermal sterilization is of interest to sterilize liquids that can be pumped, compared to techniques such as radiation, high pressure filtration, and the addition of chemicals that inhibit microorganisms or viruses. The reasons for this are manyfold, including the ability to measure temperature and to provide a device to control the time of exposure, plus the significant backlog of information that has been developed over the years for the inactivation of numerous bacteria, spores, yeast, molds, and viruses, using heat; thermal sterilization can be done in simple systems. To discuss this subject intelligently, some definitions and background are necessary.

In-line (thermal) treatment was first used in the dairy industry with milk and cream, which was used to make butter, because the organism that caused tuberculosis in humans could originate in the cow producing milk. If the cow tested positive for brucellosis, it most likely had the organism *Brucella abortus* and *Mycobacterium bovis* in its system, and these organisms were transmitted to the milk produced. If the organisms were not inactivated, the individual consuming the milk stood a good chance of contracting tuberculosis (Traum 1924).

When public health investigators first determined what time and temperature was required to inactivate this organism, D values, F values, Z values, and others were not yet defined. In fact, many of the tests were somewhat elementary, although they were very effective. A milk sample infected with tubercle bacilli was heated to various temperatures, held, and then injected into guinea pigs. If the guinea pig survived the time-temperature condition, the heat treatment was adequate. If the guinea pig contracted tuberculosis, it was proof that the heat treatment was inadequate (Traum 1924). These studies lead to various states requiring the pasteurization of milk in the early 1900s. It was found that this organism could be inactivated by heating the milk to 143°F, which was later increased to 145°F to inactivate the rickettsia that causes Q fever (*Coxiella burneti*) (Frobisher 1957), and holding it for 30 minutes (vat pasteurization). This was the common pasteurization technique in the 1930s. By 1950, the use of high-temperature short-time (HTST) pasteurization, where the temperature was increased to 161°F and the time was 15 seconds, was approved by many states.

This was continuous in-line thermal treatment where the milk was heated and held in a closed system (vats were open); this technique is still used today. Because vats can be opened, an air space heater that uses steam to maintain 5°F above the 145°F in the space between the milk and the vat cover is required. Vat pasteurization is still used for certain dairy by-products. Both methods are specified with practically all methods defined in the PMO (Pasteurized Milk Ordnance) or in supplementary flow diagrams and process descriptions.

Another "legal" term applied to milk is *ultrapasteurization*, which is defined as treating the dairy product at 280°F and holding it for 2 seconds with the purpose of extending the refrigerated shelf life.

As late as the early 1950s, the situation was summarized by Ben Freedman in his *Sanitary Handbook:* "In the period from 1938 to 1950 milk was eight (8) times more powerful in causing illness than were water borne diseases" (FDA 1982).

When the viscosity of the product increases (sweetener is added) or the fat content is 10 percent or more, the temperature is increased 5°F for vat or HTST pasteurization. When egg nog and frozen dessert mixes such as ice cream are pasteurized, the temperature and/or time increases again (see Table 5.1).

The pasteurization of milk employs a number of times and temperatures when plate-type heat exchangers are used, which are classified as ultrahigh temperature (UHT) pasteurization time-temperature standards, which range from 191°F to 212°F (Read et al. 1968).

## Pasteurization/Sterilization of Foods Other Than Dairy Products

With food products (not milk where pasteurization is a legal term), pasteurization consists of "heating a product to a temperature below the boiling point of water in order to destroy certain spoilage organisms in the product" (Carlson 1972). This definition is one used by the canning industry for many acid products, such as fruit juices, tomato products, nectars, and purees. In certain cases, the temperature actually used is above the boiling point of water (pressurized) to inactivate enzymes.

In late 1996, a strain of *Escherichia coli* 0157:H7 contaminated apples. This strain is not normally associated with fruits or the juices made from them. However, this organism is very common in the soil, water, and air. Hence, if a fruit is grossly

Table 5.1. Pasteurization Time/Temperature Requirements as Specified by the U.S. Public Health Service / U.S. Food and Drug Administration

| Product | Vat | | HTST | | HHST | |
|---|---|---|---|---|---|---|
| | Time | Temp. | Time | Temp. | Time | Temp. |
| Whole Milk | 30 min | 145°F | 15 sec | 161°F | 1.0 sec | 191°F |
| | | | | | 0.5 sec | 194°F |
| | | | | | 0.1 sec | 201°F |
| | | | | | .05 sec | 205°F |
| | | | | | .01 sec | 212°F |
| Milk Products with increased viscosity, added sweetener, or fat content 10% or more | 30 min | 150°F | 15 sec | 166°F | Same | |
| Egg Nog, Frozen Dessert Mixes | 30 min | 155°F | 25 sec | 175°F | Same | |
| | | | 15 sec | 180°F | | |

Note: Those products pasteurized at 280°F for 2.0 sec are to be labeled as "ultrapasteurized."

contaminated because it has fallen to the ground, it is extremely difficult to wash the fruit and completely remove every cell. If a few remain, entire lots are contaminated, and consumers can become ill if the juice is not pasteurized. Because of the severity and frequency of illness that has been attributed to *Escherichia coli* 0157:H7, it would not be surprising if juices were required to be pasteurized, as milk and egg products are, for public health and safety reasons.

One step above pasteurization is *commercial sterilization*, which is a term used in the canning industry (Ito 1970). *Sterilization* is an absolute term. The treatment of food or nonpharmaceutical products is based on statistics. Certain organisms may be present after thermal sterilization; however, they cannot grow or develop, do not produce toxins that are harmful, or do not change the quality of the product. Many of the terms that have developed come from in-container sterilization in the canning industry. In this process, the food product is put in a can or jar, and the lid is secured. The container is then subjected to high temperatures and pressures in a pressure cooker, or it is filled hot, held, and then cooled. The pressure cooker or hot fill system may be either a continuous or batch system. The terms used are logical for continuous in-line thermal sterilization. Even though the mathematics to convert heating/cooling rates in a retort (in-container) to a continuous in-line process are somewhat complicated, the following apply:

- *D value:* The time at a constant temperature to destroy 90 percent of the organisms present. Any D value stated must also indicate what temperature is applicable, such as $D_{280}$ (Ball and Olson 1957).

- *Z value:* The temperature (°F) required for the thermal death time curve to traverse one log cycle (change the time by tenfold) (Pflug 1987).

- *F:* Number of minutes to destroy organism at 250°F (Ball and Olson 1957).

- $F_o$: The time in minutes required to inactivate a population of organisms in a specified environment at 250°F when Z = 18 (Ito 1970).

- *Thermal death time:* The time it takes at a certain temperature to kill an organism (spores) under specified conditions (Ito 1970).

Usually, in the canning industry, the D value for a process for nonacid products at 250°F is established at 12. This will inactivate $1 \times 10^6$ spores of *Clostridium botulinum* per mL with a safety factor of $1 \times 10^6$, or the safety factor will vary with the number of spores present (Pfeifer and Kessler 1995). This is commonly referred to as the 12D process, as $(1 \times 10^6) \times (1 \times 10^6) = 1 \times 10^{12}$. If the initial load of *Clostridium botulinum* spores is 1,000/mil, or $1 \times 10^3$, then the safety factor of the process will be $(1 \times 10^{12}) - (1 \times 10^3)$ or $1 \times 10^9$, or the probability of the process failing will be 1 in $(1 \times 10^9)$. It is estimated that the actual probability of a botulism incident is of the order of $10^{-11}$ to $10^{-13}$ (Pflug 1987).

If the product is in a nonacid medium, the types of organisms that will grow are different than those that will grow in an acid medium (Fernandez et al. 1995). Acid acts as an adjunct to the time and temperature when inactivating organisms in the liquid. If the liquid is not acidic to provide this additive effect (the liquid can support the growth of practically any type of organism, e.g., bacteria, yeast, or mold), the sterilization process must be at a much higher temperature to obtain commercial sterility. Commercial sterility (in the food industry) relates to the inactivation of vegetative cells and spores of microorganisms that will *grow* and *develop* in the product. Also, yeasts and molds are a consideration with certain products (primarily acidic liquids). However, sterilization processes of foods are not defined to inactivate other entities that should be of concern (e.g., viruses) (see Table 5.2). The reason is twofold: (1) Many viruses come from bacteria and if the bacteria are inactivated, then the ability to produce the virus is eliminated; (2) Most viruses have fairly low thermal process inactivation requirements and can be easily inactivated with heat. However, considering the viruses that are being discovered today, coupled with the fact that many of these do not have established thermal death times and the importance of disease-causing abilities are not completely known, it may be worthwhile to develop in-line thermal sterilization procedures that are adequate to cope with the viral worst-case situation.

**Table 5.2. Viruses—Thermal Resistance and Inactivation**

| Product | Processing Condition | Effect of Processing on Virus | Reference |
|---|---|---|---|
| Whole Milk | 110°C/30 sec | – | DeLeeuw and Van Bekkum (1979) |
|  | 148°C/2 sec | + | Cunliffe et al. (1979) |
| Cream | 93°C/15 sec | – | Hyde et al. (1975) |
| Cultured Butter | 93°C/15 sec Fermentation | Minimum of 4 months | Blackwell (1978a) |
| Casein | 72°C/0.25 min Isoelectric pt, pH 4.6 | –42 d at 25°C | Cunliffe and Blackwell (1977) |
| Cheese | No Heat | –90 d, +120 d curing | Blackwell (1976) |
| Cheddar | 63°C/10 sec 67°C/15 sec | –, processing, but not 30 d curing | Blackwell (1976) |
| Camembert | 72°C/15 sec | –21 d, +35 d curing | Blackwell (1976) |
| Mozzarella | 72°C/15 sec | + processing alone | Blackwell (1976) |
| Whey—Acid | 72°C/15 sec, pH 4.6 | + processing alone | Blackwell (1978b) |
| Whey—Sweet | 72°C/15 sec, pH 5.2 | – | Blackwell (1978b) |

– Virus survived processing conditions, + Virus was inactivated by processing conditions

Many viruses originate from organisms within foods. Diseases that are caused by viruses include foot-and-mouth disease, smallpox, poliomyelitis, rabies, yellow fever, and hepatitis (A and B) (Frobisher 1957).

> *Heating denatures capsid (viral surface) protein, thus preventing specific attachment of the virion to the cell receptor site, inactivating virion-associated enzymes needed for synthesis of progeny virus and interfering with removal of coat protein and release of viral nucleic acid. At temperatures above 70°C, unwinding of double-stranded viral nucleic acid, and random breaks in the sugar phosphate backbone are part of the irreversible changes that occur in the viruses. With a liquid containing organism(s) which can excrete viruses, or which contain excreted viruses, heat for a specified time can be used to inactivate the viruses. At temperatures above 70°C (158°F) unwinding of double standard viral nucleic acid and random breaks in the sugar phosphate background are part of the irreversible changes that occur in the viruses.*

This statement was made in an article titled *Food Borne Viruses: Their Importance and Need for Research* by John Blackwell and colleagues (Blackwell et al. 1985). Also in the same article, it would appear that 69°C (156.2°F) is a critical temperature in the thermal processing of animal products because ASF (African swine fever), SVD (swine vesicular disease), HC (hog cholera), and FMD (foot-and-mouth disease) viruses are inactivated at this temperature. Blackwell et al. (1985) also indicated FMD virus recovered from lymph nodes of infected cattle after heating for 2 hours at 69°C, 1 hour at 80°C, and 0.25 hour at 90°C. There is at least one important exception; the virus homologous serum jaundice withstands boiling for 7 minutes (Frobisher 1957).

It is also possible that certain viruses can take temperatures above boiling to as high as 138°C (280°F). The holding period evaluated was 2 seconds, and the virus tested causes foot-and-mouth disease (Blackwell et al. 1985). This is a defined pasteurization process in the PMO, published by the U.S. Public Health

Service (now part of the FDA). Because certain processes are defined legally for processing food products, it does not mean that viruses cannot survive these processes. The same logic can be applied to nonfood products (e.g., water). Other factors must also be considered, including pH, and whether other processes are occurring at the same time as heat treatment, such as radiation or pulsed light (refer to other chapters in this book).

## THERMAL PROCESS DETERMINATION

An additional term that needs defining when discussing what D value to use when treating a specified material is water activity. *Water activity* $(A_w)$ is the vapor pressure of the solution (solutes in water in most foods) divided by the vapor pressure of the solvent (usually water) (Frazier 1958). With many liquid pharmaceutical products, $A_w = 1$. Water activity usually does not have an effect on the restriction of bacterial, yeast, mold, or viral growth in the cells of organisms until $A_w = 0.80$–$0.85$ or below. Unless the solvent is a syrup with considerable amounts of sugars, salts, or some other solute, it will not have a preserving effect. Generally, it will not have an effect on the D value that is selected for a specified solution, but it should be determined that this situation exists (Frazier 1958).

Most likely, the overriding controlling factor will be the type of organism (e.g., vegetative bacteria cell or spore, molds, yeasts, or virus) and the number of organisms present. This will then dictate the temperature required for inactivation, which in turn dictates the time for one log kill (D value). Acidity of the solution (pH below 4.6) lowers the D value (Fernandez et al. 1995), but, with the esception of some dialysis solutions, most drug solutions are neutral.

The pressure required for the inactivation of vegetative cells is 90,000–100,000 psi. Pressures must be even higher to inactivate spores or alter enzymes; however, high pressures when used synergistically with other treatments can be reduced to obtain equal inactivation levels. High pressures (>3000 bar) are effective to inactivate the viruses tested (Vardaq 1995).

Generally, most pharmaceuticals cannot have bactericidal agents, such as halogens or halogen-containing compounds, added to them. A material such as glutaraldehyde, which is poisonous and toxic, cannot be used.

The D value selected would normally be based upon a liquid pharmaceutical, such as Water for Injection (WFI), infant formula, dietary feeding solution (oral), gavage solution (tube feeding), and so on. The D value would not be applicable to an ointment or salve, which probably would have a very low water activity and may incorporate certain preservatives.

## Time-Temperature to Select

As previously indicated, the first criteria to be established is the temperature of sterilization. A temperature should be selected that is reasonable and will obtain the D value desired with a rational hold time. Generally, hold tubes are less expensive than heat exchangers; hence, holding periods of at least 30 seconds at the rate desired are common, and as long as 5 minutes is possible with reasonable costs for the holding tube. The holding tube should be designed so it is configured in an insulated coil and provides an *r/D* (ratio of the centerline radius of the coil to the diameter of the tubing making up the coil) of a maximum of 8–10. If the holding tube is designed accordingly, and the velocity is 1 ft/sec or more, then secondary flow can be assured, and the hold of all portions of the solution will be for the period defined. If a straight holding tube with 180° return bends is used, then the product may be held for variable times, depending on the velocity and viscosity of the product. The product (if viscous and flow is laminar) may be "theoretically" held for as much as twice as long as the mean velocity calculation would indicate. However, the actual time for a portion of the liquid may be considerably more because the velocity through the straight section of the tubing may be 0 at the interface between the wall of the tube and the product. This means that the velocity of some portion of the product is greater than twice the average mass flow. Whether or not this occurs is a function of the viscosity of the product (VRC 1995).

## Calculation of D Values

The determination of the D value is a function of the safety desired, as well as the type and number of organisms or viruses present. If the type and number of organism requires a sterilization temperature of 70°C for 15 sec to provide a D value of 1, then a D value of 6 or 8 can be provided by increasing the hold time 6 or 8 times, to 90 or 120 sec. If the solution being treated thermally contains a total number of a specific type of organisms of $1 \times 10^1$ per mL, the safety factor of the process is somewhere between 5 and 7. Theoretically, the actual D value required to inactivate 90 percent of the 10 organisms (i.e., 1 organism will survive) is 15 sec at 70°C. As time increases to 90 or 120 sec, this equals 6 or 8 times the inactivation of 90 percent of the organisms present. If the decimal reduction value is subtracted from the 6 or 8, a theoretical D value of 5 or 7 exists, and the 1 surviving organism will be subjected to moist heat for 5 or 7 times the time required to kill it (Evans and Hawkinson 1970).

The safety of a process, or increasing the D value, can also be accomplished by elevating the temperature and maintaining the hold time. Often this is not desirable as more energy must be used to increase the temperature (at higher costs); if a lower temperature can be used without negatively affecting the product, the operational costs are less. However, with certain products, it is desirable to heat to as high a temperature as possible and hold them for a minimal time to obtain the D value desired. This is possible because the inactivation of bacteria (and viruses) is logarithmic and can be calculated if the contributing factors are known. The thermal death time curve must have a constant slope in the temperature range that is selected. See Chapter 2 for the particulars.

## Hold Time

The system must be interlocked to ensure that the hold time is not reduced. The conventional way of checking to verify holding time with milk pasteurizing systems is with a salt solution. The salt solution is injected at the beginning of the holding tube, and electrical sensing and indicating devices are located at the end of the holding tube to verify how much time it takes to travel from the beginning to the end. Some products may actually flow through the holding tube faster than the 15 seconds detected by electronic sensors; some products, particularly viscous products, do not flow at all at the wall of the tube. When an HTST process is tested, the tests are performed using water as the product to determine the system design, including the hold tube design. After the system has been tested and is satisfactory, it is then used on the various dairy products for which it has been designed. All commercial products are refrigerated (less than 45°F), hence, survivors are inhibited by the refrigerated temperatures that are required for pasteurized dairy products in the United States. If food or other products are not adequately pasteurized, refrigeration may not prevent the growth of organisms that may be in milk. Hence, the organism or virus could survive the process and possibly cause a problem by contaminating other materials that are made from the water or solution (Goff and Davidson 1992). By maintaining water at 80–82°C, many organisms and viruses that are typically in water will be inactivated by the temperature; however, certain viruses can survive at this temperature.

## SYSTEM DESIGN—CONCEPTS WITH PHARMACEUTICAL GRADE WATER AND WATER FOR INJECTION

Mechanical designs can be provided to heat to 150°C using conventional boilers, sanitary heat exchangers, and system designs. Approximately one-half of the heating from an incoming temperature of 15°C to 70–80°C can be done by the hot product in a safe manner. In this process, the hot product is cooled to 55–65°C, which renders it at 85–95°C. This is at a high enough temperature to inactivate many of the viruses or vegetative cells, yeasts, and molds that are of commercial significance to producers/ users of WFI and pharmaceutical grade water (PGW). Hence, many commercial WFI systems utilize storage at these temperatures to prevent growth as these temperatures can be easily maintained. The storage tank is insulated to prevent heat loss to the atmosphere and for safety. Another approach is to heat room temperature water to approximately 80°C prior to use; this inactivates many bacteria and viruses. It may or may not be adequate.

If there is any chance that the water can be contaminated, either from the atmosphere or from components that may exist downstream, the tank and piping system should be designed so that such contaminants cannot enter the WFI system. A PGW system such as this is shown in Figure 5.1. Figure 5.2 shows a PGW regenerative sterilization system that cools water to 80°C (approximately 176°F). The product or water can be diverted from these systems to a use point, such as a filler, instead of a tank.

## Indicator Tests or Procedures to Verify the Process

When milk is pasteurized and the time-temperature specified is met, the enzyme phosphatase is inactivated. Tests for the presence of this enzyme can be performed quickly to verify that the pasteurization process has been performed properly. In addition, regulators from cities, counties, states, and federal agencies check to ensure processes are being operated properly and records maintained to indicate that adequate temperatures are being met; whenever a deviation occurs (temperature is lost), safety devices, such as flow diversion valves, operate. The operation of such flow diversion valves is indicated on a chart with a mark indicating that an event has occurred. The same general concept is followed for the thermal sterilization of food products that are aseptically packaged and labeled pasteurized or ultra-pasteurized. However, when a diversion occurs because of low temperature or pressure, the system must be resterilized before the flow diversion valve can move to the forward position.

The process for sterilization specifies the time-temperature must be filed with the FDA if the product being processed is nonacid (pH 4.6 and above, and an $A_w$ above 0.85). There are many other specifics that are required, including how the system will be sterilized initially. How the sterile product will be

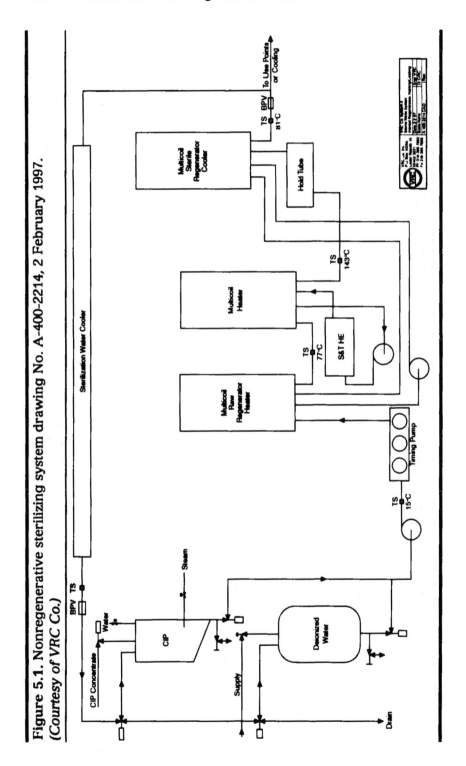

Figure 5.1. Nonregenerative sterilizing system drawing No. A-400-2214, 2 February 1997. *(Courtesy of VRC Co.)*

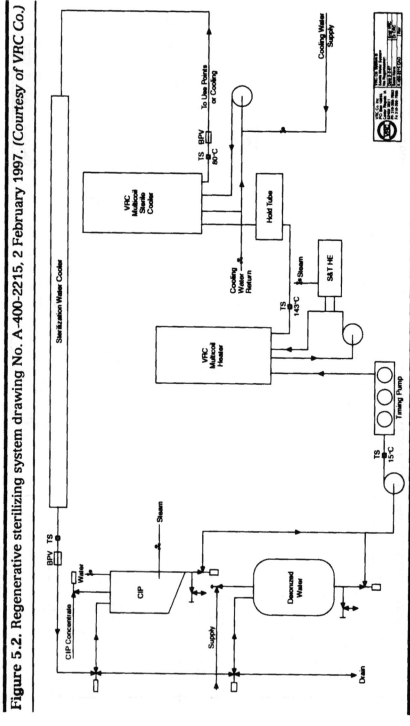

**Figure 5.2.** Regenerative sterilizing system drawing No. A-400-2215, 2 February 1997. *(Courtesy of VRC Co.)*

protected from nonsterile pieces of equipment or surfaces, how the system is to be cleaned, what happens if sterility is lost, also must be defined. Because a commercially sterile product (shelf stable at room temperatures) is being processed, as opposed to a pasteurized product that will be refrigerated at a later date, conventional flow diversion valves (which are not sterile) are not used. The general concept, which may be interlocked in the programmable logic controller (PLC), is that if temperature is lost, the system is switched to water (which will not be packed because a diluted product would result), and the system will be flushed and cleaned if necessary. After cleaning, the system will be resterilized, and the product will be processed.

Because the product is normally packed in an aseptic packaging system following processing, the product will be commercially sterile with an indefinite shelf life at any reasonable temperature. This allows it to be transferred across state lines, which involves the FDA. The inspector will verify that the system is operated according to the process filed with the FDA. The manufacturer's records will show that no deviations have occurred; or if deviations have occurred, that the product has been stored in a warehouse until cleared by the FDA.

## Suggested Procedures for Pharmaceutical Products

The same basic procedure will be logical for pharmaceutical products, even though they may not be aseptically packaged. It may be used in the preparation of other products, and may be stored in an aseptic or sanitary tank. The system should be designed and controlled, and records kept during processing should be produced. The system should be supplied with a timing pump that is interlocked to ensure the flow of product has not exceeded the rate the process specifies. Holding tubes should be designed as to guarantee that all portions of the product are subjected to the hold desired. The process (time-temperature) should be specified so the desired D value is obtained and the thermal inactivation of microorganisms (vegetative cells, spores, yeast, molds) and viruses is accomplished.

If the process involves products that are regulated by the FDA or the USDA, the proper procedures specified by the agency must be followed. If the product is "nutritional," it may be necessary to file a process with the FDA Food Group to allow processing of this product in the manner defined. Certain "pharmaceutical" products are legally considered "food" products, and the proper process must be filed with the FDA. If the product is a naturally acid product, or if its acidity is adjusted to a specified pH, records showing the pH are required by FDA. A continuous in-line pH sensor and meter with recorders, or pH records of each batch must be provided so the FDA can examine these records and verify the product was actually acidified as specified (CFR 1988).

Nonacid food products are normally sterilized at 250°F for a minimum of 3 minutes. An equivalent process that will provide the same sterilization values may also be used. The process at 250°F for 3 minutes provides an F of 3, commonly considered the minimal process required for most nonacid food products. This is based on a D value of *Clostridium botulinum* of 0.21 min (12.6 sec) at 250°F. For a 12D process, the theoretical time is 151.2 sec or just over 2.5 min. When lag times for heating and cooling a fluid product in the tubes (used in the laboratory where the D value is determined) are added, the time equivalent is approximately 3.0 min (Pflug 1987).

The thermal death time for organisms, including *Clostridium botulinum*, is determined by heating known suspensions in special capillary tubes that are heated to various temperatures, held for various times, cooled below lethal temperatures, and the survivors noted. With this information, a thermal death time curve indicating inactivation at different times and temperatures can be prepared.

In an "in-line sterilization system," time is measured through the holding tube for a fluid product, such as water, that is doubled for a viscous product, such as pudding or cheese sauce. When the product is viscous and the mass volume is doubled, some of the product is held for 6.0 min, while some of the product may be held for 3.0 min. In this manner, a safe product from a public health standpoint is produced. In actual practice, much higher temperatures are used (e.g., 284–290°F [140–143°C]).

However, as indicated earlier, this doubling of the hold time could be eliminated by the proper design of the hold tube (coiled with the proper *r/D* dimensions, and insulated).

## ELECTRONIC CONTROLS AND THE USE OF PLCs

Advancements in the design of electronic controls, particularly PLC instruments, have progressed so that many of the operations in the systems designed today can be automatically sequenced and interlocked to provide proper operations easily. If the operation is not proper—including temperatures, flows, pressures, etc.—interlocks can prevent moving from one phase to another; however, the system must be designed to use a PLC to obtain full advantage.

Advancements have also been made in sensors and instruments, including simple PID controllers and cascade, feedforward, and fuzzy controls. Such controls can be used for controlling product temperature, pressure, flows, and other processes. If incorrect conditions exist, all operations can be stopped.

The proper temperature and the time for pressures and flows must be controlled exactly. Flowmeters may not be as accurate as desired with certain products, so, in certain situations, they should be used as a check only to verify flows, which require the use of exact, positive pumping methods. If the product does not contain particles, and will always be at the same temperature and viscosity, flowmeters can be used to sense the flow through the system.

For homogeneous fluid products that do not contain particles, a mechanically driven piston pump similar to a homogenizer is ideal. These pumps have been used for years. Flow from the pump will not vary unless mechanical problems exist. Problems could be inadequate suction pressure, worn belts (which may slip), sticking valves (in the piston timing pump), worn or leaking gaskets, and so on. Because failure can occur, a monitor that verifies the operation of the pump (i.e., an ammeter with an indicator) may be installed on the drive motor. If amperage is not adequate, or is too high, the operator should make corrections. This technique is commonly used to verify the pressure (homogenization) of the pump. It can also be used to verify that the output is reasonably close to that desired.

Obviously, a physical check of the flow through the system should be made with a flowmeter on a periodic basis. Certain flowmeters can determine variations in density, so a product change, or a switch from product to water, can be verified from density values. Flowmeters should be used as checking devices— not as controlling devices.

## SANITARY DESIGN

Equipment must conform to sanitary design recommendations or regulations. The 3A Symbol Council has developed recommendations for a wide variety of equipment common to the dairy industry. E3A recommendations are used by the egg industry for liquid egg products such as liquid whole eggs, yolks, whites, salted yolks, and sugared yolks. Many other food industries (e.g., the baking and beverage industries) have recommendations for design purposes. The American Society of Mechanical Engineers (ASME) has a set of standards that have been adopted by the American National Standards Institute (ANSI) (ANSI/ASME F2.1 1986). These standards are used for traditional canning equipment; they incorporate general specifications with recommendations for equipment falling outside the definitions or recommendations issued by 3A, BISSC (Baking Industry Sanitation Standards Committee), E3A, or other standards.

In essence, equipment must be designed to be cleaned and to maintain sterility after it has been cleaned and sterilized. Initial sterilization must be with heat, so the equipment must be designed to withstand high temperatures (150°C, 302°F) without undue wear or potential contamination problems (Pfeifer and Kessler 1995). Usually, inert materials, such as 304, 316, and 316L stainless steel (or other noncorrosive alloys) must be used. When plastic or rubber goods are used, they must be noncontaminating, nontoxic, and of no public health concern. Finishes must be smooth so that the surface can be cleaned easily and thoroughly.

Equipment must be designed so that it can be inspected either visually (manually) or with instrumentation (e.g., a flexible borescope that can be used to examine pipelines and certain types of fabrication methods [e.g., welding]), and to determine if cleaning has been effective.

## PUMPING AND TIMING

Pumping the product through the system is extremely critical and must be done at a constant rate. With most products, a piston pump (Figure 5.3) that is mechanically or hydraulically powered must be used. Slippage in a piston pump is practically nonexistent, and the product flow rate is constant. This assumes the pump is fed adequately and has a net positive suction head required in order to pump to its full capacity.

Another advantage of a piston pump is that, assuming the total pressure against which it is pumping is not excessive, it will produce a constant flow continuously, regardless of the downstream pressure. Many times, heat exchangers will tend to foul; valves may be opened or shut that will change the flow pattern and the pressure of the system; the filler will actuate fill valves; or the filler may be equipped with a surge bowl where the level or pressure changes, thereby changing the pressure that the pump must overcome. Rotary pumps or progressive cavity pumps tend to vary in output based on pressure. The amount of variation will depend on many factors, including wear.

An additional problem with rotary or progressive cavity pumps is that they will pump at one rate because the pressure is lower when the system is sterilized with water (at a viscosity of 1 cps or below), and at another rate when the pump is pumping product, which might have a viscosity of several cps. Attempts can be made to even this variation by using either a multistage progressive cavity pump that is grossly oversized for the pressure it will be pumping against, or to arrange two or more rotary

**Figure 5.3.** Gaulin homogenizer (mechanical piston pump) used as a timing pump. *(Homogenizer manufactured by and photo courtesy of the APV Homogenizer Division.)*

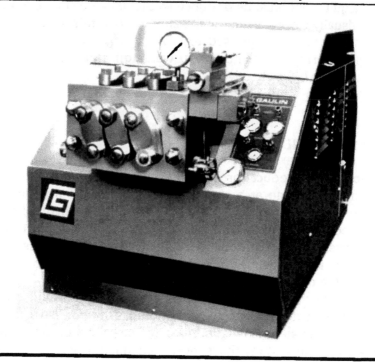

lobe pumps in series, thereby reducing the pressure differential of the last pump.

The suction and discharge pressure at the pump, which is acting as a timing device for the system, must be measured and possibly interlocked. If the pressure should fall below a predetermined minimum, diversion could occur; if the pressure increases to a predetermined maximum, the flow through the system may not be at the rate desired for the inactivation or kill desired (D value). Diversion can be done either automatically, semiautomatically, or manually. For fluid products, a piston pump (i.e., a homogenizer), equipped with larger pistons as the pressure is reduced, may be used. Most mechanical piston pumps that are equipped with either poppet or ball valves can transfer water used for sterilization and CIP solutions without significant wear. When pumping water for sterilization, wear will be nominal; most importantly, the flow will be constant.

Often, if a rotary positive or progressive cavity pump is used, it is bypassed during the sterilization or CIP process, as the temperatures used may be quite high and wear will be excessive. With certain rotary pumps (Figure 5.4), care should be taken depending on where they are located in the system to ensure that the proper clearance (hot) between the rotors and case is provided so the pump will not seize. Obviously, the pump pressure must not exceed the pressure capabilities of the piping, fittings, and valves that may be in the circuit downstream from the pump. As piston pumps develop high pressure, they may cause fittings or valves to leak or even burst.

**Figure 5.4.** Waukesha universal rotary pump used as a timing pump particularly in pasteurizing systems. *(Photo courtesy of Waukesha Cherry-Burrell.)*

# HOLDING

A critical phase of an aseptic or in-line sterilization process that has not received proper consideration is the holding of the product at the sterilization temperature. When continuous pasteurization of milk was first developed, a major concern was holding all portions of the milk for the minimum period of time required. There were designs and patents on holders that contained as little as one quart. The quart was put into a cup and held at the pasteurization temperature of 161°F for 15 seconds by conveying it in an atmosphere that maintained this temperature. At the end of the hold, the product was discharged into a tank, from which it was transferred through a cooler for filling and packaging.

Following this, it was determined that in a continuous-flow holding tube, if turbulent conditions existed, all particles would be held for the period of time for which the hold tube was designed. This was verified by many tests, such as timing the passage of an injected salt solution through the length of the holding tube. Turbulent flow was easily achieved with a fluid product like milk. When more viscous products, such as cream and ice cream mixes were pasteurized, the holding time was increased to 25 seconds and the temperature increased to 175°F. The theory was that these products may contain bacteria that are more heat resistant, and the higher product viscosity further protected them from heat; hence, longer times and higher temperatures were used. Certain ingredients, such as eggs and chocolate, may contain spores, which increases the heat resistance of the microflora in these products.

## Flow in Conventional Holding Tubes

A conventional holding tube consists of straight sections connected by 180° bends. Laminar flow in the straight section typically exhibits a range of velocities from the tube wall to the center. In the bends, a swirling motion called *secondary flow*, caused by centrifugal forces, is initiated. This secondary flow, depending on its intensity, acts much like turbulence in reducing velocity variations in the tube. However, this advantage is quickly lost in the next straight tube section, where the original flow pattern resumes.

A condition that may develop with products containing particles (which may be very small, such as starch granules) or some viscous products flowing in straight tubes is adhesion of some of the particles to the tube wall, effectively reducing the area available for flow. While these adhering particles will be held for more than twice the designed hold time, other particles and portions of the liquid may be going through the tube in less than one-half of the calculated hold time. This phenomenon has been seen with chocolate syrup. At the start of production, the specified hold is achieved. As time elapses, the chocolate coats the wall, gradually reducing the cross-sectional area of the holding tube. At the end of the day, the effective diameter of the hold tube may be only one-half of the tube inside diameter (d), which cuts the holding time to a quarter of the specified time for which the holding tube was designed. Bacteriological analyses through the day correlate well with the reduction in the time of hold; as the holding time decreases, the bacterial count increases.

The phenomenon of progressive reduction of holding time in an aseptic process has not been recognized by the FDA as a problem with homogeneous liquid products, in that the overall product is "safe." However, the work of certain investigators with more viscous products, such as ice cream mixes that tend to coagulate on the walls of the holding tube, has suggested otherwise (Goff and Davidson 1992). Considering some of the questions that have come about regarding conventional holding tubes, a holding tube should be designed in such a manner that everything possible is done to "stack the cards in favor of the processor," which will render a product bacteriologically inactive.

## Proper Hold Tube Design

To take advantage of the velocity-leveling effect of secondary flow in curved tubes, where secondary flow exists, the holding tube should be designed as one continuous helical coil, with an adequate r/D ratio. Secondary flow is also effective in mixing the phases and keeping the particles in uniform suspension. Further, the holding tube must be insulated from the environment to minimize heat loss and drops in temperature. This approach eliminates the need to oversize the holding tube to account for the wide variation in straight tubes (depending on the viscosity). A properly designed holding tube will produce a uniform sterilized product, which will be of the same quality at the beginning and end of the day.

# HEATING

## Plates

Plate heat exchangers were tried many years ago with fluid products (i.e., milk and juices), an extension of commercial pasteurizing operations at lower temperatures. However, because of the difficulties related to their use, plates in aseptic (or in-line sterilization) systems should be used only in heating temperatures to 90–100°C. One of the inherent problems with plates is their ability to be cleaned using CIP. Plates were originally developed for use in food applications as they could be disassembled, manually cleaned, inspected, and reassembled into the plate pack. With the advent of higher capacity systems, plates were cleaned by washing with various solutions (rinse, clean, rinse); consequently, they were not hand cleaned or inspected. Because of the higher temperatures used in aseptic systems and burning of product to the plate, or precipitation of salts on the plate, CIP sometimes did not remove all of the soil or salts. If the heat exchanger was dirty, the question was, "How dirty?" Another problem with plate cleaning occurred when juices containing pulp were processed. The pulp tended to gather in the low-velocity section of the plate and was difficult to remove using CIP. After this, soil tended to build up over a period of days, weeks, or months. The initial sterilization cycle may not have been adequate to sterilize the plates, thus introducing contaminants to the commercially sterile cold product. Spoilage then resulted. The same situation can develop when plates are used to heat solutions containing salts or minerals that precipitate when heated.

There is a basic question about plate sterility, because on one side of the gasket, the cool product is sterile. On the other side of the plate and gasket is potential contaminating air or media. If the plate does not have significantly greater pressure, or if the pressure in the plate should vary for whatever reason, contaminants can literally be sucked into the sterile zone and cause contamination. This can be a problem because plates develop stress cracks or pinholes since the material used to make the plate may be quite thin (0.020 in.).

Also, there was a mechanical problem. When the plates were sterilized with heat, they expanded. However, the frame and tie bolts that held the sections of the frame together (follower and fixed end) did not expand. This put enormous stress on the gaskets, and failure was at an accelerated rate.

This was further complicated as the elastomeric seals softened and weakened at the elevated sterilization temperatures; hence, leakage sometimes occurred during sterilization. Efforts were made to put more pressure on the gaskets by tightening the plate pack; however, this put more stress on the gaskets to the point where they would shear and fail.

Since plates were initially used, other heat exchangers have been designed and perfected to take their place. These heat exchangers not only withstand sterilization operations but also operate in a reliable manner, so sterility does not have to be a question. Plates are of interest from a historical standpoint but probably have little application for sterilizing liquids.


## Tubes

In the very early systems (1940s), when aseptic processing was first developed, commercial lines were installed, and capacities were quite low. Simple, tubular heat exchangers were used. A section of small-diameter tubing was formed into a helical coil, which then was placed in a casing with flanges on both ends. The coil entered the casing through a packing gland and discharged in the same manner. Inlets and outlets were provided for steam or water. On heaters, steam was used; as the product went through the small-diameter coiled tubing, it was heated by the steam. Condensate was trapped off the bottom and this sufficed quite well. Coolers were made so that cold water could be introduced, generally in counterflow to the product. In other words, the product may have entered a tube at the bottom and the water would enter at the top. For low-capacity systems (3–4 gpm; 12–16 Lpm) and easy-to-heat-and-cool products (i.e., milk or juices), this was perfectly adequate. However, as the filling equipment improved and container size increased (for industrial markets), the production rate of the systems had to increase; thus, this design no longer did the proper job. The number of tubes became excessive, and mechanical problems (i.e. vibration, water hammer, etc.) developed; design modifications, obviously, were in order.

One of the designs that resulted from this basic concept was to place a core inside the tube to displace volume. The tube was tightly wrapped around the core, and the shell was then wrapped around the tube to force the water or steam to flow in a helical manner countercurrent (when a liquid coolant was used) to the product in the tubes. This also resulted in certain problems in that the vibration of the tube, particularly with heaters using steam, was excessive, leading to the failure of a large number of tubes. This was accentuated with heaters or coolers when salts precipitated from the steam or water chemically eroded the tubing and caused pinholes or stress cracks. In one case, the design was changed to use straight tubes with 180° return bends, and supports in the tube were bundled to counteract vibration. The vibration was accentuated because a piston pump with harmonic vibrations was used. Pressures would be quite high, as homogenization pressure would be transferred back through the tubes.

## HydroCoil®

The next design of tubular heat exchangers (coiled) was the HydroCoil®, which incorporated many design advances that were not present in the earlier designs. First, if the $r/D$ ratio is in a certain range, *secondary* flow develops. This flow is caused by centrifugal forces that are present if the flow through the tube (velocity) is above a minimal level. By arranging the tube in this manner, the movement of the carrier and particles in the tube is much greater than it is in a straight tube. If products with particles (which may be extremely small) are processed, this secondary flow causes the particles and carrier to move. Heat transfer between the carrying liquid and the particles (or within the liquid) is very high. In essence, the carrier and a particle will move to the wall of the tube, and a second particle will try to displace the first particle, causing movement of the carrier and particles in the product stream. Movement can be controlled by adjusting velocity so that the ideal compromise between the heat exchange rate, the pressure that will be developed, and the damage to the particles or gel can be made (Carlson 1995).

In commercial plants, damage does not occur, and the heat exchange rate may be two to four times as much as in a straight tube; however, the pressure loss will be higher than in a straight tube. Heating and cooling is accomplished by putting hot water for heating, and cold water for cooling, on the outside of the tube. Large quantities of water are circulated, eliminating a limiting factor of the amount of energy available for heating or cooling.

If a heat-sensitive product is heated, such as low-cholesterol liquid eggs, the velocity may be increased significantly, which will result in very high heat transfer rates. Turbulence will be so high that heating can be done quickly, fouling will be reduced, equipment's operating time increases, and CIP of the equipment can be done.

HydroCoil® heat exchangers use pressurized hot water for heating, and tower water or refrigerated water for cooling. The water tends to act as a cushion, and tube failure is eliminated. In essence, the liquid acts as an absorber of the energy that may emit from the piston-type timing pump.

Most mechanical-type timing pumps have pressure/discharge curves that are a series of high and low points. The value is high when the piston is operating at its normal speed. However, when the piston stops at the end of a stroke, pumping ceases and the volume decreases to zero. By overlapping this pressure-volume with three, five, or seven pistons, and by operating the pump at a high speed, or by using a pulsation dampener (high pressure), the pressure can be evened to a degree. However, pistons can operate only at a maximum speed for a set period of time before problems result relating to the feeding of the pump or excessive seal wear. To have a pulsation dampener that is sanitary and rated at high pressures is not practical. Most pulsation dampeners rated at this pressure are industrial types that cannot be cleaned. A more practical approach is to use a liquid, such as water, as the heat exchange medium, and let it absorb the energy differences caused by the pump pulsations.

## Tube-in-Tube Heat Exchangers

Tube-in-tube heat exchangers have been used to heat and cool certain fluids without particles, or with very small particles. For these applications, mechanical devices have been employed to develop turbulence, to increase the heat transfer rate to minimize the number of tubes, and to heat and cool difficult viscous products. When the product is somewhat viscous and fairly uniform (i.e., tomato puree) or fluid (e.g., juices), tube-in-tube heat exchangers have done a fairly good job. However, if the product contains particles (i.e., pulp in orange juice, crushed pineapple, yogurt fruits, etc.), problems with particles clinging to the mechanical devices have resulted. It is very difficult to clean tube-in-tube heat exchangers.

Particle integrity is important. In essence, the particle is damaged when the product is heated or cooled. This damage has been most noticeable with products containing fragile particles (i.e., sliced strawberries, raspberries, and cubed potatoes).

When tube-in-tube heat exchangers are used, certain mechanical actions must be known. Most heat exchangers of this type are straight lengths of tubing, surrounded by other straight lengths of similar tubing of a greater diameter. When the heat exchangers are sterilized with heat, they become longer. After sterilization, and when they are cooled, the heat exchangers return to their normal length. When the heat exchanger tube resumes its normal length, it can draw contaminants into the sterile zone.

If tubular heat exchangers are used, they should be either circular, to expand or contract much like an accordion, or straight lengths of tubing, with only water or steam on the outside of the tube, and product on the inside of the tube. The outer tube should be fitted with an expansion joint so undue stress is not developed during the sterilization cycle. Another approach is to have the inner tube free (not attached) of the outer tube so that stress does not build up.

Another consideration, both with tube-in-tube and plate heat exchangers, is the limited amount of heat transfer medium that is on the outside of the heating/cooling tube or between the plates. Designers have tried to overcome this limitation by using steam at pressures that will release latent heat as the steam condenses. However, the amount of latent heat that can be released is limited, once again, by the volume between the inner and outer tube or between plates.

## VRC® MultiCoil Heat Exchangers

VRC® MultiCoil heat exchangers are similar to HydroCoils® except they have two or more coils in each shell (Figure 5.5). The coils are of the proper dimensions so that the *r/D* ratio provides maximum secondary flow and heat transfer. The advantages of putting multiple coils in one shell include the following:

- All connections (product and media) are convenient—1–2 ft off floor.

- Less headspace required—heat exchange shell more compact.

- Connecting pipe is a heat exchanger.

- Reduced temperature loss.

- Proper pressures at discharge of heat exchange tube.

- Lowest cost per square foot of heat transfer surface.

- Heat and cool fast.

**Figure 5.5.** Cross-section of a VRC® MultiCoil heat exchanger.

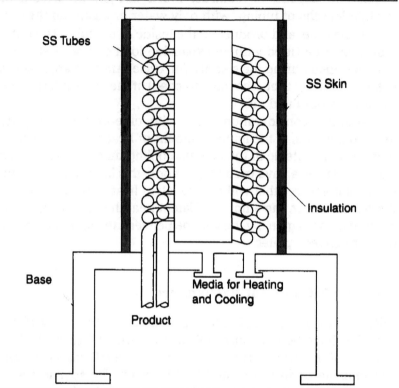

- CIP capability.

- No maintenance costs—no gaskets.

- Flexible—can be used as heaters or coolers.

- Minimum floor space.

One major advantage of having two or more coils in one shell is that the flow of product can be in series or parallel, and it can be either counterflow or parallel to the media flow. This offers much more flexibility as far as providing the ideal arrangement of the heat exchange surface for the product that may be heated or cooled. This may not be as great an advantage with a fluid product of low viscosity (e.g., PGW), but it is important when products are viscous or contain pulp (orange juice concentrate), or large particles (pie fillings). By arranging the surface in the proper manner, advantage can also be taken of the proper temperature splits between the heating medium (pressurized hot water) and the product, or the cooling medium (tower, chilled water, or glycol) and the product. By making the temperature splits ideal, the amount of heat transfer surface needed will be less, which means the initial cost of the equipment will be less, and heating/cooling will be faster. This is important with heat-sensitive products to minimize flavor degradation in products such as milk or "natural" fruit juices. Fouling of products, such as liquid eggs when heated, is dependent on the rate of heating and the temperature split. The best compromise for the longest operation is needed, and the VRC® MultiCoil heat exchangers provide this (reduced fouling).

By making the units more compact, which is a trend with all items today, heat exchangers occupy less space—both on the floor and overhead. By using this design, no sacrifices are made from a mechanical standpoint, and the efficiency is actually improved. Hence, there is an advantage to the user as less space is required, the heat exchangers are more user-friendly, and more efficient means less cost.

## CONTINUOUS STERILIZATION OF LIQUIDS— SYSTEM(S) DESIGN

There are many designs of systems that can be provided for the in-line sterilization of liquids. The following considerations are applicable:

- The physical nature of the liquid

- The temperatures that are to be used

- The hold time

- The type of timing pump to be used

- What utilities are available (steam, water, air, electrical)

- What degree of automation desired

- Identifications of the critical points

- The usage of automation or recording devices and interlocks

When the general specifics are identified, a very detailed flow schematic should be developed to verify that all of the desired operations can be accomplished. This flow schematic should then be discussed with the major individuals or departments within a company—including production, research and development, quality assurance, maintenance, engineering, and general management. The quality of the product, initial production rates, and expansion capabilities should be defined (McWilliam 1995).

## Regenerative Sterilizing System

Of primary concern is the ability of the system to heat the liquid to the temperature desired, hold it for a constant period of time to affect sterilization, and cool the product to the temperature desired. If the quantity to be processed is significant, it may be desirable to consider regeneration. Regeneration is using the hot sterile product to heat the cold incoming product and using the cold incoming product to cool the hot sterile product. Sometimes, the cold, nonsterile product may contain undesirable elements that must not contaminate the sterile product. This would include bacteria cells or spores, yeast, molds, or viruses. To prevent contamination of the sterile product, regeneration is normally done with a sterile medium (e.g., water). The water is heated by the hot sterile product, then cooled with the cold incoming raw product. The two are separated by the wall of the heat exchanger. If the heat exchanger is made from materials that are very substantial, contamination is not likely. As an example, coils of the VRC® MultiCoil heat exchanger usually use stainless steel tubes that are approximately 0.109 in. to 0.120 in. thick. As a comparison, some plate heat exchangers use plates that are 0.020 in. thick that have been stressed with physical alterations, such as corrugations or dimples being pressed into the plate. When the plates are stamped to form these corrugations or dimples, the metal is stretched and stressed. When it is stressed, a condition known as "stress corrosion" can develop. This is accentuated if the product being processed contains either active chemicals (e.g., halogens), or is used to heat or cool a pure liquid, such as deionized water, that acts as a solvent and extracts various elements from the steel.

A "typical" system to be used with various liquids is shown in Figure 5.2 (see page 121). This system incorporates regeneration as it is designed to heat some product—possibly WFI or PGW—to 77°C with indirect regeneration, and further heat this product to 143°C. The reason for heating to such a high temperature is to ensure that no microbial contaminants survive. Vegetative bacteria cells and spores, yeasts, and molds will be inactivated at the temperature that has been selected.

The system uses indirect regeneration, where water is transferred continuously from the regenerative heater to the regenerative cooler. At the point where the product enters the regenerative cooler, it is commercially sterile. Even though some heat exchangers, such as VRC® MultiCoils, utilize thick wall tubing, it is also possible to incorporate in the media pipe line filters that will remove any particle larger than 1 micrometer.

After the liquid to be processed is heated regeneratively to 77°C, it will be heated in the VRC® MultiCoil heater to 143°C. Water at 150°C is used to do this heating. Water that is pressurized is a much better heat exchange medium than steam, and its temperature can be better controlled. One of the reasons for better temperature control is that steam valves and traps are not opening and closing as the process is being accomplished. In addition, more energy is available and the heat exchange surfaces are always being washed.

Following heating, the product is held for the desired period of time. The hold should be in a coiled, insulated tube to ensure that all portions are held for the desired length of time and that temperature loss is minimal. The timing pump for the system that indirectly controls the hold is a piston pump that will not be affected by varying pressures as long as the pump is properly fed and the supply of product to the pump is maintained. For fluid product, a centrifugal pump and a meter with proper controls and records can be used, if the liquid does not foul the pipe or pump and develop back pressure.

From the holding tube, the product is cooled to 81°C by regeneration in the VRC® MultiCoil regenerative cooler. When the product discharges from this regenerative cooler, it can be directed to an aseptic tank and held at this temperature, which may be desirable if it is PGW or WFI. If the product should be further cooled, it can proceed to an additional VRC® MultiCoil cooler with cooling done to the desired temperature. The cooling medium can either be tower water, chilled water, glycol, or some other medium. It is common to use a back pressure valve as the product discharges from the cooler to ensure the pressure on it is greater than the pressure of the cooling medium. This is done to prevent cooling medium contamination from entering the sterile product. Actually, if a pinhole exists in the heat exchanger in the right place, venturi action will "suck" the cooling medium into the sterile product, thus contaminating it regardless of the pressure used. For this reason, the heat exchangers on the sterile side of the system should be thick and regularly pressure tested to ensure that pinholes do not exist. However, the use of elevated pressures on the sterile side is one additional safety factor that is commonly used to help prevent contamination (and may be legally required).

## Nonregenerative Sterilizing System

The VRC® nonregenerative sterilizing system (see Figure 5.1 on page 120) is less expensive to purchase for comparable capacities than a regenerative system. Because it does not incorporate regeneration, it probably would be used where the demand for sterile water is less. It may be used for a few hours during a day, or direct the sterile water or product to an aseptic holding tank that would then supply each use point.

The system should be sequenced and interlocked to guarantee that it is properly cleaned, initially sterilized, and operated properly; when switching from the product being processed to water, proper idling must occur. The system should be interlocked so that the desired temperatures are obtained and maintained for a specified period; if temperatures are lost during processing, the system diverts or causes distilled water to be directed to a noncontaminated area. This can be done at the start of the system by switching to distilled water that will not be packed or used, or a flow diversion valve can be installed after the final cooler. If a flow diversion valve is installed, it must not contaminate the distilled water; if it actuates the system, it must be resterilized before "forward flow" is possible. In addition, other items may be installed at this point (e.g., flowmeters to verify the output of the timing pump, and metal detectors to ensure the distilled water contains no metal prior to use). Flowmeters and metal detectors are available that will not contaminate sterile water.

## Modified-Simplified System

If the regenerative heater and cooler are removed, the entire heating load must be accomplished with one heater. The same thing is true with the cooler; however, depending on the final temperature, it may be desirable to have two cooling sections. One would be to use cooling water from an evaporative cooler (tower water); the other may use water from a well or city water source, or a refrigerated source. If the sterilized solution is cooled with tower water, it may be considered by "regeneration," as no energy is expended to cool the tower water. Another possibility would be to heat the water to about 70–80°C and deposit it directly into a tank.

## ASEPTIC TANKS

Aseptic tanks should be avoided if possible. They do not add value to products, they can be contaminating devices, and they cause certain mechanical problems. The mechanical problems relate to the fact that many products that are stored in tanks, such as water, act as a solvent for various minerals in the metal, causing pock marks to be etched into the surface. For this reason, 316SS ELC minimum should be used in the manufacture of aseptic tanks, as this material will reduce or eliminate this defect. Water and other products can cause discoloration, resulting in the tank appearing "dirty" after a period of time. Depending on the discoloration, it can be very difficult to remove; literally, the tank must be polished and the material must be removed to the point where bare, unstained metal again exists. This can be a very exasperating experience because the problem keeps reccurring with no apparent easy answer.

## CLEANING OF THE SYSTEM

The systems depicted in Figures 5.1 and 5.2 are capable of CIP. They can be automated, and CIP concentrate can be delivered to the water tank, which also functions as a CIP tank; optionally, the cleaning concentrate can be added manually. The temperature of the solution in the CIP tank will be automatically controlled, and flow from the tank will be programmed by the PLC to provide a RINSE, WASH, and RINSE cycle. The temperature of the rinse and wash cycles is automatically controlled. Temperature increases should be through the direct steam sparger in the water tank—not through the heat exchangers in the system. If the product being heated (sterilized) is sensitive and will foul heating surfaces, or if it contains various minerals, vitamins, or other additives that can adhere to heated surfaces, the use of heat during CIP will cause these materials to be "baked" to the surface; minerals and salts can be precipitated onto the surface and be very difficult to remove.

The steam that is used in the water supply tank, which is also functioning as a CIP solution rinse tank, must be "clean" steam. It must be produced using the techniques and recommendations required for steam of this quality, and not contain any deleterious materials—chemical or physical.

Another variation that would simplify the system would be to eliminate heating of the rinse or cleaning solutions in the supply tank; however, this is not recommended. The heating of cleaning solutions should be accomplished in the circulation tank. If this is not done, the product can foul the heat exchange surface in the system, or minerals and salts can be deposited on the hot surfaces and be extremely tenacious and difficult to remove.

If the system is initially sterilized with hot water, and undesireable salts do precipitate on the surfaces, they can be removed with a mild food-grade acid that can also be certified for pharmaceutical products. An acid, such as citric acid, will dissolve the salts, is nontoxic, and compatible with most products in the event of failure, and should become part of the product being processed. By circulating the acid in the manner of the solution being sterilized, salts and minerals will be removed from the heat exchange surface, and mechanical efficiency will be restored.

## Clean-in-Place (CIP)

Even though the sterilizing temperature may have been obtained, if the system is not properly cleaned, it may be impossible to sterilize. The CIP system does not have to be sophisticated and interlocked; however, if cleaning efficiency is determined through manual inspection, or by observing charts, thermometers, gauges, and so on, then an exact procedure must be established, followed with daily records to ensure cleaning is accomplished properly. If cleaning is done by a sophisticated system that delivers cleaning solutions at desired temperatures and pressures to the process system to be cleaned, the system should be inspected on a periodic basis after the CIP operation has been completed. This inspection should be made for the most difficult-to-clean components in the process system. During this inspection, the equipment, pipelines, and valves can be examined to ensure that they are operating properly, and do not have any cracks or crevices or gaskets that should be replaced.

On a periodic basis, gaskets should be taken from joints, inspected, cleaned if necessary, and replaced if bad. If the gasket is in good condition, it is logical to use it again after cleaning and treating it with a bactericidal solution. Automatic controls with sensors can indicate, control, and record items (i.e., pressure, temperature, and pH); however, they cannot look at a gasket and determine whether it needs replacing or cleaning, whether it is hard and brittle, and so on. Hence, it is important to establish a program to inspect critical points at regular times to verify that they are correct.

## CIP Solutions

CIP solutions are normally made by adding certain chemicals to well or city water that may or may not be filtered or treated. If the water is very hard or contains numerous minerals, excessive cleaning compounds must be used. If the strength of the chemicals is excessive, reaction between these chemicals and the stainless steel of the equipment will proceed at a much faster rate. The equipment will deteriorate and will need to be replaced at more frequent intervals. Also, many cleaning chemical companies use chlorine to enhance the ability of the chemicals to clean. Chlorine reacts with soil and does an excellent job of cleaning. Unfortunately, it also causes stainless steel to corrode and fail. This is accentuated at the elevated temperatures that are often used in the CIP cycle. If difficult-to-clean products are being processed, such as liquid eggs or dairy products, the CIP cycle may need to be longer at elevated temperatures. This further suggests the need for stronger CIP solutions, which, if used, will attack the stainless steel and cause it to fail that much faster. Probably the best approach is to prepare CIP solutions from high-quality water that is low in mineral content. This may mean using distilled or reverse osmosis–treated water. Often, the cost of supplying water of this type will be returned manyfold.

The use of water that has been magnetically treated often reduces the effect of "hard water"; a more thorough CIP process is accomplished and chemical corrosion of metals (stainless steel) does not occur. Various companies, such as Descal A Matic, provide equipment of guaranteed performance.

If the equipment is not properly cleaned, the question of how dirty it is and whether the initial sterilization cycle is adequate for sterilizing dirty equipment must be addressed.

Therefore, equipment should be examined after cleaning to verify that the CIP cycle has been adequate. The design of the equipment should be assessed to verify it can be cleaned by CIP, the CIP cycle should be performed at regular intervals, and the CIP solution should be as weak as possible to do the job. Heating of the CIP solution should be done by the CIP system, not the heat exchangers in the "sterilization" system. Heating with the equipment in the system may cause films or precipitated minerals to adhere more tenaciously to the heat exchangers and cause cleaning to be less effective. This means that the CIP cycle must be longer, which will deteriorate the equipment, or the CIP solution must be stronger or hotter, which will do the same thing.

The subject of CIP solutions—the type of solution, the duration of use, and the temperatures of use—is very involved. CIP merits a book by itself. The quality of water cannot be emphasized enough; if steam is added directly to the CIP vats, it will have a significant effect on the CIP solutions and cycles required.

## ASEPTIC BARRIER SEALS AND VALVES

### Barrier Seals

Generally, barrier seals have space sufficient to allow the flow of a sterilizing media. If the item to be sealed is the plunger of an aseptic homogenizer (pump), or a valve stem, the sealing distance (barrier) should be greater than the stroke of the pump or the valve. In this way, no portion of the plunger or valve in the sterile zone ever moves to the atmosphere and becomes contaminated. Pump plungers and valve stems can move at speeds faster than microbial contaminants can be sterilized; hence, the seal area must have a greater length than the stroke. Any portion of the plunger or valve stem that moves must travel from the sterile processing zone into the seal area only.

The plunger or valve seal area cavity must initially be sterilized with heat that can be conducted to all portions—including the areas between gaskets and sealing flanges, cracks, and crevices (Pfeifer and Kessler 1995). Heat can be conducted to these areas to sterilize them. After heat from steam or water is used to sterilize the item the area can be maintained in a sterile condition by the use of low temperature chemicals (Figure 5.6). This technique is used with heat-sensitive products (i.e., liquid eggs and dairy products) that tend to burn or denature when heat is applied.

## Aseptic Valves

Aseptic valves have "evolved" as aseptic processing and packaging in the food industry have developed. Initially, aseptic processing systems were used with Dole Aseptic Canning Systems® (DACS®), and the aseptic valves that were used were furnished with the DACS®. The "Christmas tree" (fill manifold) was a

**Figure 5.6.** Cross-section of barrier sealed valve.

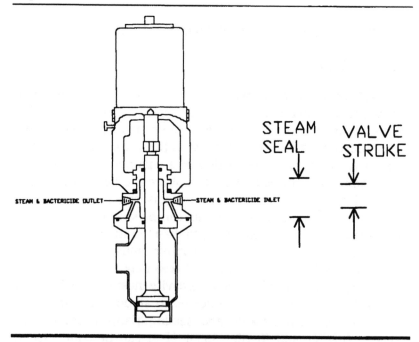

STEAM SEAL

VALVE STROKE

STEAM & BACTERICIDE OUTLET

STEAM & BACTERICIDE INLET

combination of hand-actuated piston-type valves that were maintained as sterile by enclosing them in the filling chamber, which was maintained at 265°F (130°C) with superheated steam. Hence, the main requirement of the valves at that location was an ability to withstand this temperature. Because of this requirement, most of these valves had metal-to-metal seats.

The DACS® was limited to liquids (milk, drinks, ice cream mixes, creams, and soups). Back pressure was maintained with a limiting orifice just prior to filling. If the product viscosity was different from the viscosity of the previously processed product, the orifice was changed. The flow was constant; hence, back pressure could be maintained satisfactorily.

Routing valves, in either tee or tee-tee configurations, were piston valves (Figure 5.6). These valves directed the product to a filler, aseptic tank, or drain supply. The concept was that the piston was in a barrier atmosphere, such as steam, and the barrier was of greater length than the stroke of the valve. Hence, no portion of the sterile valve ever moved to the contaminating atmosphere. These valves were available with barrier seals on the connections; if the valve was used in a vacuum, any contaminant drawn into the sterile area would have to move through the barrier. The steam was sealed from the atmosphere with an O-ring; hence, a double seal existed and leaks did not occur.

A disadvantage of a steam barrier was that slow flow of the product caused the product to increase in temperature by several degrees. Slow-flowing product would burn at the hot barrier surface. With heat-sensitive products, such as liquid eggs or milk, the barrier area was first sterilized with heat, and the barrier material was changed to a cool liquid sterilant (Figure 5.7). This sterilant could not sterilize between gaskets and the sealing surface or cracks/crevices. As systems became more complex and more sterile valves were required, the number of barrier seals became excessive and complicated. Also, for integrity to be guaranteed, the failure areas had to be interlocked. This complicated the piping in the system and the control system. Hence, efforts were made to develop designs that could be used without barrier seals.

Piston valves were equipped with diaphragms. As the piston moved up or down, the diaphragm flexed when the stem

**Figure 5.7.** Aseptic barrier seal piping schematic.

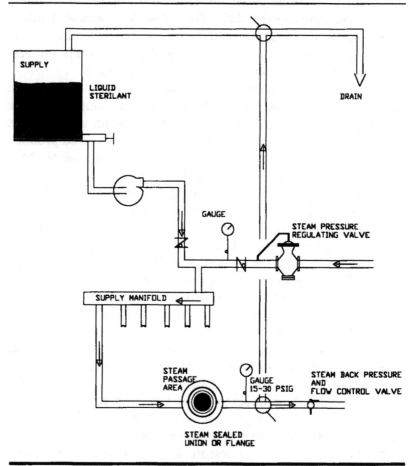

moved, and no barrier seal was required. The materials initially used for valve construction were not satisfactory. New materials and designs resulted in reliable valves at present.

Initially, the major manufacturers of diaphragm valves did not make stainless steel versions with diaphragms approved for use with food. Today they do. The diaphragms are reliable and withstand a number of cycles and varying temperatures; however, the diaphragms should be changed on a regular basis.

Certain plants use sterilizing solutions on top of the diaphragm to guard against failures such as pinholes or cracks. If a failure occurs, then leakage will be from the sterilizing liquid, such as oxonia or iodine. The sale of valves and fittings to the food industry has been limited; hence, manufacturers have not made a significant effort to upgrade designs. In-line sterilization has escalated in the pharmaceutical industry. Valve manufacturers have redesigned products to satisfy this demand. Piston-type valves are available in 1.5–3 in. sizes. Diaphragm valves are available in 0.5–4 in. sizes.

## CONTROLS

Control interlocks must be incorporated (Figure 5.8). The final temperature to which the product is heated must be indicated and recorded continuously, either on a round or continuous strip chart. The product temperature is sensed with a high-quality sensor, which must be standardized.

The temperature at the end of the holding tube must be indicated and recorded continuously on either a round chart or continuous strip chart (Carlson 1995b). Control of the sterilizing temperature (the temperature to which the product is heated at the discharge of the final heater) can be by an electronic or pneumatic instrument. Control valves are pneumatic due to moisture. The controlling instrument may be electronic; however, a transducer is used to convert the electronic output of the instrument to the pneumatic actuation of the valve. The electrical lead should be an extension wire of the type that is compatible with the sensor and the controlling instrument. This wire is available in either conventional extension or "standardized" wire; it is important to use the proper type. Transmission wire (e.g., like that used in conventional wiring) is likely to produce errors with thermocouple sensors.

Sterilization must be with heat. It cannot be with heat and chemicals (e.g., acidified water for acid products) because potential contaminants exist between cracks, crevices, flanges, and gaskets. If heat is not conducted into the area between the flange and the gasket, or in a crack or crevice, cryptic microbes can migrate to the sterile product (Pfeifer and Kessler 1995).

**Figure 5.8. VRC Co. in-line sterilization control system.**

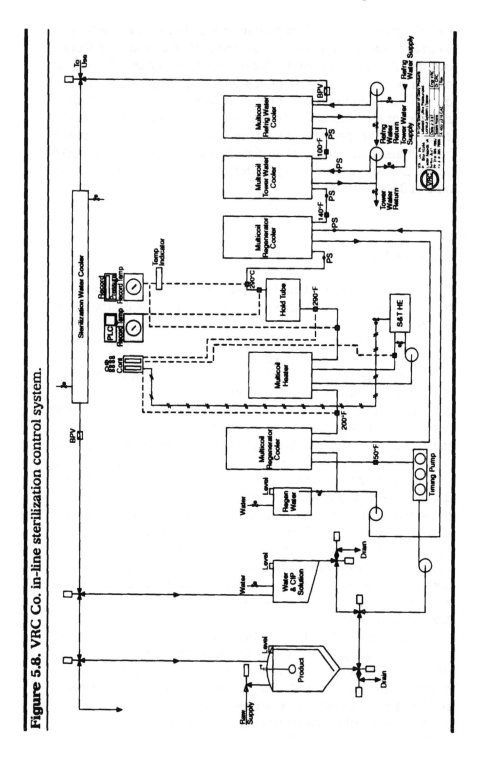

It may be desirable to control preheat temperatures automatically so that the desired preheat temperature is obtained and the temperature of the product entering the final heater will be consistent. Oftentimes, the preheat temperature will govern the water or steam temperature used for heating.

If the product is an acid product, and a system incorporates a provision for idling, the water used must be acidified to the pH of the product or lower (preferable). If this water is not of the proper pH and the switch is made from product to plain well or tap water, some contaminants in the water may not be killed. Hence, the pH of the water that will be used must be recorded and controlled to the proper level. If it is not to the set pH, the system should divert and be interlocked so that product cannot enter the system (Fernandez et al. 1995).

It may be desirable to provide an interlock on the rate at which product moves through the system. If there is any chance the rate can be increased to the point that sterility may be lost, flow can be diverted.

PLCs have been programmed for controlling other operations in systems used to sterilize fluid products. They are particularly well adapted to controlling the sequence of opening and closing the tower and refrigerated water valves used with aseptic surge tanks. They control the operation of air-gas sterilizing systems used in conjunction with aseptic surge tanks during sterilization and cooldown. They control the pressures and temperatures within the tank and maintain the desired pressure to vent a tank and unload excessive pressure. Pressure unloading where the pressure is maintained constant is common with sterile tanks.

Other operations in plants using in-line sterilization techniques are also commonly sequenced and controlled with PLCs. Besides providing sequencing of valve opening/closing and motor starting/stopping, PLCs can be arranged with temperature control loops to provide simple PID, cascade, feed-forward, or fuzzy control. Oftentimes, a plant will utilize a PLC to perform all operations, so a complex control system that is performing many functions utilizes only a small amount of space.

## SUMMARY

In-line thermal sterilization of liquids has been discussed. Terms developed include D value, F value, and water activity. Thermal sterilization of liquids has been used by the food and dairy industry for many years. This was in response to public pressure for safe, nonspoiling consumables. By retarding spoilage, products could be distributed widely.

With pasteurization used by the dairy industry, and hot fill and retort techniques used by the canning industries, considerable technical information concerning the inactivation of bacterial spores, enzymes, and toxins was developed. It was determined that the use of thermal inactivation is predictable and can be measured (time and temperature). To establish a process (time and temperature), a D value must be determined which is affected by many factors. It is possible for pharmaceutical operations to determine D values for the products that can be in-line sterilized. The exact time and temperatures determined will also be affected by mechanical and economic factors.

A considerable amount of work has been done relative to the inactivation of viruses that most processors (pasteurizers, hot fill, and retort) do not consider. Some of the recommendations relate to enzyme inactivation and the prevention of enzyme regeneration after thermal exposure.

To perform in-line sterilization, it is necessary that the time of exposure and the temperature at which exposure occurs be exactly controlled. Time can be significantly affected by the design of the pump and the holding tubes. The temperatures possible and time can be affected by the type of heat exchanger. The heat exchanger must be capable of CIP, should be sterilized and maintained in a sterile condition, and should not alter product characteristics (flavor, color, odor, viscosity changes; precipitation of salts; fouling; etc.).

Because systems can easily be designed to heat products to sterilization temperatures and hold them without adding significant costs to the operation, they are worthwhile to consider. Various designs of components, such as tanks (aseptic should be avoided), valves, and controls, as well as pumps and heat exchangers, were discussed.

The timing pump should be a true positive type, and the heat exchangers should be tubular without gaskets. The effect of secondary flows was indicated, and designs that develop these should be used when possible.

The design of systems and equipment that are used for in-line sterilization are now being used in the bioengineering field to produce pharmaceutical and food products. Bioengineering has been practiced for a number of years in the food industry to produce fermented products, varying from alcohol to wines, and has included products as diverse as cheeses and vinegar, and in the pharmaceutical industry products such as penicillin. However, in-line sterilization is becoming a science of its own. Technology, equipment, and systems exist today that allow the use of in-line sterilization by the pharmaceutical industry.

## REFERENCES

ASME/ANSI F2.1. 1986. Food, drug, and beverage equipment. New York: The American Society of Mechanical Engineers.

Ball, C. O., and F. C. W. Olson. 1957. *Sterilization in food technology*. New York: McGraw Hill Book Co., Inc. p. 189.

Bartlett, F. M., and A. E. Hawke. 1995. Heat resistance of *Listeria monocytogenes* Scott A and HAL 957E in various liquid egg products. *Food Protection* 58:1211–1214.

Blackwell, J. H. 1976. Survival of foot and mouth disease virus in cheese. *J. Dairy Sci.* 59:1574–1579.

Blackwell, J. H. 1978a. Persistence of foot-and-mouth disease virus in butter and butter oil. *J. Dairy Res.* 45:283–285

Blackwell, J. H. 1978b. Potential transmission of foot and mouth disease in whey constituents. *J. Food Protection* 41:631–633.

Blackwell, J. H., D. O. Cliver, J. J. Collis, N. D. Heidelbaugh, E. P. Larkin, P. D. McKercher, and D. W. Thayer. 1985. Food borne viruses: Their importance and need for research. *Food Protection* 48:717–723.

Carlson, V. R. 1972. Pasteurization. Delavan, WI: Cherry-Burrell Corp.

Carlson, V. R. 1995a. Air sterilization. Cedar Rapids, IA: VRC Co.

Carlson, V. R. 1995b. Processing product with particles, delicate gels, and sensitive flavors. Paper presented to the AIChE Conference on Food Engineering.

CFR. 1988. *Code of Federal Regulations*, Title 21 Part 113.40: Thermally processed low acid foods packaged in hermetically sealed containers. Washington DC: U.S. Government Printing Office.

Cunliffe, H. R., and J. H. Blackwell. 1977. Survival of foot-and-mouth disease virus in caesin and sodium caseinate produced from milk of infected cows. *J. Food Protection* 40:389–392.

Cunliffe, H. R., J. H. Blackwell, R. Dors, and J. S. Walker. 1979. Inactivation of milkborne foot-and-mouth disease virus at ultra-high temperature. *J. Food Protection* 42:135–137.

DeLeeuw, P. W., and J. G. Van Bekkum. 1979. Report of the session of the research group of the Standing Technical Committee of the European Commission for the Control of Foot-and-Mouth Disease. 12–14 June, in Lindholm, Denmark.

Evans, D. A., and D. J. Hawkinson. 1970. Heat resistance of certain pathogenic bacteria in milk using a commercial plate heat exchanger. *J. Dairy Science* 53: (12): 1659–1665.

Fernandez, P. S., F. J. Gomez, M. J. Ocio, T. Sanchez, and A. Martinez. 1995. D values of *Bacillus stearothermophilus* spores as a function of pH and recovery medium acidulant. 58 (June): 628–632.

FDA. 1982. *Milk Pasteurization Controls and Tests*. Rockville, MD: Food and Drug Administration, State Training Branch, HC-61, pp. 11–15.

Frazier, W. C. 1958. *Food microbiology*. New York: McGraw Hill Book Co., Inc., pp. 6, 226.

Frobisher, M. 1957. *Fundamentals of microbiology*. Philadelphia: W.B. Saunders Co., pp. 61–75, 575–587.

Goff, D. H., and V. J. Davidson. 1992. Flow characteristics and holding time calculations of ice cream mixes in HTST holding tubes. *Food Protection* (Jan): 34–37.

Hyde, J. L., J. H. Blackwell, and J. J. Callis. 1975. Effect of pasteurization and evaporation on foot-and-mouth disease virus in whole milk from infected cows. *Can. J. Comp. Med.* 39:305–309.

Ito, K. 1970. Aseptic processing bacteriological aspect. Paper presented to AIChE NFPA Western Research Lab, Dublin, CA.

McWilliam, A. J. 1995. The design of high purity water distribution systems. *Pharmaceutical Engineering* (Sept/Oct): 54–71.

Pfeifer, J., and H. G. Kessler. 1995. Heat resistance of *Bacillus cereus* spores located between seals and seal surfaces. *J. Food Protection* 58:1206–1210.

Pflug, I. J. 1987. Factors important in determining the heat process value, $F_T$, for low acid canned foods. *J. of Food Protection* 50:(6) 528–533.

Read, R. B. Jr., R. W. Dickerson, Jr., and H. E. Thompson, Jr. 1968. Time-temperature standards for the ultra-high temperature pasteurization of grade A milk and milk products by plate heat exchange. *J. Milk and Food Tech.* 31 (3):72–74.

Three A (3A) Sanitary Standards, 3A Accepted Practices, and E3A Sanitary Standards. Des Moines, IA.

Traum, J. 1924. *The thermal death points of* tubercle bacilli *and their bearing on pasteurization of milk.* Albany, NY: J.B. Lyon Co Printers, pp.174–203.

Vardaq, T. 1995. High pressure: A real alternative in food processing. *Int'l. Food Marketing and Technology* (Feb). 42–47.

VRC. 1995. *Holding Tubes.* Cedar Rapids, IA: VRC Co.

# 6

# Sterilization of Medical Products Using Gas Plasma Technology

*Phillip A. Martens, Ph.D.*
*Victoria Galliani*
*Gary Graham, Ph.D.*
*Ross A. Caputo, Ph.D.*

AbTox Inc.
Mundelein, IL

Increasingly stringent rules governing the use of ethylene oxide have created an interest in developing alternative low-temperature sterilization technologies. One such technology uses low-temperature gas plasmas to kill microorganisms. Although much of the technology was developed for other purposes in the 1970s, it was not used extensively in the sterilization industry until the 1990s. This chapter describes low-temperature plasmas, basic plasma chemistry, the equipment used to create plasmas, and the performance of present-day plasma sterilizers.

## PLASMA

Plasmas have been studied for over 100 years. These studies have been important in the development of quantum mechanics, modern physics, and chemistry, and they have played roles in the development of devices as diverse as lasers, nuclear fusion reactors, light bulbs, automobile engines, and computer chips. Plasmas are enormously complicated, and they are being studied more intensely today than ever before.

In extremely general terms, *plasma* refers to a gas in which some of the gas particles are electrically charged, with positively and negatively charged particles present in approximately equal numbers. Furthermore, these charged particles are present in sufficient quantities to affect the overall behavior of the gas. In a normal gas, in which virtually all of the constituent particles are uncharged, the forces between the particles are quite weak. Particles move independently of one another until they collide. Particle motion is determined by these collisions.

In contrast, the movement of a charged particle is affected by the presence of other charged particles at a relatively large distance. Its motion will be determined not only by collisions with other particles (charged or uncharged) that are directly in its path, but also by attraction to oppositely charged particles and repulsion from particles with the same charge that may be very far away.

In an ordinary gas, such as the air we breathe, there are usually a few charged particles, but they are so sparse and so far apart that they have virtually no effect on the behavior of the gas. An opposite extreme is the interior of the sun, where almost all of the particles are charged, and electrical interactions between the particles dominate the behavior of the gas. In the man-made plasmas typically used in sterilization, only a small fraction of one percent of the gas particles are charged, yet the forces between these particles are strong enough to alter the overall behavior of the gas substantially.

Physicists have formalized the concepts underlying the behavior of plasmas, and they have devised several criteria for determining whether or not a substance is a plasma, based on the density (or concentration) and energies of the charged particles. Applying these criteria, they have determined that plasmas include the interiors and atmospheres of stars, the earth's ionosphere, lightning, the aurora borealis, and man-made phenomena such as welding arcs, fluorescent lights, and neon lights. In fact, it has been estimated that 99 percent of the matter in the universe is in the form of plasma.

The man-made plasmas used for sterilization are in the category known as low-temperature glow discharges. Fluorescent lights and neon lights are also in this category. These plasmas are formed by applying an electric field to a gas. Under the right circumstances, this will lead to events that cause the neutral particles in the gas to dissociate or split into other neutral species. In other cases, an atom or molecule may be split into two charged particles, one with a positive charge and one with a negative charge. (These reactions are discussed in more detail below.) Overall, there are as many positive charges as there are negative charges in the plasma, and one might expect that the effects of the positive and negative charges would cancel each other out. However, on a microscopic scale, there are regions in which there are excess positive particles and regions in which there are excess negative particles. The localized charge imbalances are responsible for the behavior of the plasma.

## PLASMA FORMATION

There are several physical and chemical reactions that are quite important in the study of glow discharges. Consider a very simple system in which two electrodes are separated by a volume of an ordinary gas, such as air. One of the electrodes has a positive voltage applied to it, and the other has a negative voltage applied. The difference in voltage creates an electric field between the electrodes. As mentioned above, there are a few charged particles in just about any sample of gas one might care to examine. In the simple system considered here, such naturally occurring charged particles will move in response to the electric field. The negatively charged electrons will be repelled from the negative electrode and attracted to the positive electrode. Conversely, the positive ions will be repelled from the positive electrode and attracted to the negative electrode. The particles will continue to pick up speed and energy until they collide with an electrode or another gas particle, charged or uncharged.

If a low-energy particle collides with another low-energy particle, the two generally just "bounce off" each other. However, if the particles collide with sufficient energy, an electron can be completely dislodged in the collision. This reaction is called *ionization*. In glow discharges, this usually happens when an energetic electron collides with an atom or molecule. The minimum electron energy necessary for the reaction to occur is called the *ionization energy*. The ionization reaction may be written as follows:

$$e^- + A \rightarrow e^- + e^- + A^+$$

where $e^-$ represents an electron, A represents a neutral atom or molecule, and $A^+$ represents the positive ion formed by the collision. (Usually, ionization is caused by collision of an energetic electron and an atom or molecule; ionization caused by the collision of an ion and an atom or molecule occurs but is relatively rare.) Note that another free electron and an ion are produced in the collision. Both electrons will be accelerated in the field and may interact in other ionization reactions. This sets up a reaction in which neutral particles are converted to ions and electrons in rapidly increasing numbers.

The production of charged particles is limited by the reverse reaction, which is called *recombination*. In this reaction, two oppositely charged particles combine to form one electrically neutral particle. The reaction may be written as follows:

$$e^- + A^+ + M \rightarrow A + M$$

where M represents a third body. The laws of physics require a third body in this reaction to absorb some of the energy and momentum when the positive and negative particles join to form one particle. (There are other recombination mechanisms that do not require a third body, but their occurrence is relatively rare.) The third body may be another gas particle or a surface. In many low-pressure plasma systems, recombination at surfaces is much more common than recombination involving three gas particles.

As mentioned previously, the positive particles tend to migrate toward the negative electrode and the negative particles migrate toward the positive electrode. The movement of charges to the electrodes perturbs the electric field; a charged particle between the two electrodes will be affected not only by the electric field from the voltage applied to the electrodes but also by other charged particles in the gas. This has the effect of reducing the overall electric field in the volume between the electrodes and increasing the field at the electrodes and other surfaces in contact with the plasma. This phenomenon is responsible for establishing the shape and uniformity of the plasma. It is also self-limiting: If the field strength at a particular point is too low, ionization will drop, charging at the electrodes will drop, and the resultant field strength will increase. If the field strength is too high, ionization will increase until the charged particles produced form their own counterbalancing field.

The plasma reaches a steady state when the rate of formation of charged particles matches the rate of their destruction. Gas particles move fairly quickly, so it takes very little time to establish a steady state once a field is applied to the gas.

When the external electric field is removed (that is, when the voltage is removed from the electrodes), the fields caused by the charged particles remain. These fields cause the particles to rearrange themselves; ultimately, they bring about the destruction of the plasma. This happens in the following manner: The electrons, being thousands of times lighter than ions, diffuse much more rapidly than ions. Because of this great mobility, electrons tend to migrate quickly to the surfaces of the plasma system, and an electric field is established between the ion-rich gas, which has a positive charge, and the electron-rich surfaces, which have a negative charge. The positive ions are repelled from other positive ions in the gas and attracted to the electrons at the surfaces. They therefore move toward the surfaces to recombine with electrons and form electrically neutral particles. As the number of charged particles in the gas dwindles, the associated electric fields also decay. The localized fields quickly reach a point where they are no longer strong enough to promote ionization, and the production of charged particles ceases. Recombination continues until virtually all of the ions and electrons have formed neutral particles. The few remaining charged particles are spread so far apart that they do not have a pronounced effect on the behavior of the gas, and the plasma ceases to exist.

The times required to establish a plasma and to allow it to decay once the external field is removed depend on a number of factors, such as the physical size of the system and the pressure, temperature, and composition of the gas. In most man-made systems, these times are very short. For example, the plasma in a camera's electronic flash tube is created and destroyed within a few ten thousandths of a second. Similarly, the spark in an auto ignition system is a plasma that has (and must have) a very short lifetime.

At this point, it is worthwhile to consider some factors that limit plasma formation, since they pose constraints on the design and use of practical plasma sterilization systems. First, the creation of plasmas requires a strong external electrical field. Without such a field, electrons cannot be accelerated to sufficiently high energies to ionize the gas molecules with which they collide. Without ionization reactions, gases cannot accumulate enough charged particles to become plasmas.

Gas pressure is also critical. Pressure depends on the gas density (the number of gas particles per unit volume) and on the gas temperature. At a given temperature, pressure is proportional to density. Gas density is also related to the distance a particle can travel before it collides with another particle. If an electric field is applied to a gas at a relatively low pressure, an electron in the gas will be able to travel far enough (or long enough) between collisions to pick up enough energy to cause ionization when it finally does collide with an atom or molecule. If the gas pressure is too high, an electron being accelerated by the same field will experience a collision and dissipate its energy before it has gained enough energy to cause ionization. Therefore, for a given gas composition and a given electric field, there is an upper limit to the pressure at which plasmas can be formed.

There is also a lower pressure limit for plasma formation. If the pressure is too low, an electron might collide with a surface before it collides with another gas particle. To create a plasma, gas ionization must occur, so there must be enough gas particles between a typical electron and a surface to make the likelihood of an ionizing collision reasonably high. By the same token, if a plasma is to form between two surfaces, the surfaces must be far enough apart that a sufficient number of gas molecules can fill the space between them. These requirements must be seriously considered in the design of all plasma systems, and they have had an enormous impact on practical plasma sterilization processes.

Pressure also has an effect on the "shape" of the plasma. At relatively high pressures, man-made plasmas tend to resemble arcs, such as welding arcs and spark plug sparks. The charged particles tend to move along a fairly narrow path between electrodes, constrained by frequent collisions with other gas particles. At lower pressures, collisions are less frequent, the charged particles can spread out more, and the plasma is correspondingly more diffuse. Fluorescent lights and neon lights are examples of low-pressure plasmas.

## OTHER IMPORTANT PLASMA REACTIONS

The preceding discussion focused on the production and behavior of the charged particles that differentiate a plasma from an ordinary gas. There are many other reactions that also occur in a plasma that have a relatively small impact on plasma formation but have an enormous impact on plasma sterilization.

One such reaction is called *excitation*. This occurs when an electron strikes a gas particle, an atom for instance, and causes one of the electrons in the atom to jump to a higher energy level. The high energy electron is still bound to the atom, although it is (on the average) farther from the nucleus. An atom with an electron in a high energy state is said to be excited. The same process can also happen when an electron strikes a molecule.

It takes less energy to excite an electron to a high energy state than it does to separate an electron from a gas particle completely. In a plasma, some electrons have enough energy to cause ionization; many more have enough energy to cause excitation. It follows that if there are electrons with sufficient energy to sustain the ionization reaction necessary for plasma formation, there are electrons capable of causing excitation.

A process that may be thought of as the reverse of excitation is *relaxation*. In this process, an excited electron drops to a lower energy state. This is accompanied by the emission of a light photon that has an energy equal to the difference between the energies of the initial and final states of the particle. The visible light emitted in relaxation is what causes the glow of a glow discharge. The light from neon lights, sparks, camera flashes, and lightning is generated by this process. The light can span wavelengths from infrared to visible to ultraviolet, depending on the composition and pressure of the gas and the strength of the electric field. Germicidal lamps utilize a plasma to produce their microbiocidal light.

Relaxation is often a very fast process. Once an electron jumps to a high energy state, it usually takes less than a microsecond (and often about a nanosecond) to relax to a lower state. This makes plasmas very useful in camera flash units. However, there are important exceptions to this rule in which the excited electrons are prevented from dropping to a low-energy state immediately, but instead drop back after intervals of time from milliseconds up to several seconds.

If an energetic electron collides with a gas molecule, it can cause the molecule to break into pieces. This process is called *dissociation*. For instance, if an electron collides with an oxygen molecule, it can break the molecule into two oxygen atoms.

$$e^- + O_2 \rightarrow e^- + O + O$$

This type of reaction is extremely important in many plasma processes and particularly in plasma sterilization processes. The molecular fragments created in dissociation reactions often have unique and desirable chemical properties. In many instances, plasmas are employed solely to produce these fragments. With the proper plasma reactor configuration, the uncharged fragments may be separated from other reactive components of the plasma (such as the charged particles and ultraviolet light) and may be used for reactions that would be quite difficult to accomplish by other means.

The fragments produced by dissociation may react with other fragments, or with other gas particles, to either reform the starting molecules or to form new compounds. This is called *recombination*. (It is similar to, but not to be confused with, the recombination of oppositely charged particles to produce neutral particles.)

For instance, consider the example of a plasma with oxygen and hydrogen in the gas. Dissociation would result in the formation of oxygen and hydrogen atoms:

$$e^- + O_2 \rightarrow e^- + O + O$$

$$e^- + H_2 \rightarrow e^- + H + H$$

Recombination could produce the following:

$$O + O \rightarrow O_2$$

$$H + H \rightarrow H_2$$

$$O + H \rightarrow OH^1$$

$$O + H_2 \rightarrow H_2O$$

$$H + OH \rightarrow H_2O$$

In the above example, two relatively unreactive gases, hydrogen and oxygen, are converted into hydrogen and oxygen atoms in a plasma. $O_2$ and $H_2$ are chemically quite benign. O and H atoms are quite reactive. O and H atoms can also recombine to form the hydroxyl radical OH, which is also quite reactive. O, H, and OH can also recombine with each other to form $H_2O$, $O_2$, and $H_2$.

The impetus for developing plasma sterilization processes has been to replace ethylene oxide with a more benign sterilant that would work at low temperatures. The above example shows

---

1. Hydroxyl radical, not to be confused with hydroxyl ion.

one way in which this can be done. Gases that are relatively safe are converted into highly reactive chemicals that interact with microorganisms and disrupt their essential functions. Further reaction causes these chemicals to return to their original form or to form other relatively safe chemicals. In an ideal plasma sterilization system, microbes are exposed to lethal chemicals; patients and healthcare personnel are exposed to harmless chemicals.

One final and rather obvious plasma reaction should be mentioned: heating. The energy the electric field imparts to the charged particles may be dissipated as heat when the particles collide with other gas particles and with surfaces. This results in the heating of the gas and of the surfaces. The electrical power put into the system is ultimately converted into chemical energy (through the formation of new compounds), light, and heat. This must be considered in the design of a plasma sterilization system intended to sterilize delicate, heat-sensitive instruments.

The discussion above is intended to be a simple description of some of the most basic plasma processes. However, as mentioned at the beginning of the section, a plasma is an enormously complicated entity. The ions and electrons found in all plasmas may be formed by several mechanisms not discussed here, and the formation of light and reactive chemicals are topics that have been barely touched. Furthermore, there are exceptions to virtually all of the rules presented here, and there are several industries that have been based on these exceptions. Interested readers, preferably with an inclination toward mathematics, are referred to the following books for more complete descriptions of plasma phenomena: *Glow Discharge Processes* (Chapman 1980) or *Fundamentals of Gaseous Ionization and Plasma Electronics* (Nasser 1971).

## PLASMA REACTORS

All low-pressure plasma reactors have several common features: a vacuum chamber and a vacuum pump, a power source to supply the requisite electrical field, and a source of gas.

## Vacuum Systems

As noted above, when a discharge is initiated in a gas at atmospheric pressure (about 760 torr or 1013 mbar), the plasma tends to take the form of an arc. High-pressure discharges pose a number of engineering challenges, particularly heat dissipation and material compatibility. Therefore, practical plasma reactors utilize gases at reduced pressures where the plasma tends to spread out and cover a wide area. Consequently, the plasma is formed in a leaktight chamber capable of withstanding the forces encountered when the gas inside is evacuated. Also, a vacuum pump is required to remove the gas from the chamber.

The vacuum system represents a significant portion of the cost of a plasma reactor. The vacuum chamber must be made of mechanically strong materials, such as stainless steel or aluminum (although thick glass or quartz have also been employed). Special care must be taken in designing and fabricating the system to ensure that it is leaktight.

The vacuum pump must be chosen according to the volume of the system and the flow of gas required. Large vacuum chambers generally require large vacuum pumps, and it is not unusual for the pump to be the most expensive component of the entire plasma system.

One consequence of using low-pressure plasma systems for sterilization is that liquids cannot be sterilized. In fact, nothing that is wet should be placed in the plasma reactor. There are three consequences to putting water in a vacuum system. First, the water will slowly turn into water vapor and make it very difficult to evacuate the system. The evolving vapor may make it appear that the system has a leak even though the vacuum integrity of the apparatus is satisfactory. Second, part of the water may freeze, cooling its surroundings and thereby slowing chemical reactions nearby. Third, the water (liquid or ice) may block the chemicals in the plasma from reaching the surfaces underneath and thereby prevent the desired chemical sterilization reactions.

## Power Sources

One of the major considerations in designing a plasma reactor concerns the way in which the necessary electrical field is introduced and the nature of the electrical field itself. The field can be produced from a direct current source or from an alternating current source; in the latter case, the frequency of the alternating current source can vary from a few Hz to a few GHz.

The frequency of the source can have a large impact on the plasma. In a direct current plasma, electrons migrate toward the positive electrode and positive ions migrate toward the negative electrode. The charges that build up near the electrodes tend to cancel out the field in the regions between the electrodes, and so they affect the behavior of the plasma between the electrodes.

If an alternating current voltage is applied to the electrodes, the charged particles will be attracted to one electrode half of the time and to the other electrode the other half of the time. If the field oscillates fairly slowly, the resulting plasma resembles a series of direct current plasmas. For example, household electricity oscillates at a frequency of 60 Hz, and the voltage varies sinusoidally. Let us imagine an example in which power is applied to a neon lamp at an instant when there is no voltage difference between the power wires (or the electrodes in the lamp). Within a few milliseconds, one electrode (for example, the left electrode) will acquire a substantial positive charge, and the other (right) electrode will acquire a negative charge. A glowing plasma will form, with electrons migrating left and positive ions migrating right. The line voltage will peak after 1/240th of a second, and will then begin to fall. Soon, there will not be enough voltage between the electrodes in the lamp to sustain the plasma, so the plasma will decay, charged particles will recombine to form neutral particles, excited particles will relax, and the lamp will go dark. After 1/120th of a second, the line voltage will be back to zero. The voltage on the electrodes will then increase again, but with the positive charge on the right electrode and the negative charge on the left electrode. Another plasma will form, with charged particles moving in directions opposite the first plasma. The voltage will then drop again, the plasma will be extinguished, and the system will return to its starting state (no voltage difference between the electrodes) 1/60th of a second after the power was initially applied.

Now consider what will happen if the frequency of the applied voltage is increased. Since the electrons are much lighter than the ions, they travel faster than the ions and they build up a substantial charge around the positive electrode more quickly than the ions build up a charge around the negative electrode. As the electrodes change polarity faster and faster, there will come a point at which the ions will not move quickly enough to form a substantial accumulation at the negative electrode before it changes polarity, although the electrons will still be able to charge the positive electrode. In this case, the plasma will be relatively rich in positive ions, which will cause it to behave quite differently than the direct current plasma. As the frequency continues to increase, it will reach a point where neither electrons nor ions will have enough time to cluster around the electrodes in substantial numbers; most of the particles will just oscillate in the space between the electrodes. This plasma will behave differently than either the direct current or intermediate frequency plasma.

In practice, alternating current power sources operating at radio frequencies and above are preferred over direct current sources because they are more efficient at producing the plasma components useful in sterilization. Alternating current fields can also pass through insulators, and plasma reactors can be built utilizing insulating walls with electrodes placed outside the reaction chambers. These are called capacitatively coupled systems because the electrodes act in much the same way as plates in a capacitor. Also, with alternating current fields, the two electrodes can be replaced by a coil. Current through the coil generates an electromagnetic field inside the coil, and this field produces the plasma. These systems are said to be inductively coupled.

Plasma power systems are usually operated at either 13.56 MHz or 2.45 GHz. Both frequencies have been designated for industrial use, which means that they do not interfere with radio communications. The first frequency, 13.56 MHz, is considered a radio frequency; it lies between the highest AM radio frequency and the lowest VHF television frequency. The second frequency, 2.45 GHz, is in the microwave band, well above the highest UHF television frequency. Home microwave ovens operate at a frequency of 2.45 GHz.

## Gas Sources

The chemical species generated in the plasma depend on the composition of the gas used to generate the plasma. In most cases, compressed gases are simply fed into the plasma reactor via a controlling mechanism, such as a needle valve or mass flow controller. In some cases, the desired gas is formed by evaporating a liquid. The liquid is usually evaporated by heating, and the flow may be regulated either by controlling the flow of the liquid into the heating device or controlling the flow of the vapor produced. When liquid sources are used, it is important to take appropriate measures to prevent the vapor from condensing in the reactor.

## THE DEVELOPMENT OF PLASMA STERILIZATION

The development of plasma sterilization has followed the development of plasma reactors reasonably closely. This section will be devoted to describing the construction and operation of various low-temperature plasma reactors, and the sterilization studies performed in each type of reactor.

Two of the earliest reports of plasma sterilization came from Menashi (1968) (Figure 6.1) and Ashman and Menashi (1972) (Figure 6.2). In the first patent, Menashi described a system in which a container, such as a plastic bottle, could be sterilized in a plasma formed at atmospheric pressure. A coil was wrapped around the container, and gas was introduced into the container through a tube inserted into the mouth of the container. When power was applied to the coil, a plasma was formed. This plasma impinged on the interior surfaces of the container and caused sterilization. The system was designed to create a very hot plasma and expose the containers to the plasma for a very short time (one-tenth of a second or so). In the words of the later patent, the method relied "primarily on a momentary flash of

**Figure 6.1.** Treatment of surfaces. *(Adapted from Menashi [1968], U.S. Patent 3,383,163)*

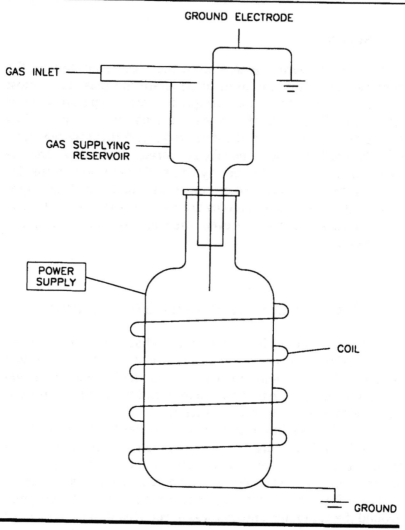

intense heat" to raise quickly the temperature of the container's surface to a lethal level. The plasma was to be turned off after the surface became hot enough to destroy microorganisms but before the container was damaged by the heat.

**Figure 6.2.** Treatment of surface with low-pressure plasmas. *(Adapted from Ashman and Menashi [1972], U.S. Patent 3,701,628)*

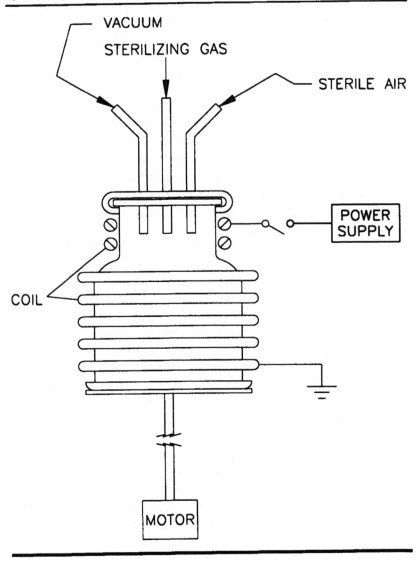

This system had a few drawbacks. First, it required careful timing to create and extinguish the plasma before the plastic being treated was damaged. Second, because the process was carried out at atmospheric pressure, the plasma resembled an arc, and it was hard to avoid the formation of "hot spots." These problems were addressed in the second patent, in which the plasma was formed at reduced pressure. Again, the second patent was directed toward sterilizing containers. In one embodiment, a container was placed in a vacuum vessel made of a nonconducting material. A metal coil was wrapped around the vessel. To sterilize the container, the vessel was evacuated by a vacuum pump, and gas was introduced into the system. Radio frequency power was applied to the coil, and a plasma was formed. The container was exposed inside and out to the glowing ionized gas. After a short period of time, the power was turned off, the plasma ceased to exist, and the vacuum chamber was vented. (In the example cited, chlorine gas was used at a pressure of 2 torr [2.7 mbar], the power was characterized as being 6 MHz and 1000 volts [power not given], and the exposure time was 3 seconds.) In another embodiment, the coil outside the vacuum chamber was replaced by two electrodes inside the vacuum chamber. One electrode was a rod that went into the container being sterilized through its mouth, and the other was a screen surrounding the outside of the container.

The Menashi (1968) patent suggested that one should use argon gas to form the plasma, since it is nontoxic and inert. The Ashman and Menashi patent (1972) suggested forming the plasma from chlorine, bromine, hydrochloric acid, or iodine, and also mentioned trials with plasmas formed from hydrogen, water vapor, oxygen, and nitrogen.

Further tests of these systems were performed at A. D. Little under the sponsorship of the U.S. Army Medical Research and Development Command. The results of these tests were rather disappointing, and corrosion problems reportedly resulted in the termination of further tests. A summary of this work was presented in an article by Boucher (1985).

At about the same time, two patents were issued to Brumfield et al. (1970a,b) (Figure 6.3). The plasma reactor described by Brumfield et al. was designed to treat either a packaged or an

**Figure 6.3.** Microwave reactor and process. *(Adapted from U.S. Patent 3,551,090)*

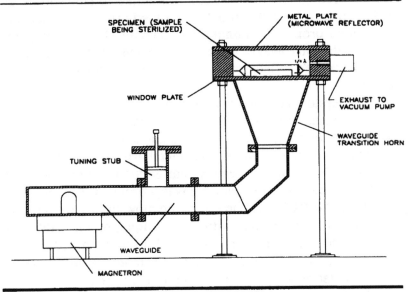

unpackaged product. In the case where the product was packaged, asepsis was accomplished by creating a plasma inside the package. Their reactor utilized a microwave power source. The power was introduced into a vacuum chamber containing the product to be treated through a window made of a nonconducting material, such as glass or quartz. The vacuum chamber was fashioned to be part of the microwave waveguide, and the product had to be positioned in such a manner that it was located at a point where the microwave field was the strongest. Once the load was properly positioned, the vacuum chamber was sealed and a vacuum was drawn, gas was introduced at a proper pressure (air at 0.05–10 torr was suggested). The microwave power source was turned on and microwave energy was applied. Power levels of 800–1000 watts at 2.54 GHz were suggested. After a few seconds, the power was turned off, and the product was recovered.

The inventors also devised a special package for use in their system as described in U.S. Patent 3,490,580. The package incorporated a nonconductive porous material that allowed air to be exchanged and acted as a barrier to microorganisms. It also featured conductive reflectors designed to concentrate an applied microwave field onto the packaged product. As stated in this patent, proper positioning was accomplished by the package, and the microwaves were concentrated inside the package by the package's reflectors instead of by reflectors built into the reactor.

This system also had several flaws. First, only nonconductive items could be treated, since metal objects would perturb the microwave. Just as one cannot utilize metal dishes in a home microwave oven, one cannot expose metal articles to a strong microwave field in the described system. Second, as in the Ashman and Menashi system, temperature was a problem. The inventors stated that "at 0.05 [torr] pressure, the glow discharge can proceed for up to 10 seconds without package ignition." Although the dressing they tested survived the 2–3 second exposure, it is clear that many articles would not withstand the heat, particularly if longer exposures were found to be necessary.

In 1976 and 1977, Tensmeyer of Eli Lilly obtained patents for killing microorganisms using plasmas created from high power laser beams. The first patent described a system in which a pulsed laser beam was focused inside a gas-filled container. In the example cited, glass vials were filled with air at atmospheric pressure, though other gases and pressure ranges were also mentioned. When the laser was activated, the light pulse interacted with the gas at the focal point to cause a plasma "spark." The patent described using light pulses from 0.1 to 300 nanoseconds long; 15 nanosecond pulses were used in the example given. The sparks generated were said to exist from about 5 nanoseconds to about 5 microseconds. Presumably, products from the plasma spark (including ultraviolet light) traveled from the interior of the vial to the walls of the vial to inactivate microorganisms on the walls. Tests were run in which up to 1600 sparks were generated in each vial, resulting in a 10–100 fold reduction in contaminating *Bacillus subtilis* spores.

The second patent (1977) described a system in which the pulsed laser was supplemented with a microwave or radio frequency electromagnetic field. Since the pressures used were relatively high, an enormous field would be required to initiate a plasma. However, it was found that much less power was required to sustain a plasma once it was initiated. Thus, the pulsed laser was used to initiate a plasma "spark" in a container. Once started, the plasma was sustained by the electromagnetic field. The field was applied for a relatively long time (up to a second or so), which allowed the plasma to expand into the containers. In the example given, glass vials were filled with argon at atmospheric pressure. The plasma was sustained by a 2.45 GHz microwave field; about 200 watts of power was absorbed by the plasma. The length of the plasma treatment was 1 second. The treatment reduced the *Bacillus subtilis* spore population by at least 100 fold.

However, these processes still suffered from some of the drawbacks of the previous processes: overheating and penetration into narrow recesses were problems, and it was difficult to sterilize wrapped items. In addition, special consideration had to be given to opaque and metallic items to ensure that the laser light and the electromagnetic field, respectively, could penetrate to the interior of the item.

Thus, it became clear that the exposure of articles to the plasma had several drawbacks. Charged particles in the plasma bombard surfaces of articles placed in the plasma. This heats the surfaces at best; at worst, the bombardment can break chemical bonds on the articles' surfaces, thus changing their characteristics. Nonuniformity of the plasma can also create problems: the processing must be long enough that the coldest spots are sterilized, at which point the hottest spots may be damaged.

As a practical example, consider plasma sterilization of a metal hypodermic needle fitted with a plastic hub and wrapped in a plastic blister pack covered with paper. This assembly is subjected to a strong electric field and a plasma is formed. The external electric field creating the plasma cannot penetrate the metal or travel very far down the bore of the needle, so there will be no plasma generated inside the needle. Charged particles created outside the needle will not diffuse very far into the interior of the needle, since in the absence of the external field they recombine very quickly. Very little of the ultraviolet light generated in the plasma will illuminate the inner walls of the needle. The inner surface of the needle will only be subjected to those uncharged chemical species present in the plasma that are capable of diffusing to the center of the needle.

While those chemicals are diffusing, the exterior of the package will be bombarded by charged particles and ultraviolet light, as well as uncharged particles which may be too reactive to play a role in sterilization of the needle's interior. The charged particles will also heat the surfaces they strike, and it is unlikely that the heating will be uniform. If the dimensions of the package are large enough to allow plasma to be formed inside, the package will be exposed to charged particles and ultraviolet light from both sides, and the plastic hub and the outer surface of the needle will also be subjected to all the components of the plasma. Thus, the interior of the needle will be in a relatively benign environment, while the outside of the needle and the package will be in an extremely harsh environment. In the time it takes to sterilize the interior of the needle, the hub and package may be destroyed.

Two approaches have been used to circumvent these problems. In one approach, measures were taken to remove the most aggressive components of the plasma and expose the articles to be sterilized to less reactive components. In the other approach, plasma was used in conjunction with another antimicrobial agent.

## Removing Aggressive Plasma Components

In the mid 1970s, Fraser et al. (1974, 1976) of the Boeing Company investigated plasma sterilization technology for spacecraft applications. They used a plasma system in which gas flowed through a chamber where it was exposed to an electromagnetic field, and then into a chamber containing articles to be sterilized. The samples were kept away from the intense field, and the excited gas flowed to the region where the test samples were mounted. The charged particles in the plasma recombined, and the plasma decayed as it traveled from the electrode region to the test samples.

Plasmas were generated from argon, helium, nitrogen, and oxygen at pressures from 0.2 to 1.0 torr (0.26 to 1.3 mbar). The electromagnetic field was supplied by an radio frequency generator operating at 13.56 MHz, with power levels from 50 to 300 watts. They found that a plasma of helium at 0.2 torr, 300 watts power, was most effective.

These investigators looked into several possibilities for the lethal effects they observed and concluded that the primary mechanism for killing microorganisms was ultraviolet light. Samples that were separated from charged particles in the gas by quartz or sapphire windows, which transmit ultraviolet light, were sterilized. Much less lethality was exhibited when glass windows, which did not transmit ultraviolet, were used.

The Boeing group also obtained two patents on plasma sterilization (Fraser et al. 1974, 1976). These patents discuss systems in which the plasma is created in a chamber separate from the chamber containing the articles being sterilized (Figure 6.4). The inventors indicated that the sterilizing chamber could be made of metal or encased in metal, which would presumably reduce the amount of stray electromagnetic energy that could enter the sterilizing chamber. The basic plasma reactor described is similar to the reactor used in the previously discussed Boeing monograph. Other designs adapting remotely-generated plasma to sterilization of blood oxygenators, implements used on spacecraft, and spacecraft themselves are also discussed.

In the mid 1970s, a new type of plasma reactor was developed that had a major impact on plasma sterilization. In the production of semiconductor electronic devices, there are processing steps in which organic films must be removed from silicon substrates. In many instances, it is advantageous to use oxygen atoms and other active components in an oxygen plasma to perform these steps, but if a strong field is present during the processing, the electronic devices may be ruined. A reactor to fulfill these requirements was designed by Bersin and Singleton (1975) (Figure 6.5). In this system, the vacuum chamber was constructed from a nonconducting material, such as quartz. The electromagnetic field was created by placing the chamber between two curved electrode plates and applying radio frequency energy to the plates. The gas entered the chamber through a manifold at the top of the chamber and flowed out to a vacuum

**Figure 6.4.** Sterilizing process and apparatus utilizing gas plasma. *(Adapted from Fraser et al. [1976], U.S. Patent 3,948,601)*

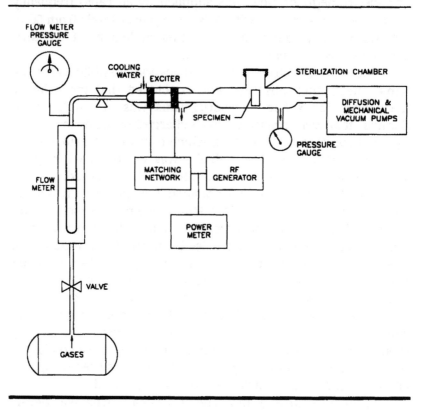

pump through a port or ports in the bottom of the chamber. Inside the chamber was a cylinder formed from a perforated conducting material, such as aluminum. The cylinder excluded the electromagnetic field from its interior, so no plasma was generated within the cylinder.[2] Instead, the plasma was generated in

2. This device is sometimes called a Faraday Cage. In the early 1800s, Michael Faraday discovered that electric fields would not penetrate an enclosure made of metal or metal mesh, provided the mesh size was sufficiently small. He demonstrated this in a spectacular experiment in which he sat in a large metal mesh cage surrounded by machines capable of generating large electric fields. When the machines were activated, sparks flew to the cage; yet Faraday was unharmed, and sensitive measuring instruments within the cage failed to detect the presence of an electric field.

**Figure 6.5.** Plasma etching device and process. *(Adapted from Bersin and Singleton [1974], U.S. Patent 3,879,597)*

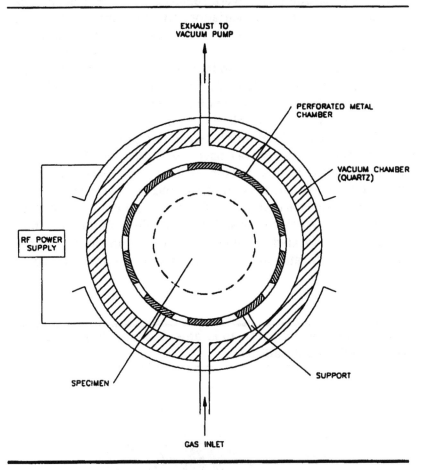

the space between the cylinder and the walls of the vacuum chamber. Virtually all charged particles in the vicinity of the cylinder would recombine on its metal surface, and the cylinder's interior was essentially free of both fields created by the external electrodes and fields created by charged particles. However, uncharged gas particles were free to flow through the perforations and interact with material inside the cylinder.

This reactor became very popular in the semiconductor industry, and a number of variations were devised. In one variation, the electrode plates were replaced by a coil (Jacob 1982), (Figure 6.6). In another variation, the vacuum chamber was made of metal. The chamber wall served as one electrode and the perforated cylinder served as the other. It should be noted that in most of these designs, the voltage on the cylinder will oscillate and may reach very high levels; however, everything within the cylinder will be at the same voltage. Therefore, no significant electric field will exist within the cylinder, since an electric field requires a voltage difference between two points.

A similar reactor was used by Bithell (1982) to sterilize wrapped items. In this reactor, articles were placed in an aluminum cage that had slots on its top and bottom to facilitate gas flow. The cage was mounted inside an aluminum vacuum chamber. This cage shielded the articles held within in the same manner as the perforated metal cylinders in the previous examples. Slotted or perforated metal electrodes were mounted above and

**Figure 6.6. Gas discharge apparatus.** *(Adapted from Jacob [1982], U.S. Patent 4,362,632)*

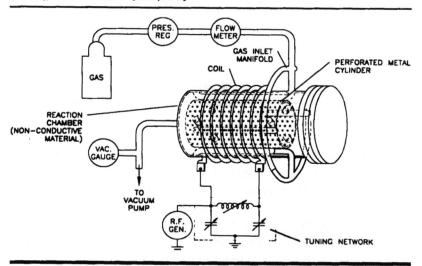

below the cage and were connected to each other electrically. The gas system was designed to permit gas to flow through the top electrode, through the top of the cage, around the articles being sterilized, through the bottom of the cage, through the bottom electrode, and out to the vacuum pump. A radio frequency voltage was applied and plasma was generated between the electrodes and the cage.

Tests performed by Bithell showed that it was possible to sterilize wrapped items using components made in an oxygen plasma. Active components created in the plasma were able to diffuse through the wrapping material and interact with the items inside. The positioning of the articles within the chamber was not important, and he was able to stack packages with little impact on processing time. Since charged particles were excluded from the cage, ion bombardment and associated uniformity and heating problems were eliminated. Thus, by eliminating the electromagnetic field and the associated charged particles, he was able to sterilize packaged items without damaging the packaging. The active gas components created in the plasma (sometimes called "active species" or "active agents") behaved much like ordinary gases instead of like charged particles.

## Combining Antimicrobial Agents with Plasma

In 1980, a patent titled "Seeded Gas Plasma Sterilization Method" issued to Boucher (Figure 6.7). It described a process in which a plasma was formed from a gas mixture containing at least one aldehyde; "seeding" referred to the introduction of the aldehyde into the gas. Preferably, the mixture would also contain oxygen or an oxygen-containing gas, such as air, $CO_2$, or $N_2O$. The articles to be sterilized were immersed in the plasma. The resulting sterilization process was more effective than sterilization with aldehyde vapor alone or with plasma formed from gas not containing aldehydes.

There are several explanations for this effect. First, the aldehyde may be converted into a more active compound, or more active compounds, in the plasma. For instance, the aldehyde could be dissociated into active free radicals by electron impact

**Figure 6.7.** Seeded gas plasma sterilization method. *(Adapted from Boucher [1980], U. S. Patent 4,207,286)*

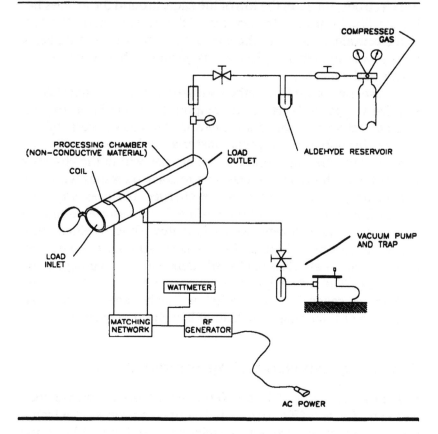

reactions. Second, the aldehyde could react with other components formed in the plasma (such as oxygen atoms) to form a very active sterilant. Third, the aldehyde could interact with a microorganism to alter an essential component in a nonlethal manner; a plasma component could then act on this in a separate reaction to kill the microbe. Fourth, the plasma could alter an essential component of the microorganism, and the aldehyde could react with the altered component to kill the cell. It is not difficult to imagine still more variations. Given the complexity of the plasma and microorganisms, it is conceivable that all of the mechanisms contribute to the lethality of the process, and it would be difficult to determine which mechanism is dominant.

Boucher used two types of plasma reactors for his experiments. One was an inductively coupled system operating at 13.56 MHz, similar to those described above. The second system operated in the microwave band (2.45 GHz) and utilized a microwave chamber resembling a home microwave oven. Within the microwave chamber was a vacuum container holding the samples being sterilized. In both systems, the items being sterilized were directly exposed to the plasma. The test items were unwrapped AOAC (Association of Official Analytical Chemists) carriers, although it was reported that spore strips wrapped in surgical gauze were also sterilized.

A very similar approach was utilized by Jacobs and Lin (1987). They used an inductively coupled reactor operating at 2.49 or 3.89 MHz. The plasma was generated from a vapor formed by evaporating a hydrogen peroxide solution. Whereas Boucher flowed gas through his reactors while the plasma was being generated, Jacobs and Lin employed a static system. After the samples were loaded and the sterilizer chamber was evacuated, the vacuum pump was valved off, and hydrogen peroxide was evaporated into the system. The hydrogen peroxide was allowed to diffuse for a period of time. Then radio frequency power was applied to the chamber and a plasma was generated around the samples from the pretreatment hydrogen peroxide vapor. After a prescribed period of time, the power was turned off and the samples were recovered.

In addition to ions, electrons, and ultraviolet light, the plasma of the Jacobs and Lin system contains many neutral active species, such as atomic oxygen, atomic hydrogen, hydroxyl radicals, and hydroperoxyl radicals. Many of these components are created in other sterilizing systems (gamma irradiation in particular). Some of the components, particularly the hydroxyl radical, also arise in the body from normal oxygen metabolism, and they have been studied extensively for their roles in causing cell damage, aging, and so on. In many respects, the chemistry of the interactions between microorganisms and the neutral components of a hydrogen peroxide plasma is very similar to the chemistry of vapor-phase hydrogen peroxide sterilization and the water radiolysis reactions of gamma sterilization.

Jacobs and Lin reported that they were able to sterilize *Bacillus subtilis* spores inoculated on paper carriers and packaged in spunbound polyethylene (Tyvek®) packages. They reported that the plasma had to be generated in the immediate vicinity of the article being sterilized in order for the process to be effective. They stated that if the article was shielded from the electromagnetic field creating the plasma (a wire cage was used as an example of shielding in the patents), the sterilization efficiency was reduced to levels seen without plasma (i.e., using hydrogen peroxide vapor alone).

To summarize, first Boucher and then Jacobs and Lin found that a plasma formed from a gas containing an antimicrobial agent could kill microorganisms efficiently. However, both used reactors in which the articles being sterilized were exposed to the electromagnetic field used to generated the plasma. This is undesirable because of the attendant shielding and uniformity problems discussed above.

## MODERN PLASMA STERILIZERS

There are two plasma-based sterilizers being used in hospitals today. One is the AbTox Plazlyte® [3] sterilization system, and the other is the ASP Sterrad® [4] sterilization system. Both incorporate the use of antimicrobial agents and both protect the articles being sterilized from the electromagnetic field used to generate the plasma. However, there are substantial differences in the operation of the two systems.

---

3. Plazlyte is a registered trademark of AbTox Inc., Mundelein, Illinois. The AbTox system is protected by several U.S. and foreign patents, for example, Moulton et al. (1992).

4. Sterrad is a registered trademark of Advanced Sterilization Products, Irvine, California.

## The AbTox Plazlyte® Sterilizer

A simplified cross-sectional view of the AbTox Plazlyte® sterilizer is shown in Figure 6.8. For purposes of discussion, the sterilizer may be divided into two main sections: The plasma generation section and the sterilization chamber section. In Figure 6.8, the plasma generation section is above the sterilization chamber. For clarity, only one plasma generator is shown; the Plazlyte® system actually incorporates three identical plasma generators that are connected to a single gas distribution manifold.

**Figure 6.8.** Plazlyte® sterilization system cross-sectional view.

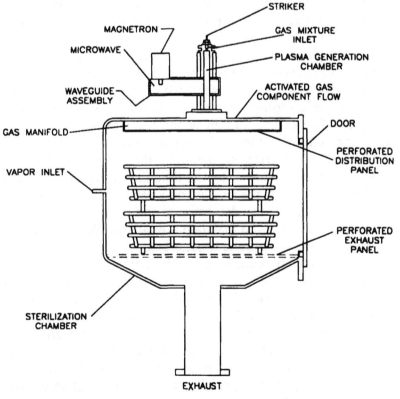

As the name implies, plasma is generated in the plasma generation section. When plasma is required, a microwave electromagnetic field is produced by a magnetron operating at 2.45 GHz. The microwave energy is transmitted through a waveguide and focused on a quartz tube. The tube contains the gas from which the plasma is generated, a nonflammable mixture of argon, oxygen, and hydrogen. The gas and the microwave field interact to form a plasma.

The gas flows continuously while the plasma is being generated. The gas enters the system through an inlet, passes through the microwave waveguide (where it is converted into a plasma), through an outlet, and into a gas distribution manifold. The flowing gas enters the sterilization chamber through an array of holes in a perforated distribution panel that are sized and spaced for uniform delivery. The gas then interacts with the articles being sterilized. Finally, the gas flows from the sterilization chamber through an exhaust line and is removed by a vacuum pump.

The plasma generator is designed to keep the microwave energy confined within the bounds of the waveguide. The waveguide is separated from the gas distribution manifold by a metal support piece designed to inhibit microwaves from traveling through it. The quartz tube in which the plasma is generated passes through this support. As the gas travels from the waveguide to the gas distribution manifold, the microwave field becomes weaker and weaker; by the time the gas reaches the gas distribution manifold, the field is too weak to form a plasma. A gas flow restrictor at the downstream end of the quartz plasma generation tube serves to maintain a relatively high pressure in the region where the plasma is generated.

As discussed previously, when the field used to generate a plasma is removed, the charged particles recombine very quickly to form neutral particles. In the AbTox system®, the field is attenuated to negligible levels as the gas flows from the waveguide to the gas distribution manifold, and there is a corresponding decrease in the concentration of charged particles. The gas also spends a considerable amount of time in the manifold (many milliseconds) during which most of the few remaining charged particles recombine. By the time the gas reaches the sterilization chamber, virtually all of the charged particles have recombined to form neutral particles.

Most of the excited particles formed in the plasma relax before they reach the sterilization chamber. Excited particles generally have a lifetime on the order of a microsecond or less, so they barely enter the manifold before they relax. The manifold is also designed to block ultraviolet light created in these relaxation processes from reaching the sterilization chamber. Finally, the gas is heated by its interactions with the microwave field in the waveguide. When it flows into the gas distribution manifold, it interacts with the walls and cools.

In brief, the system is designed so that charged particles and short-lived excited atoms and molecules present in the plasma and the ultraviolet light generated in the plasma do not reach the articles being sterilized. The lethal agents in the system are "relatively" nonreactive, that is, reactive enough to inactivate microorganisms but sufficiently stable to diffuse through wrapping materials and over the surfaces of the items being sterilized.

The sterilizer also incorporates a vaporizer. This is used to evaporate an antimicrobial solution of peracetic acid; the purpose of the peracetic acid vapor is described below. The outlet of the vaporizer is connected to the sterilization chamber via a valve, and the inlet is connected to a valve system that introduces measured amounts of the peracetic acid solution. The peracetic acid is stored in a reservoir.

## The Plazlyte® Sterilization Process

The Plazlyte® process utilizes two antimicrobial gases, peracetic acid vapor and the active species of the plasma previously described. The gases are alternated several times during a process cycle.

After loading the sterilizer, the door is sealed and a valve is opened to connect the sterilization chamber to the vacuum pump. The air in the chamber is withdrawn until the pressure reaches a predetermined level. The valve is then closed, and the sterilization chamber is isolated from the vacuum pump.

Next, the valve between the vaporizer and the sterilization chamber is opened, and peracetic acid solution is introduced into the vaporizer. The resulting peracetic acid vapor enters the sterilization chamber. Solution is introduced into the vaporizer until the chamber pressure reaches a predetermined value. Once the prescribed pressure has been achieved, the process timer starts. The peracetic acid is held in the chamber for 20 minutes.

At the end of the 20-minute time period, the valve between the vaporizer and the sterilization chamber is closed, and the valve between the sterilization chamber and the vacuum pump is opened. The peracetic acid is withdrawn from the chamber and the pressure drops to a predetermined level.

At this point, valves open to allow the argon-hydrogen-oxygen gas mixture to flow into the three plasma generators. Once the flow is established, the magnetrons are activated and plasmas are established in each of the plasma generators. The gas flows through the plasma generators, the distribution manifold, the sterilization chamber, and out through the vacuum pump. The gas flows continuously during the plasma phase, which lasts 10 minutes. At the end of the plasma phase, the magnetrons are turned off, the gas flow is terminated, and the remaining gas is withdrawn from the chamber.

When the pressure has fallen to a predetermined level, a second exposure to peracetic acid vapor is initiated. This treatment also lasts 20 minutes. This is followed by a second plasma treatment. This pattern is repeated until the articles being sterilized have been exposed to 6 rounds of peracetic acid vapor and 6 rounds of plasma. The exposure times and operating parameters for the second and subsequent exposures are the same as those for the first exposure for both the peracetic acid treatment and the plasma treatment.

When the last plasma treatment has ended, the chamber is evacuated, the vacuum pump is valved off, and an "air wash" procedure is initiated. This procedure is used to reduce the levels of residual peracetic acid vapors absorbed by the articles being sterilized. Sterile air is introduced into the chamber through a valve until the pressure reaches a level slightly lower than atmospheric pressure. The sterile air valve is then closed, the valve between the sterilization chamber and the vacuum pump is opened, and the air is withdrawn from the sterilization chamber. This is repeated several times. After the final evacuation, sterile air is introduced into the chamber until the sterilization chamber reaches atmospheric pressure. The door is then opened and the sterile load is removed.

## The ASP Sterrad® Sterilizer

The Sterrad® sterilizer utilizes the design represented in Figure 6.9. The vacuum chamber is constructed of metal in a generally cylindrical shape. Within this chamber is mounted a cylinder of perforated metal that is slightly shorter than the interior length of the vacuum chamber. The inner, perforated cylinder is concentric with the outer vacuum chamber, and the inner cylinder

**Figure 6.9.** ASP Sterrad® sterilizer front view.

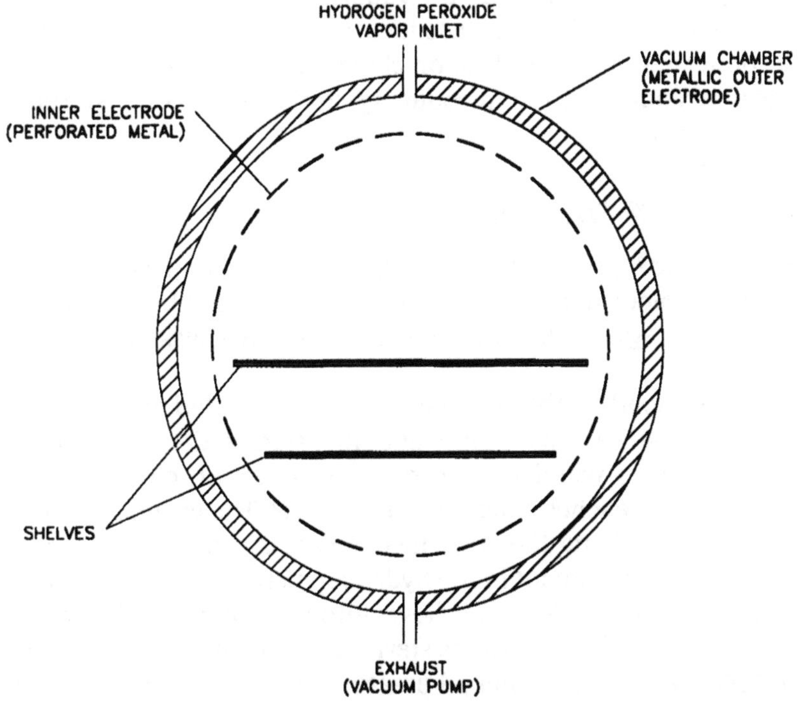

does not touch the outer chamber at either end. Shelves to hold the items being sterilized are mounted within the inner perforated cylinder and are electrically isolated.

The wall of the vacuum chamber and the inner cylinder are used as electrodes to generate the plasma. The vacuum chamber is grounded and the inner cylinder is connected to the output of a radio frequency generator operating at 13.56 MHz. As discussed above, the inner cylinder acts as a Faraday shield to prevent the articles being sterilized from interacting with the electromagnetic field used to generate the plasma.

The Sterrad® system also incorporates a vaporizer to convert an antimicrobial solution (in this case, hydrogen peroxide) into a vapor. The hydrogen peroxide is stored in sealed compartments in a cassette holder; each compartment contains a fixed volume of hydrogen peroxide for one sterilization cycle. At the appropriate time, a hollow lance pierces a hydrogen peroxide compartment. The contents of the compartment flow through the lance and into the vaporizer.

The Sterrad® system also incorporates a vacuum pump that is connected to the sterilization chamber via a valve, and a filter/valve arrangement for introducing sterile air into the sterilization chamber.

## The Sterrad® Sterilization Process

The Sterrad® process uses a single gas consisting of hydrogen peroxide vapor and water vapor. This gas is used both as an antimicrobial agent and as the gas from which a plasma is generated.

After loading the chamber, the door is sealed and the valve between the vacuum pump and the sterilization chamber is opened. Air is withdrawn from the sterilization chamber until the pressure reaches a predetermined level. The vacuum pump is then valved off and the lance mechanism is activated. The lance pierces a compartment in the hydrogen peroxide holder, and the hydrogen peroxide solution flows into the vaporizer. The resulting vapor then flows into the sterilization chamber. Once the vapor reaches a predetermined pressure, an exposure timer starts. The vapor is held in the chamber for approximately 44 minutes.

At the end of the hydrogen peroxide exposure period, the valve between the sterilization chamber and the vacuum pump is opened. When the pressure reaches a predetermined level, the vacuum pump is again valved off. At this point, the radio frequency generator is activated, and a plasma is formed from the remaining hydrogen peroxide vapor. The plasma is formed in the annular region between the inner perforated metal cylinder and the outer metal cylinder, which serves as an electrode and as the wall of the vacuum chamber. Plasma products can enter the interior of the perforated metal cylinder by diffusion.

Since the electromagnetic field in the interior of the perforated cylinder is extremely weak as a result of the Faraday shielding, no plasma is formed within the cylinder. For the reasons discussed above, virtually all charged particles that diffuse to the interior of the perforated electrode quickly recombine to form electrically neutral chemical species. Thus, the primary microbiocidal components in the interior of the perforated electrode are neutral active species from the plasma, ultraviolet light from the plasma, and residual hydrogen peroxide. Since most sterilization wrapping materials block ultraviolet light, wrapped articles are exposed mainly to neutral active species from the plasma and any remaining hydrogen peroxide.

The atoms and radicals created in the plasma do not usually recombine to reform hydrogen peroxide; pathways leading to the formation of water and oxygen are preferred. The plasma ultimately decomposes the hydrogen peroxide vapor, and by the end of the plasma exposure period, virtually all of the hydrogen peroxide in the plasma is converted to water and oxygen.

At the end of the plasma exposure, the valve between the vacuum pump and the sterilization chamber is opened and the pressure is reduced to a predetermined level. The pump is valved off, and the chamber is brought to atmospheric pressure with sterile air. The door may then be opened and the sterile articles may be retrieved.

## APPLICATIONS OF PLAZLYTE® STERILIZATION

AbTox has made its technology, the Plazlyte® Sterilization System, available to the healthcare and other industrial markets. It has been cleared by the U.S. Food and Drug Administration (FDA) for acute care applications, meeting or exceeding regulatory criteria.

Utilizing established methodologies, the Plazlyte® sterilizer has met the requirements for safety and efficacy as reported by Caputo et al. (1993). The half-cycle overkill approach, utilizing resistant biological indicator spore carriers, has been used to validate delivery of $10^{-6}$ sterility assurance levels to representative loads. A variety of studies demonstrated that no toxic residuals were deposited on articles following sterilization. These tests included cytotoxicity, sensitization, irritation/intracutaneous toxicity, acute systemic toxicity, hemocompatibility/hemolysis, ocular irritation, and muscle implantation.

Sterilization tests verified sporicidal, bactericidal, tuberculocidal, fungicidal, and virucidal efficacy, including meeting the sporicidal requirements described in the AOAC.

Just as sterilization is a critical component of providing conscientious medical care in hospitals, sterilization is also important for certain applications in the industrial market. The design of the Plazlyte® sterilizer allows for greater flexibility in developing and customizing sterilization cycles appropriate for the products and materials being processed. Plazlyte® sterilization of products in areas as diverse as medical device, pharmaceutical, biotechnology, food, and aseptic manufacturing illustrate its broad applicability.

A medical device manufacturer received FDA market clearance for Plazlyte® sterilization of polyethylene components used in orthopaedic implants. The Plazlyte® process was also successful in sterilizing musculoskeletal tissue used to repair or replace bone damaged by trauma or disease. Biological materials, such as bioresorbable polymers used for surgical implants, have posed challenges for sterilization in the past. Because of their unique structure and sensitivities, previous sterilization options have negatively impacted their functionality. In a joint effort with

AbTox, a sterilization process was developed that successfully sterilized samples for clinical trials. As reported by Brekke (1995), little degradation of the polylactic acid and hyaluron samples was experienced following sterilization in the Plazlyte® system.

## ACKNOWLEDGMENT

The authors gratefully acknowledge the technical assistance of Dennis Hess, Ph.D., Georgia Institute of Technology.

## REFERENCES

Ashman, L. E., and W. P. Menashi. 1972. Treatment of surface with low-pressure plasmas. U.S. Patent No. 3,701,628.

Bersin, R. M., and M. Singleton. 1974. Plasma etching device and process. U.S. Patent No. 3,879,597.

Bithell, R. M. 1982. Package and sterilizing process for same. U.S. Patent No. 4,321,232.

Boucher, R. M. G. 1980. Seeded gas plasma sterilization method. U.S. Patent No. 4,207,286.

Boucher, R. M. G. 1985. State of the art in gas plasma sterilization. *Medical Device and Diagnostic Industry* 7 (2):51–56.

Brekke, J. 1995. Architectural principles applied to three dimensional therapeutic implants composed of bioresorbable polymers. In: *Encyclopedic Handbook of Biomaterials and Bioengineering*, edited by D. L. Wise et al. New York: Marcel Dekker, Inc., pp. 689–731.

Brumfield, R. C., J. T. Naff, and A. T. W. Robinson. 1970. Containers and process for asepsis. U.S. Patent No. 3,490,580.

Brumfield, R. C., J. T. Naff, and A. T. W. Robinson. 1970. Microwave reactor and process for asepsis. U.S. Patent No. 3,551,090.

Caputo, R. A., J. Fisher, V. Jarzynski, and P. Martens. 1993. Validation testing of a gas plasma sterilization system. *Medical Device and Diagnostic Industry* 15 (1):132–138.

Chapman, B. 1980. *Glow Discharge Processes.* New York: John Wiley.

Fraser, S. J., R. L. Olson, and W. M. Leavens. 1975. Plasma sterilization technology for spacecraft applications. NASA Report NASA-CR-146314.

Fraser, S. J., R. B. Gillette, and R. L. Olson. 1974. Sterilizing and packaging process utilizing gas plasma. U.S. Patent No. 3,851,436.

Fraser, S. J., R. B. Gillette, and R. L. Olson. 1976. Sterilizing process and apparatus utilizing gas plasma. U.S. Patent No. 3,948,601.

Jacob, A. 1982. Gas discharge apparatus. U.S. Patent No. 4,362,632.

Jacobs, P. T., and S. M. Lin. 1987. Hydrogen peroxide plasma sterilization system. U.S. Patent No. 4,643,876.

Menashi, W. P. 1968. Treatment of surfaces. U.S. Patent No. 3,383,163.

Moulton, K. A., B. A. Campbell, and R. A. Caputo. 1992. Plasma sterilizing process with pulsed antimicrobial agent treatment. U.S. Patent No. 5,084,239.

Nasser, E. 1971. *Fundamentals of Gaseous Ionization and Plasma Electronics.* New York: Wiley Interscience.

Tensmeyer, L. G. 1976. Method of killing microorganisms in the inside of a container utilizing a laser beam induced plasma. U.S. Patent No. 3,955,921.

Tensmeyer, L. G. 1977. Method of killing microorganisms in the inside of a container utilizing a plasma initiated by a focused laser beam and sustained by an electromagnetic field. U.S. Patent No. 4,042,325.

# 7

# Sterilizing Filtrations: Process

*Wayne P. Olson*

Oldevco
Beecher, IL

## UTILITY OF STERILIZING FILTRATION IN PROCESS

Over 75 percent of the drugs produced in the United States and Europe are processed aseptically. That is to say, solutions of the drugs are filtered through sterilizing-grade membrane filters that remove all bacteria and fungi. Of necessity, all therapeutic proteins and peptides are cold sterilized by filtration (Olson 1983); steam sterilization destroys the structure and efficacy of peptide drugs. Some polysaccharides may be autoclavable (Lou et al. 1994) and some drugs, like succinylcholine chloride, are heat stable (Schmutz and Muhlenbach 1991). But heat sterilization of relatively simple glucose-containing fluids for peritoneal dialysis gives rise to acetaldehyde, glyoxal, methylglyoxal, formaldehyde, 5-hydroxymethylfurfural, and 2-furaldehyde (Nilsson-Thorell et al. 1993). Sterile filtration followed by aseptic filling results in none of these. Provided the filling system can be validated to a sterility assurance level (SAL) of 6 logs of heat-resistant spores, aseptic processing is the preferred method.

A substantial proportion of conventional drugs of relatively low molecular weight ($M_r$) (e.g., less than 2 kDa) exhibits heat lability (e.g., phenylmercuric nitrate, Parkin et al. 1992), hence the widespread use of the aseptic process. The view of the pharmaceutical industry is that the process that least degrades the product, yet results in a safe, efficacious, sterile, and pure product in the final container at a competitive cost, is the method of choice. The U.S. Food and Drug Administration (FDA) presently accepts sterilization by filtration with reluctance because traditional filling systems have not been equally as effective as sterilization in the final container. Aseptic filling usually has been shown to have an SAL of 3 logs (0.1 percent, no more than 1–3 sterility failures in a media fill of 3,000 final containers).

## HISTORY AND THEORY OF STERILIZING FILTRATIONS

A useful starting point for those highly involved in filtrative sterilization is Timothy Leahy's dissertation (1983).

### Adsorptive Mechanisms

Until approximately 1976, the method of choice for cold sterilization of bulk liquids and gases was filtration through fine fibers of asbestos. At about that time, the FDA promulgated a regulation requiring that pharmaceutical firms using asbestos as a filtration medium seek replacement materials or methods. When inhaled, asbestos induces abnormal states in the lung, including cancer (see Fiore et al. 1980 for discussion and references). The fear was that the inadvertent injection of chrysotile asbestos fibers with a parenteral drug might have the same effect. Many millions, perhaps billions, of gallons of wines were (and are) filtered through chrysotile fiber filters with no apparent increase in stomach cancers, but ingestion is quite different from inhalation.

The initial response of manufacturers of injectable drugs sterilized by filtration was to place membrane filters (293 mm diameter discs, rated at a pore size of 0.2 micrometer) downstream of the asbestos depth filters housed in plate-and-frame presses. The objective was to scavenge asbestos fibers released into the otherwise essentially particle-free filtered liquid.

Chrysotile asbestos fibers have a hollow, cylindrical structure, hence a very high surface area per unit weight. The isoelectric point is 8.3 (attributed to $Mg^{2+}$ that encompasses the fibrils); therefore, chrysotile has a positive charge in neutral aqueous solutions. Bacteria and most other particles are negatively charged in neutral suspensions. It is likely that asbestos and other positively charged filter matrices remove small microbes and other particles by adsorption and ionic interaction (Fiore et al. 1980) in addition to sieving, although, to the best of our knowledge, the proof of mechanism has been indirect.

Pall Corporation developed filters of submicroscopic fibers of potassium octatitanate held together with a potting glue. Like asbestos, the fibers presented a very high surface area per unit weight and a positive charge that encompassed the fibrils. The filter mats were integral and could be bubble-point tested (see Nondestructive Tests) prior to pleating. However, pleating of the matrix, to form a high surface area cartridge, ruptured the matrix at the pleats and destroyed the bubble-point integrity of the filter. Rupture of the matrix at the pleats also made possible the release of fibers. Titanium microfibrils cause mesotheliomas, and production of the microfibrils was discontinued by DuPont soon after their cancer-causing potential became known. However, they remained available from Kerr-McGee. The titanate fiber cartridges were discontinued by Pall Corp. when a pleated nylon membrane became available.

## Sieving Mechanisms

Membrane filters usually are taken as polymeric sheets containing voids (spaces) that penetrate from one surface to the other. The voids are usually interconnected, in which case the void

volume may be as much as 98 percent of the total volume; or un-
connected (as in track-etch filters, see below), in which case the
void volume may be 10 percent of the total, or less. Polymeric
composition has ranged from the highly nonpolar [poly(tetrafluoroethylene), also known as PTFE or Teflon™], sheets that are
stretched to produce oblong, slitlike pores (Figure 7.1); to the po-
lar poly(ethersulfone) that is cast by the traditional phase-
inversion process (Figure 7.2). Most membrane filters are 150–
200 micrometers thick and, for those filters rated at a pore di-
amter of 0.2 or 0.3 micrometer, bacteria are retained on and
within the upstream side of the filter matrix (Figure 7.3).

## Adsorption by Membrane Filters

In the filtration of liquids, particles larger than the largest pores
of a membrane filter remain on the filter surface. Particles and
organisms smaller than the pore size *may* pass into the void volume below the filter surface; many fine particles, on impacting
the filter matrix, are retained. If the microbes or particles are retained, the mechanism is some form of adsorption. For example,
bacteriophages sorb to membrane filters cast from mixed esters
of cellulose, although the viruses are far smaller than the filter
pore size. Presumably, adsorption occurs because mixed ester
membranes contain hydrophobic patches, and the capsule proteins of many viruses contain hydrophobic amino acid residues
(e.g., tyrosine). Nylon, which is notorious as an adsorptive filter,
contains $-(CH_2)_6-$ (hydrophobic) as well as $-NH_2$ (cationic) and
$-C(O)OH$ (anionic) regions.

Membrane filters that are rated at 0.2 micrometer on the
basis of bacterial retention often do not retain polystyrene latex
beads larger than 0.2 micrometer in diameter. Many of the pores
in membrane filters are greater than the stated pore size. Where
the filter is intended to render a fluid bacteriologically sterile,
there is no misrepresentation of the functional filter pore size.
However, where particle retention (e.g., for the removal of inorganic matter) is the objective, such filters may not perform as
implied in the designation since rigid inorganic particles are often less adsorptive, especially to hydrophobic filter matrices.

**Figure 7.1.** Poly(tetrafluoroethylene) (PTFE) filter rated at 0.2 micrometers on the basis of particle retention. *(Electron micrograph courtesy of Gelman Sciences, Ann Arbor, MI)*

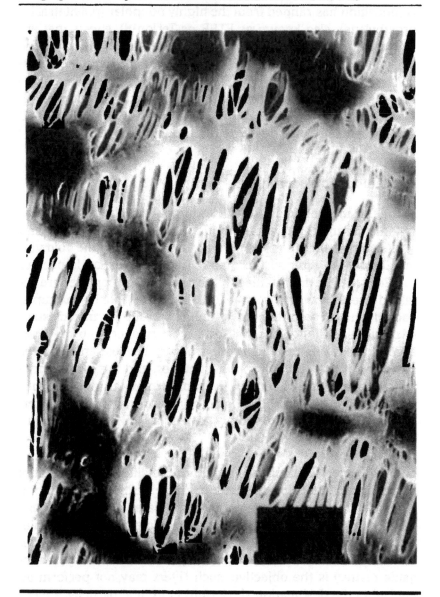

**Figure 7.2.** Phase-inversion membrane pore structure of poly(ethersulfone). *(Electron micrograph courtesy of Gelman Sciences, Ann Arobor, MI)*

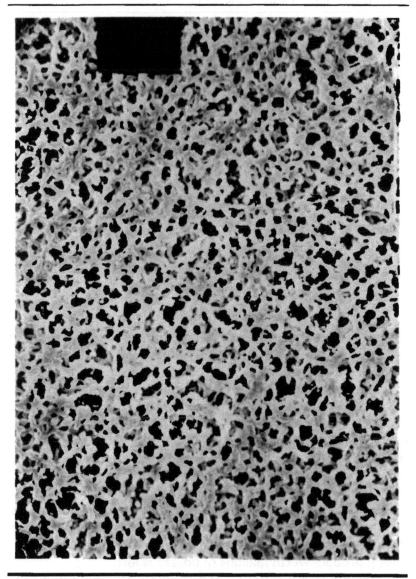

**Figure 7.3.** Organism retention as a function of phase-inversion membrane filter depth, for filters of various pore sizes. *(From Fifield and Leahy 1983)*

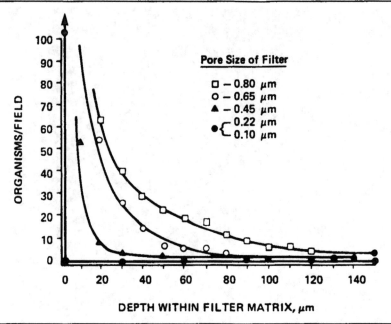

Although adsorptive retention often is advantageous (e.g., the retention of viruses by oversized pores), it can be extremely disadvantageous as well. For example, an attempt to sterile filter human clotting factor VIII (hF-VIII) through 0.2 micrometer-rated nylon 66 membrane filters results in irreversible adsorption of the product to the filter matrix and loss of the product. This occurs despite the presence of a far higher concentration of a relatively inert background protein. The background protein may adsorb reversibly and the hF-VIII irreversibly, a point of more than academic interest to a producer of human blood fractions if product worth millions of dollars is lost.

For the sterile filtration of human source or recombinant human proteins for injections, several rules of thumb apply:

- The more highly charged the membrane (e.g., with carboxyl or amino groups or both), the more adsorptive to proteins.

- The more hydrophobic the membrane, the more adsorptive to proteins.

- The smaller the pore size rating of the filter, the more adsorptive to proteins.

- The higher the $M_r$ of the protein, the greater the tendency of the protein to adsorb irreversibly to surfaces.

- Protein adsorption by membrane filters is a saturable process and eventually ceases with large volumes containing more than 1 percent protein.

Most of these points have been made in Nema and Avis (1993). However, those authors found an anomalously high adsorption of lactate dehydrogenase (LDH) with poly(vinylidine)difluoride (PVDF) membranes of a pore size of 5 micronmeters.

PVDF is fairly closely related to PTFE and, like PTFE, is hydrophobic. Millipore renders PVDF hydrophilic by coupling to the surface a polyol of moderate $M_r$. As a result, hydrophilic Durapore (as the polyol-coupled PVDF is designated) has very low protein binding. This author wonders if the unmodified 5 micrometer PVDF inadvertently had been substituted for the material of choice in the Nema-Avis work.

Filter manufacturers are most aware of the tendency (or not) of their filters to wet with water and often include a wetting agent (surfactant) in the formulation of the more hydrophobic membrane filters to promote the imbibition of water (see Olson et al. 1980). This still is done with many filters. Usually, hydrophilic membrane filters for process contain no such wetting agents, or, if they do, the manufacturer makes a point of the need to rinse the filter prior to use. Hydrophobic filters are good protein adsorbers. Most cartridge filters (but not all) are rinsed with high-purity water before drying and packaging, ergo the extractables from many cartridge types are very low. But low extractables does *not* translate to low adsorptivity, which is a function of filter matrix composition.

## Comparison of Adsorption and Sieving

Adsorptive retention of microbes and particles is a saturable process. If all of the adsorptive sites are occupied, pass-through may not occur because the extent of particle/macromolecule adsorption has decreased mean pore size. Sieving, by way of contrast, should be absolute in the retention of organisms, depending on the diameters of the organisms and the pore size distribution of the membrane filter. Filtrative removal of microbes by a sieving mechanim appears to be a probabilistic event, as is the killing of organisms with moist or dry heat. This is dealt with under Validation of Filters and Tests.

## The Special Case of Filtering Air and Gases

The gold standard for the filtration of air (e.g., vents for sterile product tanks) is 0.2 micrometer PTFE. First, a hydrophobic filter is preferred as it is unlikely to absorb moisture and ultimately become blocked by water. When a hydro*philic* membrane filter with small pores becomes wetted with water, whether from steam, Water for Injection (WFI), or another source, free water in the filter blocks the pores and air (or the passage of another gas) is diffusion limited, an extremely slow process. The imbibition of water by the vent filter(s) can lead to tank collapse during pump out and, in fact, such events have occurred.

Vent filters usually operate at atmospheric pressure, which, for most hydrophilic membrane filters with small pores, is insufficient to displace free water in the pores. Therefore, for membrane filters in venting applications on tanks, the filter matrix is a hydrophobic polymer that wets with water very, very poorly if at all. If the matrix is hydrophilic, the filter housing must be sufficiently heated to prevent condensate from accumulating in the filter pores.

Fibrous filters contain a very wide range of pore diameters and, although the smallest pores may clog readily, larger pores tend to remain open. Fibrous filters usually are called *depth filters* because particle retention is throughout the depth of the matrix as contrasted with membrane filters that tend to trap particles on and immediately beneath the surface of the filter.

The most common fibrous vent filters are of fine borosilicate glass. They are somewhat hydrophilic and are kept dry by heating the filter holders, usually with high resistance heat tapes.

Air passing through a dry fibrous or a membrane matrix imparts a high static electric charge to the filter. Fixed charges operate over longer distances in air than in water, and since the particles impinging on a filter actually are a mosaic of charges, ionic interactions can and do occur. The particle removal efficiency of a matrix, depth or membrane, is about 4 times greater in air than in water. Presumably, a hydrophobic vent membrane filter rated at 0.8 or even 1.0 micrometer pore diameter would prove to be as efficient as a filter rated at 0.2 micrometer per the data of Liu and Kulmey (1971). The advantages of using the higher pore size–rated filters are that smaller surface areas are needed to filter the same volume of air, and blockage by moisture becomes even less likely. However, most vent filters in use in pharmaceutical manufacturing are rated to 0.2 micrometer.

## Filtration as a Statistically Driven Event

Sterilization by steam autoclaving is taken as the gold standard for sterilization, but it is not an absolute event at any given time. Even with a microbial population that appears to be in every way uniform, in an autoclave chamber that appears to be uniform in temperature and moisture, organisms are destroyed at a determinable *rate*. For example, in the first minute at temperature, 90 percent of the organisms may die, and 90 percent of the survivors may die in the second minute at temperature, and so on. That is to say, in this example, 1 log of the test organism dies during each minute of exposure to the given temperature. The D value is 1 minute, and 6 logs of the test organism will be killed in 6 minutes at the given temperature.

A filter has a pore size distribution on the filter surface and throughout the depth of the matrix. Track-etch filters are an exception, but their use in process is minimal. Because there is a pore size distribution, a possibility exists that a viable particle might traverse the 150 to 200 micrometer depth of a membrane filter (or the much greater thickness of a fibrous depth filter) via the successive pores though which a fluid must pass in going from the upstream membrane surface to the downstream surface. The number of pores through which a particle must pass in transit from surface to surface will range from 150–1,000 depending on pore geometry. Therefore, even particles significantly smaller than the pore size rating have a severely limited probability of passage.

A case in point is the rating of sterilizing-grade membrane filters. Pore size ratings of membrane filters are established on the basis of microbial retention. For example, a filter that retains $10^7$ *Pseudomonas diminuta*/cm$^2$ filter area is said to have a pore size rating of 0.2 micrometers. Individual *Pseudomonas diminuta* grown in limiting medium (saline-lactose broth, unstirred) have a diameter of approximately 0.3 micrometers. However, despite the rating of 0.2 micrometer **absolute** placed on their filters as a (brilliant) marketing ploy by Pall Corp., and later copied by the competition as a matter of necessity, almost all of the filters available in the commercial marketplace allow the passage of 0.2 micrometer diameter polystyrene latex beads (Simonetti and Schroeder 1986). Absolute was an adjective that pharmaceutical customers wanted to hear and believe, and at this writing (1997) the term is in widespread use.

However, for purposes of demonstration, events with 0.45 micrometer-rated membrane filters are easier to follow. Sterility testing usually is done with 47 mm diameter disc membrane filters (surface area, about 17 cm$^2$). When a sterile 0.45 micrometer-rated filter is challenged with 1,000 or fewer viable *Pseudomonas diminuta* (< 60 organisms/cm$^2$ filter area), the filtrate is sterile. When challenged with more than 1,000 organisms, passage occurs, and the filtrate is nonsterile. That is not to say that the pore size distribution of a membrane filter is not shifted to the left as pore size rating becomes smaller. It is. However, some degree of overlap exists because there is a range of pore sizes in such filters, hard-sell assurances notwithstanding.

There are pore-forming strategies that considerably reduce the probability that an organism will emerge on the downstream side, yet are consistent with good flow rates. Anisotropy helps, whereby the pores on the upstream side are more open than on the downstream side, and limiting pores are near the downstream side. The disadvantage of that pore-forming strategy is that the number of pores in the limiting range is reduced. Rather than small pores throughout the depth of the filter, fewer pores do the work of organism retention in an anisotropic filter.

For this reason, the filtration done upstream of the sterile, sterilizing-grade (final) filter is important. If the population of organisms challenging the final filter is small, the probability of organism passage through that final filter approaches nil. For this reason, many pharmaceutical manufacturers elect to use final 0.2 micrometer-rated membrane filters *in series*. However, for a regulatory reason, only the last of the 0.2 micrometer-rated filters is rendered sterile.

## PRACTICE

### Where to Filter in Process

#### *Into Carboys*

Human clinical reagents usually are processed as small batches. Often 10 L ($10^4$ 1 mL vials) is sufficient, in which case a clinical reagent lot can be sterile filtered into a 20 L glass carboy from which it is filled by a small filling machine. Consider, for example, that nitrogen mustard, a powerful alkylating agent, recovered from a silica column as an acetone solution, is adjusted for concentration with additional acetone and filtered through 0.2 micrometer-rated PTFE into a 20 L carboy. The carboy previously had been washed and then rinsed thoroughly with 0.2 micrometer-filtered water, followed with 0.2 micrometer-filtered acetone.

Separately, sufficient aqueous HCl per mL to precipitate from acetone solution one dose of nitrogen mustard as the HCl salt is filtered through a 0.2 micrometer-pore size membrane filter into a 5 L carboy that has been properly washed and rinsed. From this carboy, a 1 mL fill is made into each washed and dry heat sterilized 10 mL vial. This is followed immediately by a fill of one dose of nitrogen mustard in acetone into the vial, resulting in a mixing of the reagents in the final container and the precipitation from acetone of nitrogen mustard as flakes of the hydrochloride.

Individual lyophilizer trays of the filled vials, with slotted stoppers half-inserted, are gently swirled to ensure the complete mixing of the aqueous and acetone solutions, then chilled to 10°C in a lyophilizer that, when closed, pulls a vacuum of 10 torr. The acetone and water (miscible liquids) shift to the vapor phase under vacuum and sterile-filtered nitrogen mustard remains in the vials as a stable dried powder.

This clinical reagent scenario is of whole cloth. That is to say, it is a possible system for producing an injectable product but likely is done differently by manufacturers. However, it resembles the mode of production of lente insulin in that two sterile-filtered solutions are mixed in the final container and subsequently dried. It likely can be done as stated.

On 5 and 10 L carboys, the vent filters are of PTFE and 50 mm diameter vent filters are adequate. The vent filter area is doubled simply by adding another filter, if required. Filling *from* a carboy usually is done under positive air or nitrogen pressure applied through the vent filter(s), but peristaltic pumps on the outlet are also often used. Clearly, the vent filters are critical to the maintenance of the sterility of the product being filled. The sterilizing-grade filter is autoclave sterilized, together with the vent filter(s), attached to the carboy. Hose ends are covered with porous autoclave paper, usually kraft paper of the type used for sturdy grocery store bags but with fewer fibrous extractables. Any sturdy porous material will do, and it is secured with wire or tape. Secured is the operative word.

Filling *into* a carboy is done by applying air or nitrogen pressure to a closed tank so that liquid product is forced through the outlet of that tank, through the sterilizing-grade membrane filter, and into the sterile carboy. Air in the carboy is displaced through the vent filter(s). At the completion of filling, the hose from the sterilizing-grade membrane filter to the carboy is double-clamped near the filter, with a 4 cm space between the clamps; then the filter is removed for disposal. Filling sterile product into vials *from* the carboy is done by alcohol swabbing the hose between the clamps, flaming the area, cutting the flamed hose with sterile scissors, and aseptically connecting the carboy end of the hose with the filling machine. The carboy can be very lightly pressurized with air or nitrogen through the vent filters.

### Tank-to-Tank

Most filtration in the production of sterile drugs is done between two tanks. The nonsterile (makeup) tank, A, is clean. The sterile (receiving) tank, B, has been cleaned, steam or autoclave sterilized with a hydrophobic vent filter(s) in place, and kept sealed until use. As with carboys, tank B usually is sterilized with the sterilizing-grade membrane filter in place on the tank inlet, but not necessarily. (For many years, with no breaks in product sterility, one firm aseptically attached the sterilizing-grade membrane filter to tank B after the filter and the tank had been separately steam autoclaved.)

The hookup between A and B need not be done aseptically since the liquid passing through the sterilizing-grade filter into B already is on B, and product in A is not sterile. If the filter and B have been autoclaved separately, then the sterilizing filter and tank B must be connected aseptically in a small area or room supplied with high efficiency particulate air (HEPA)-filtered air, by workers who are gowned, gloved, and masked for cleanroom work. As mentioned above, this works, but it is simpler to autoclave or steam sterilize B with the sterilizing-grade filter on the tank.

### In-Line with Filling

Tank-to-tank sterilizing-grade filtration is most common because the receiving tank (B) can be stored—in the cold, if desired—until bulk testing provides proof of sterility, or a sterile filling line is available, or both. For example, if a drug solution is sterile filtered and for any reason 50 organisms pass into the filtered product, the organisms will tend to be distributed among the filled vials. If the lot size is 5,000 vials, only about 1 percent of the final containers can be contaminated (assuming no contamination during filling), and that level of contamination cannot be detected by the U.S. Pharmacopeia (USP) sterility test. Fifteen percent or more of a batch must be contaminated to ensure that the lot will fail the USP sterility test (Olson 1987). However, if the filtration is validated to produce repeatedly a sterile product, and if the sterile filling line consistently can be ready for operation shortly after product makeup, there is no reason that final filtration cannot be done in-line with filling. This is, in fact, often done.

Filtration in-line with filling usually is done with a sterile gear pump or peristaltic pump to ensure even flow rates and the proper fill volume in the filling head. However, if the filtration system is considerably oversized and there is a sterile surge tank between the filter and the filling machine, air or nitrogen pressurization of the makeup tank (A) can be suitable. The problem is that, as the final filter clogs, the flow rate through the filter slows if tank A is kept at constant pressure. What the filling machine requires is constant *rate* of feed. This is dealt with in the next section.

## Constant Pressure or Constant Rate

### Constant Pressure

The simplest systems operate under constant gas pressure and do not compensate for reduced flow rate as the filters clog. Where one is filtering between the makeup tank (A) and a sterile receiving tank (B), it is of consequence that the filtration is completed in a timely manner, but not of consequence that the filtration rate may slow throughout the process. For a constant pressure system, the makeup tank (A) must be sealed so that it can be pressurized with gas, and the sterile receiving tank (B) must be sealed so that microbes cannot gain entry to the interior of the tank or its contents prior to, during, or after filling. Therefore, one needs a means of evaluating tank and fittings integrity before commencing a constant pressure sterilizing filtration. The simplest test is a pressure-hold test of the assembled system.

If a tank is pressurized (relative to atmosphere) and sealed, and the internal pressure is monitored over time, pressure decay over time reflects the loss of gas through gaps or imperfections in a gasket, a valve, a fitting, or combinations of these. A reasonable pressure-hold test for a tank is 20 psi, with a pressure decay of no more than 2 psi over 30 minutes. Although one can calculate the size of a single gap or hole through which the pressure loss occurred, the practical objective is, in the main, to ensure that tight connections have been made (e.g., between the filter and the tank if the connection is via tubing) and that all seals and caps are intact.

If the mixing tank will hold pressure, it can be pressurized to drive the filtration. If the receiving tank will hold pressure, it will be integral and maintain the sterility of the product. In fact, the receiving tank will experience no significant pressure differential during use because tank B has a sterile vent filter(s) to ensure pressure equalization. A pressure-hold test assures that tank B meets the requirement for integrity.

Therefore, the cleaned and steam-autoclaved or steam-sterilized tanks, A and B, after closure and pressure-hold testing, are ready for use. Makeup is done in tank A and the outlet for A is connected aseptically to the oversized sterilizing filter on tank B (oversize is critical, as will become obvious later in the chapter). In some instances, the filter may be mounted separately, in which case 2 aseptic connections are made: one from tank A to the filter, and the second from the filter to tank B. Tank A is pressurized *slowly* so that filtration commences slowly and the system can be checked for leaks. The filter housing will fill and must be vented so that air in the housing escapes to atmosphere from the upstream side; a small amount of unfiltered product escapes before the vent valve is closed. This valve must *not* have a filter on it; however, the valve may be aseptically connected to a small glass sidearm flask that has a sterile vent filter, to ensure that no gross contamination enters the system upstream of the sterilizing filter during venting; this usually is unnecessary. After the filtration has commenced from a full housing and filter/connection integrity seems assured, the pressure in tank A is taken to about 1 atmosphere (1 bar, 14.69 psi). The system is left untouched until the filtration is complete, as indicated by the absence of liquid on the upstream side (one usually places a piece of glass tubing in-line with the hose between tank A and the filter; the glass tubing is filled with liquid when the filtration is in process and is readily emptied by simple movement when all of the liquid has passed through the filter). As mentioned before, the sizing of the sterilizing filter system is crucial.

Following the completion of filtration, hosing from the filter to tank B is clamped and cut between the filter and the clamp(s); many users prefer double-clamping for safety once the tubing has been cut. Tank B should be retested for pressure hold (a failed test necessitates a repetition of the sterilizing filtration) and quarantined in a cold room or at ambient temperature, depending on the stability of the aqueous product. Tank B is bulk tested for sterility and wheeled to an aseptic filling line when a clean, sterile line is available.

Often, depth filtration (Fiore et al. 1980) is needed upstream of the final sterilizing filter. Depth filters are more open than membrane filters and can trap particles and organisms throughout the depth of the matrix. Depth-filtered liquids impose a far smaller particle burden on the final filter, which clogs far more slowly as a result. If depth filtration is needed, one filters between tank A (makeup) and tank A' (depth-filtered product). *The product in tank A' is not sterile and must be sterile filtered almost immediately (usually within 1–2 hours) into tank B.* Whether a depth filtration is required is determined in sizing experiments.

## Sizing for Constant Pressure

Sizing can be done in two ways, one tedious and the other rapid. The tedious method works regardless of the mechanism of filter clog; the rapid method scales properly only for solutes that accumulate within the filter rather than as a cake on the surface. Macromolecules (usually biologics) adsorb to the interior of the filter matrix. The rapid filter sizing method described by Badmington et al. (1995) is particularly applicable to biologics and we will deal first with this.

Proteins possess charged groups and hydrophobic residues on their surfaces and so are ideal candidates for adsorption to most surfaces. As they adsorb to the polymeric framework of a membrane filter with small pores, the pores become smaller, as does the internal diameter of an artery in which arteriosclerotic plaque accumulates. As the internal pores become smaller, resistance to flow increases until the resistance becomes sufficient that flow effectively stops at constant pressure. Consequently, in this circumstance, reduction in *flow rate* is predictive of the *volume* that can be filtered.

When pores are clogged slowly, as by the internal adsorption of proteins, filtration rate and throughput can be described as:

$$t/V = (k_s/2)t + 1/Q_i$$

where $Q = dV/dt$ (flow rate per unit time) at constant pressure; $Q_i$ = initial flow rate at $t = 0$; $k_s$ = constant for filter clog; $t$ = filtration time; and $V$ = cumulative volume filtered at time $t$. From this equation we can predict the cumulative volume filtered at any time.

For the given set of conditions, $V_{max}$ is the entire volume that has been filtered when the flow rate $Q$ has decreased to zero (when the filter is fully clogged). It can be shown that

$$V_{max} = (k_s/2)^{-1}$$

$$\therefore t/V = t/V_{max} + 1/Q_i$$

where $Q_i$ is the initial flow rate. This last equation is a straight line with a y intercept of $1/Q_i$ and a slope of $1/V_{max}$ when $t/V$ is plotted on the y-axis and $t$ is plotted on the x-axis.

The same information can be obtained in another way. The flow rate at any time, relative to the initial flow rate, is $Q/Q_i$. It has been shown that

$$Q/Q_i = (1 - V/V_{max})^2$$

from which we can estimate estimate filtrate volume on the basis of flow rate/initial flow rate ($Q/Q_i$) and the volume filtered to that point ($V$). All of this assumes a linear relationship between $t/V$ and $t$, which may or may not be. If the reciprocal of rate as a function of time is linear, this is a very rapid method for sizing. However, at constant pressure, filtration until the filter clogs fully is usually highly impractical. The sensible approach is to filter until the rate is approximately 20 percent of the initial flow ($Q/Q_i \times 100 = 20$), at which time the filter will be 55 percent clogged. If one filters to 10 percent of the initial flow rate, the filter will be 68 percent clogged, but the rate, near this endpoint, will be agonizingly slow.

The older method for sizing at constant pressure involves no assumptions. One places product in a pressure vessel that contains a magnetic stir bar or is fitted with a bladed mixer. The pressure vessel is coupled at the outlet to a 47 mm filter holder containing the candidate filter (as above). The pressure vessel is pressurized slowly to the pressure that will be applied during process (e.g., 20–25 psig) and the amount of filtrate collected into a graduated cylinder is recorded as a function of time in minutes. This function of rate versus time may be linear or convex or, less often, concave.

The flow rate and the throughput (volume filtered) when the filter is 80 percent clogged become the basis of sizing. A simple proportion is set up for rate, if rate is a determinant, and another for throughput. Let us consider throughput first, and doing things backwards (without first depth filtering) so as to see the consequences of *not* doing sizing experiments systematically. The filter area of the 47 mm disc must be calculated (it may range from 13.8 cm² down to 9.6 cm², depending on the structure of the filter holder and the gaskets) and the volume to be filtered in process must be established. Let us say that the target volume is 3,000 L ($3 \times 10^6$ mL) and the throughput to 80 percent clog of the 47 mm disc is 214 mL. Then the throughput proportion is

$$\frac{214 \text{ mL}}{3 \times 10^6 \text{ mL}} = \frac{13.8 \text{ cm}^2}{y \text{ cm}^2}$$

where $y$ = required filter area in cm².

Filter areas for pleated cartridges usually are at least 2,000 cm² per 25 cm (10 inch) length. The solution to the proportion (above) is of the order of $2 \times 10^5$ cm² or the equivalent of 1,000 10-inch cartridges, which is absurd in cost, the physical manifolding that would be required, and the product holdup in the system.

What this datum likely indicates is that the liquid has a very high burden of fine or deformable particles or both, causing the filter to clog rapidly. However, if the predominant particles are dense or large, and hence tend to settle out rapidly, part of the sizing problem is in the mixing of the sample addressed to the filter. A poorly mixed sample containing far more particles/mL than the average in a well-mixed sample will result in the overly rapid clog of the 47 mm disc and misleadingly large requirements for process filter area.

If this is not a mixing problem, and even if it is, *one* relatively inexpensive depth filter usually is required upstream of the sterilizing-grade membrane filter, but which depth filter? The vendor of the membrane filter may have some candidate materials, and they should be examined for impact on the throughput of the sterilizing-grade filter. What likely would obtain in the example given is that both mixing and depth filtration of the sample are required. It is common to filter 3,000 L of an aqueous solution through 2,000 $cm^2$ of a membrane filter (one 10-inch pleated cartridge) rated at 0.2 micrometer pore size. However, it also is common to do continuous mixing as the sample is fed to the filter.

In the example cited (214 mL throughput for a 47 mm disc of 0.2 micrometer-rated membrane filter), this author would (a) modify the pressure pot to accept a simple, single-blade mixer and (b) perform a series of $Q/Q_i$ experiments to 20 percent clog whereby the membrane filter is sandwiched with an upstream depth filter. What one now will do is a sizing experiment with a 47 mm disc of the depth filter, and then a sizing experiment with the filtrate and the membrane filter of choice.

### Constant Rate

It is possible to feed a filling machine at a constant rate with a gas pressure-driven feed; this will be dealt with later in this section. The more common system is constant rate feed to a filling machine by a gear pump or peristaltic pump (filling machines with peristaltic pumps are a commonplace). If sterile-filtered product in tank B is connected aseptically to a sterile filling machine via a sterile gear pump or a peristaltic pump, liquid filtration is not a concern. However, if sterilizing filtration is done in-line with a pump, a sterilizing filter, and the filling machine, sizing and performance of the filtration system is the major concern.

With constant pressure filtration, the filtration *rate* slows as the filter clogs from particle accumulation. With constant rate filtration, the upstream pressure builds as the filter clogs until, at some elevated pressure upstream of the filter, a fitting between the pump and the filter fails, the pump fails, or the filter fails. Where elastic tubing is used for connections, the tubing tends to balloon. Usually, one prefers to size a system so that the upstream pressure on a filter never exceeds a very safe limit (e.g., 20 psi). Even so, periodic monitoring or automated monitoring of the pressure upstream of a filtration line feeding a filling line is most important. Initially, the slope of the pressure increase is low and more or less linear. As the filter becomes highly clogged, the rate of pressure increase often tends to become exponential and failure can come quickly.

Where sterilizing filtration is done in-line with a filling line, it is essential that the autoclave-sterilized final filter be preceded by an identical but *nonsterile* filter. The reason for this is that the lot size is determined by the volume that passes through the *sterile final filter.* Any filter upstream of the final filter may be changed if it commences to clog, without compromising lot integrity.

One pressure-driven system that monitors flow rate downstream of the final filter is shown in Figures 15 and 16 in Olson (1987). Where product requires an inert (e.g., nitrogen) atmosphere, such a system may be used in place of a pump to achieve constant rate to feed the filling heads. This can be, and is, done, but it adds a further complication to the system. For filtration and filling systems, wherever possible, simplest is best.

Local, Eastern U.S. FDA inspectors have required that a sterile-filtered bulk be tested for certain attributes, including sterility. Where in-line filtration/filling is done, bulk can be tested only from a surge tank installed between the sterile (sterilizing final) filter and the filling heads.

*Comparing Constant Pressure with Constant Rate*

A large-scale operation in a location where labor costs are high might best be done with a high degree of automation and integration of operations. For very large volume products, the tendency then is to dedicate one or more filling lines to that product. It is possible, even beneficial, to provide services and to supply the machines that sterilize, cool, and position the final containers and closures to the filling room, with one or two people to start, stop, take samples, and troubleshoot in a traditional filling clean room. Minimally, this halves the number of workers present in the filling area; the workers in place are skilled and perform multiple tasks.

Where a filling system and the supply services to it are dedicated to a single product, it makes sense to sterile filter in-line with filling. Most often, hopefully almost always, the elements of the system will be ready just in time, on schedule for filling, and the filtration/filling operation is routine, very safe, and highly cost-effective.

If a filling line is not dedicated to a product but must undergo changeovers in product, fill volume, and lot size, the system becomes more labor intensive than the one described above. Frequent delays may be commonplace. For such instances, filtrations between tanks A and B are simplest and most sensible, since sterile product can be stored briefly (usually less than a week) until it can be scheduled for filling. Furthermore, if any part of an integrated system (e.g., clean-in-place [CIP] and/or sterilize-in-place [SIP] system for cleaning and sterilization of the filling line) breaks down and does not have an immediate backup, product is not filtered or filled. However, if sterility of the filling line is compromised, a constant pressure filtration system between tanks A and B still can be done, and the sterile product stored until filling can be resumed. Nonintegrated systems are slower, more labor intensive, and far less cost-effective than integrated automated systems.

## Issues

### *Adequacy of System Sizing*

Filtering one-half or three-quarters of a lot, and stopping because the final filter has clogged, is unacceptable and is a measure of inadequate sizing of the system. There are three methods of sizing, two for constant pressure (by volume and by rate) and one for constant rate filtrations.

**Sizing by Constant Pressure from Volume Data.** Obtain samples of reagents from at least 3 different lots prepared by the manufacturer. Obtain makeup WFI on at least 3 different days to account, at least in part, for differences in water particle quality. Perform makeup in a 20 L glass carboy, and transfer mixed product to a 5 or 10 L stainless steel pressure vessel fitted with a 47 mm diameter filter holder on the outlet. Candidate depth filters are placed in the 47 mm disc filter holder. Mixing is achieved by repeatedly swirling the pressure vessel. This also is a tank A–tank B setup but here tank B is a 1 L volumetric flask. Volume data are taken in 1 minute increments.

The flow rate over the first minute of filtration is the baseline to which all subsequent data are compared. The **throughput** is taken as the volume filtered through the 47 mm disc until the flow rate is 20 percent of the initial rate.

### *Adequacy of Sterilization of the Filters*

A filter cannot sterilize a fluid if the filter itself is not sterile. Consequently, sterilization of final filters must be done, either with the filter connected with the receiving tank B or the line to the filling heads, or such connection must be made aseptically. The former usually is preferred as it is less amenable to error. This sterilization must be validated and reliable. It is best done in a steam autoclave, and the cold point(s) in the system determined with thermocouples during a preliminary autoclave run. It is important that the autoclave draw at least a single or a double vacuum so that steam penetration is rapid and complete. If a filter is attached to a sterile product tank with a drop line, the cold point usually is in the drop line.

## Probability of Microorganism Passage

Validation of a final sterilizing grade filter must be for the complete retention of at least $1 \times 10^7$ *Pseudomonas diminuta* (ATCC 19146)/cm$^2$ filter area when the organism is grown at 30°C as a standing culture in saline-lactose broth for 40 hours (Fifield and Leahy 1983). Invariably, so that the amount of organisms can be held to reasonable quantities, the testing is done with 47 or 50 mm diameter filter discs with a filtration area of about 17 cm$^2$. A correlation between a nondestructive test of filter integrity and pore size versus organism retention must be shown and related, at least coarsely. This is a somewhat involved and time-consuming study that may be avoided if the vendor presents a certification to the effect that *Pseudomonas diminuta* retention at a level of $10^7$/cm$^2$ filter area will be obtained with the product, provided certain conditions are met. However, filtrative sterilization is so important in aseptic processing that the demonstration of these skills is worthwhile. If it cannot be done in-house, there is a certain probability that problems, when they arise with the system, cannot be addressed in-house, and that in itself is serious.

If one is to challenge 17 cm$^2$ of filter area with at least $1.7 \times 10^8$ *Pseudomonas diminuta*, a simple and rapid method for estimating numbers of these organisms is desirable. The simplest is to raise the organisms in the indicated manner, perform standard plate counts (Greenberg et al. 1981) with 0.1 mL of logarithmic dilutions plated onto nutrient agar, and examine the light-scattering properties of dilutions of the organisms in 0.2 micrometer-filtered saline at a fixed visible wavelength greater than 500 nm (Barber 1996). The correlation of $A_{520}$ as a function of colony-forming units/mL is used thereafter to estimate the challenge to a filter. If all of the 47 mm discs perform satisfactorily, one does not have a control that fails the test. One such control is a 0.2 micrometer-rated filter that has been pinholed. That filter should grossly fail both the nondestructive test (e.g., bubble point) and *Pseudomonas* retention. Alternatively, or additionally, a filter rated at approximately 1 micrometer should pass organisms. What is *not* required is a detailed plot of $\log_{10}$ organisms retained/organisms passed through the filter (log retention value or LRV) as a function of bubble point in psi such as was done by Fifield and Leahy (1983).

## Nondestructive Tests

The most comprehensive and useful chapter on nondestructive tests for membrane filters will be found in Meltzer (1987).

### Bubble Point

The most effective, single, nondestructive test of filter integrity is the bubble point, determined by measurement of gas flow through a wetted filter over a range of upstream gas pressures. A filter is wetted with water, saline, product, or a solvent; with the exceptions noted later, one solvent system may not be substituted for another.

A modest air pressure applied to the upstream side does not displace the wetting liquid from an integral filter. As the upstream air pressure is raised slowly, at some critical pressure the menisci of the liquid in the upstream filter pores drop from the filter surface to levels in the filter where the pores are smaller. If the air pressure is again increased, liquid is expelled from the largest pores and bulk flow of gas commences through the filter. At still higher pressures, successively smaller pores unload their liquid and the unblocked pores contribute to the bulk flow of gas.

The *initial* bubble point is that at which bulk flow of gas commences and usually is shown on a graph as the intersection of diffusive and bulk flow of the challenge gas (Figure 7.4). (Note: The data for such a plot of the type shown might be taken manually. However, it is the consensus of those who perform such measurements that they are done more accurately and with greater precision by machine. Several filter manufacturers market these devices.)

The American Society for Testing and Materials (ASTM) bubble point is a higher value, recorded at 50 percent of the bulk flow of gas following the initial bubble point. The ASTM value may be more reproducible, but the initial bubble point is the

**Figure 7.4.** Bubble point determination from gas flow rate determinations; the bubble point is estimated from the intercept of the diffusion curve (A) and the curve for the bulk flow of gas through pores from which the wetting liquid has been displaced (B).

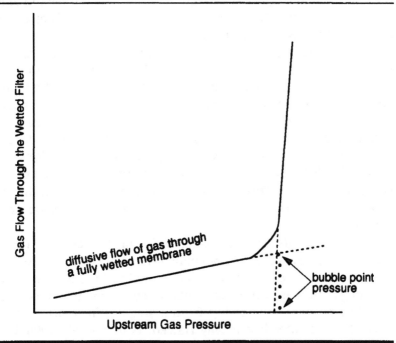

more sensitive and useful value. Either can be correlated with organism retention and used routinely as a nondestructive test before and after autoclaving of the filter and following use.

### Diffusion Only

Most often, the bubble-point test is taken to mean the ASTM test as one measure of the upstream gas pressure at which liquid is displaced from the pores of the filter. At pressures lower than the bubble point, the upstream gas *diffuses* through the wetted filter at a rate per $cm^2$ that is proportional to the thickness of the filter, the upstream gas pressure, and the solubility of the gas in the wetting liquid. Diffusion rates are also proportional to the percentage of the filter matrix that is void (i.e., not occupied by the polymer or other solid that constitutes the matrix; the assumption is that the voids in the filter are continuous) and, most dramtically and exquisitely, to temperature. The higher the temperature, the lower the diffusion rate, because gas solubility declines with increasing temperature.

The issue is that the user may not be able to control the thickness of the water layer, especially if liquid is wicked into the area between support screens and the membrane, or the ambient temperature may be significantly different from the validation temperature. A thicker water layer, like an increase in temperature, will lower the diffusion value and mask the effect of bulk flow of gas through oversized pores. If specifications are drawn tightly so as to be meaningful, such preuse test data based solely on a single, diffusive flow measurement may mean that an inappropriate filter is used and the filtered bulk may be nonsterile.

An equally likely source of error is that a filter matrix with low porosity (low void volume) may have a break through which the bulk flow of gas occurs (a bubble point, in fact). But if the overall gas flow is within the limits set by the manufacturer, the user has no way of knowing from a single diffusion measurement that the filter is crucially flawed. Waibel et al. (1996) discuss much of this. In the opinion of this author, Waibel et al., and Meltzer (1987), single point diffusion tests are contrary to good manufacturing practices and an eventual invitation to nonsterile events.

The progenitor of the diffusion rate was the forward-flow test proposed by Pall Corp. as a test for the pleated fibrous titanium fiber filter marketed to the pharmaceutical industry in the 1970s. The pleated matrix had a very low bubble point, hence the recommendation of the manufacturer to challenge the wetted filter with 5.25 psi of air. Some low gas flow rate was recommended by the manufacturer as acceptable. Subsequently, many laboratories have shown that a challenge pressure of 5 psi is far too low to differentiate a sterilizing-grade filter from a nonsterilizing filter.

However, if an integral membrane filter with an area of 90 cm$^2$ or more, and rated at a pore size of 0.2 micrometer, is wetted with water, then pressurized on the upstream side to 80 percent of the bubble point, measurable amounts of air should diffuse through the 200 micrometer-thick water layer (thickness of the filter). Pleated cartridge filters, with 20 times or more that area, are candidates for diffusion measurements, but the most conservative tests will take the upstream gas pressure to the bubble point and beyond.

## Upstream Testing of the Filter on the Tank

Where wetted filters are integrity tested on the tank prior to autoclaving, it is essential that the bubble-point test be performed, since steam will not adequately penetrate a water-laden filter. The filter must be blown down prior to autoclaving. Presteam vacuum cycle(s) help to rid the filter(s) of excess moisture, although some will remain bound. If the filters can be consistently shown to be integral before and after autoclaving and have virtually no extractables, it is acceptable not to test prior to sterilization so that the filters are dry at the time of autoclaving. However, the filters must be tested after autoclaving and prior to use with product. One way of doing this without adding filter test water (released from the filter during bubble-point testing) as a diluent to the sterile bulk tank (B) is to wet the filter with the product and to test the filter with the product.

The product, although water based, may exhibit considerably different properties in a bubble-point or diffusion test than water, saline, or saline-lactose broth alone. The differences in bubble-point pressure usually are attributed to differences in the surface tension of the liquid. Let us say that one bubble-point tests a filter with water after a WFI rinse, autoclave sterilizes the filter, and performs a second bubble-point test with sterile water. Let us say that the sterile filter now is challenged with a suspension of an organism [e.g., *Pseudomonas diminuta* (ATCC 19146) from standing saline-lactose broth for a filter with a rating of 0.2 micrometer pore diameter] in saline-lactose broth. And let us say that the effluent is sterile in every way that we examine for the passage of the *Pseudomonas diminuta*.

Now we take a fresh, clean, unused copy of the challenged filter and bubble-point test it before autoclave sterilization. However, after sterilization, we wet the filter with the product that we propose to sterile filter. Perhaps we rinse down this filter to see if we can reproduce the WFI bubble point. Our objective is to establish the ratio of bubble point with water to bubble point with product so that the equivalency is a matter of experiment. Whether the product raises or lowers the WFI bubble point is immaterial, so long as the ratio with the WFI value is established and verified.

## Tank Vent Filters

Most tank filters are of a hydrophobic polymer (e.g., PTFE) and are not readily wetted with water. On very large tanks, especially fermentation tanks of unusual size requiring extemely high volumes of sterile air, fibrous filter beds packed into silo-like structures sometimes are used, but that is exceptional; most tank vent filters in the pharmaceutical and diagnostics industries are of hydrophobic polymers as cast or stretched membranes.

Testing may be done by wetting such hydrophobic filters with a solvent of sufficiently low dielectric constant (e.g., ethanol) and blowing the filter to dryness after bubble-point testing, or the water intrusion test may be applied. The water intrusion test is a bubble point test in reverse—applying water, under increasing pressure, until the water intrudes into the hydrophobic pores of the filter. Either test can be correlated with organism retention.

It is best that the filter matrix is hydrophobic so that the matrix does not wet with water and become essentially impermeable to the free flow of air. Gas flow through a wetted filter is limited to the diffusion rate, which is always too small relative to the air volumes/unit time required.

A filter with a pore size rating of 1 micrometer in liquid testing retains particles of 0.2 micrometer diameter or less in air (Liu and Kuhlmey 1977). Therefore, although hydrophobic membranes of 0.2 micrometer pore diameter rating are considered necessary as sterile tank vent filters, the user should be aware that in these applications, filtration is very forgiving.

## Membrane Structure

Because sterilizing-grade membrane filtration is probabilistic, membrane filtration strategies are employed to reduce significantly the probability of any single microorganism penetrating the final filter into a sterile bulk tank or filling machine. The most common and most practical strategy is to use double membranes.

### Single Versus Double Membrane

Most membrane filters in cartridge configuration employ two back-to-back membrane filters. The logic is that any imperfection that may exist in a single membrane is likely to be negated by a second membrane if there is no significant mixing volume between the two. Considering that a single, straight-through pore of 0.8 micrometer diameter would not be detected in a bubble-point or diffusion test of a 2,000 cm$^2$ membrane filter cartridge, and such a pore path could readily pass a small microbe, the strategy is conservative. To further decrease the probability of microbial passage, an *unautoclaved* filter identical with the final filter usually is employed immediately upstream of the final (sterilizing) filter. That filter, should it become clogged, can readily be changed without compromising lot integrity (i.e., the lot number remains the same although flow is stopped temporarily as the unautoclaved filter is changed). Therefore, double membranes in the final filter, and doubling of the final filter are recommended strategies.

As mentioned previously, vent filters are far less demanding but still must be integral for high assurance of organism retention. Fibrous filters pose some level of risk. However, because any filter is at least fourfold more efficient at particle removal from air than from a liquid, the risk of using tight fibrous filters is modest.

### Adsorptivity

Filters that possess a high charge density (e.g., nylon) bind macromolecules by ionic mechanisms. Similarly, microfiberglass develops a high static electricity charge as a vent filter and removes particles from air by a similar mechanism (however, if the filter becomes wet it is useless in such applications). Amphipathic polymers (e.g., mixed esters of cellulose) in liquid applications bind organics, macromolecules and many particles by hydrophobic interaction. Whereas size-exclusion effects with small pore membrane filters are predictable and validatable, and avail themselves to nondestructive tests, adsorptive effects are more difficult to predict filter-to-filter.

## Issues of Filtration In-Line with Filling

Most small-volume parenteral bulks are of the order of hundreds of liters, in part because each such bulk represents many doses of, for example, 1 mL volume. Therefore, a 500 L bulk represents $5 \times 10^5$ doses that possess the same expiration date. Such a bulk can be formulated in a relatively small portable tank and sterile filtered into another such tank with ease. Such tanks fit into larger autoclaves and present the manufacturer with considerable flexibility in the handling of bulks. A sterile-filtered bulk in a sterile tank, properly vented and sealed, can be rolled into a cold room for a limited storage time until it can be filled.

Whereas the flexibility in timing for filling can be an advantage, cold room space must be constructed and maintained, and plant design must accommodate the storage, washing, and movement of portable tanks. Consequently, the design of the manufacturing area commences with a consideration of manufacturing strategies. Such methods provide flexibility in scheduling, but are not the least cost in workhours; they are labor intensive. Where products are produced in 1,000 L or greater volumes, whether small- or large-volume parenterals, the integration of formulation, filtration, and filling into a single operation becomes increasingly cost-effective. One may choose to do makeup in fixed tanks which then are cleaned and steam-sterilized in place (Seiberling 1987). Product then is pumped through filters into the adjoining filling room (all filtration is immediately outside the filling room; sterile-filtered product is pumped through a port through the wall into the filling room).

Similarly, where bulks are quite large and several activities must be integrated (vial washing and sterilization, closure washing and sterilization), integrated systems are increasingly advantageous as labor costs rise.

Where filtration is done in-line with filling, it is absolutely essential that the filtration system is oversized (50 percent or more oversized is wise), and that all filling components and machines be available and prepared for use. Failure to size the filtration system properly could result in line shutdown and a division of one lot into two, doubling quality control and other costs.

Filter lifetime is an issue with regard to grow-through. The bioburden of the fluid presented to the sterile filter should be quite modest. It would not be surprising that 40,000–100,000 L of solution that was well filtered prior to addressing the final sterile filter *could* be processed through one 20-inch pleated cartridge. If the selected filtration rate was 50 Lpm (assuming no flow decay), the filtration time would range from 13.3 to 33.3 hours, an intolerably long filtration time.

With the loosest and least adsorptive of membrane filters, the minimal grow-through time generally is taken as 12 hours, and several days may be required for grow-through with particular membranes. As a rule of thumb, sterilizing-grade filters for liquids should not be used for more than 8 hours. Therefore, in the particular application, although a single 20-inch pleated cartridge could accommodate the throughput, one would want additional filters to accommodate a higher flow rate.

## VALIDATION OF FILTERS AND TESTS

The most comprehensive reviews are those of Meltzer (1987) and Olson (1987). The most current is that of Waterhouse and Hall (1995).

### Sterilization of the Filters

The filters used as sterilizing-grade (final) membrane filters for an aseptically processed injectable drug must themselves be sterilizable, without structural collapse. The sterilization method of choice is steam autoclaving. Filter validation is done under conditions that simulate manufacturing. If final filtration is to be done between portable tanks, filter sterilization validation may be done concomitant with the validation of the cycle for the sterile bulk tank. As with the tank(s), the first step is to place many thermocouples in the tank and the filter, especially regions that are candidate cold spots that take longer to come to temperature than, for example, most surfaces in the interior of the tank. Usually, the filter is not a cold spot; steam enters the tank (under partial vacuum) via the final and vent filters. The filter for product often is attached to a dropline that extends to the bottom of the tank, and through which product passes to the filling system; such droplines may be cold spots. Usually, a separate sampling line provides access for bulk sterility samples, and this, too, must be tested for thermal profile and spore kill.

When the cold spot is known, sealed packets of *Bacillus stearothermophilus* spores ($10^6$–$10^7$/packet), referred to as spore strips, are placed outside and inside the filter cartridge and at various points inside the sterile tank and tubing. After autoclaving, the spore strips are recovered and incubated in medium to ensure that all of the spore strips are sterile. The successful cycle is repeated at least three times and must be successful on each occasion for the autoclave cycle to be considered validated.

## Bacterial Removal by the Filter

The measure of utility of a filter for cold sterilization of liquids is the removal of $10^7$ *Pseudomonas diminuta* (ATCC 19146)/cm$^2$ filter area when the organisms have been grown in a standing culture of saline-lactose broth at 30°C until the organism is in stationary phase (Fifield and Leahy 1983). Twenty-four hours to stationary phase is common. This now is part of the FDA guidelines and is expected of manufacturers (Center for Drugs and Biologics 1987). ASTM, the Health Industry Manufacturers Association (HIMA), and other standards organizations have written standards, primarily for filter manufacturers but also for users. Despite the nuisance value of such validation for the pharmaceutical manufacturer, it is important and should be done, blandishments of the filter vendor notwithstanding. However, the microbial work must **not** be done in the manufacturing facility.

Organisms that pass the putative sterilizing-grade membrane filter (usually designated as 0.2 micrometer pore diameter) are collected onto a 142 mm diameter 0.45 micrometer-rated disc filter that then is incubated on nutrient agar in a large petri dish. At first consideration, it may seem inappropriate to judge the bacterial sterility of filtrate from a 0.2 micrometer-rated filter based on the sterility of a 0.45 micrometer-rated disc filter downstream. However, as mentioned earlier, where the microbial challenge is less than 60 organisms/$cm^2$ filter area, the 0.45 filter retains all challenge bacteria. A proof of the validity of this approach has been done recently (Carter 1996). If the challenge volume of the saline-lactose broth has been 10 L or less, the filtrate (from the filter under test) can be run into a clean, sterile, 10 L glass carboy with a hydrophobic sterile vent filter. A 50 mm diameter vent filter disc of a 0.45 micrometer-rated Teflon™ filter is satisfactory.

Consider that for a 2,000 $cm^2$ membrane area pleated cartridge filter under test, one would require a challenge of at least $2 \times 10^{10}$ *Pseudomonas diminuta*, and the organisms should be dilute when the filter is challenged. Too many particles in high concentration form a filter cake on the membrane surface, actually reducing the probability of organism passage.

Ideally, the challenge concentration should be about $10^4$/mL, in which case $2 \times 10^6$ mL (2,000 L) of saline-lactose containing the organism should be used as the challenge. Considering the potential for release of the organism, in most instances, this is an unacceptable risk. Where the experiments must be done in the manufacturing plant, one might use a 142 mm or 293 mm diameter disc of the same matrix as the candidate filter, in a stainless steel filter holder. These correspond to areas of 45 and 92 $cm^2$, respectively, and to minimal challenges of 4.5 and $9.2 \times 10^8$ organisms. The most practical approach, then, is to test a disc of the filter matrix, which has been nondestructively tested prior to exposure to the test organism.

If the organisms are delivered, diluted in medium, to the surface of the sterilized filter disc(s) for which the bubble point pressure or diffusion rates have been determined, one may reasonably expect cartridge filters with the same bubble point or diffusion rate/cm$^2$ also to retain the organisms. Scientifically, this is a small reach. One must also produce evidence that a filter failing the nondestructive test is unable to retain the organisms.

Retention failure usually is shown by holing a filter with a pin after a suitable volume of sterile filtrate has been collected. All 142 and 293 mm diameter filter holders with which the author is familiar have upstream vents for the escape of trapped air when filtration commences. Pinholing with a needle can be done through the upstream vent hole. What one does is to cease filtering, seal off the sterile receiving vessel and place it in the incubator for 14 days, where it is observed for turbidity.

The pinhole is placed and the nondestructive test repeated. Of course, the pinholed filter will fail the test. Filtration now recommences into a second sterile receiving vessel. At the completion of the filtration, the nondestructive test is repeated, the second receiving vessel is sealed and incubated.

The only alternative to testing the filters with viable organisms is to accept certification by the filter manufacturer that representative cartridges of the filter cartridge lot have been shown to retain the organisms according to an appropriate protocol. Considering the potential for a microbiological accident in a pharmaceutical plant if such challenge tests are carried out routinely, the acceptance of lot certification is rational. However, the initial test of organism retention is a measure of the pharmaceutical manufacturer's competence, and proof that the nondestructive tests correlate with organism retention in the hands of the pharmaceutical manufacturer. That is to say, this is not an article of faith.

All such data must be recorded in laboratory notebooks, each page of which is signed by the investigator and countersigned by another person who understands the objectives, methods, and results. The implication of incomplete or nonexistant documentation is that the work has not been done or has not been completed. These studies should be done at the time of the sizing experiments, before manufacturing commences.

## CONCLUSION

Sterilizing-grade filtrations and aseptic methods are increasingly common for the processing of injectable solutions. However, the crucial steps in sizing, testing, and validating filtration systems are increasingly critical.

## NOTE

The interested reader may wish to obtain a copy of the following one-page article. The authors are from Millipore Corp. (Bedford, MA).

> Rowe, P., S. Walker, and D. Reader. 1997. Cartridge filter steaming-in-place allows single unit downstream sterilization. *Gen. Eng. News* 17 (15):19 (Sepetmber 1 issue).

## REFERENCES

Badmington, F., R. Wilkins, M. Payne, and E. S. Honig. 1995. $V_{max}$ testing for practical microfiltration train scale-up in biopharmaceutical processing. *BioPharm* (September): 46–52.

Barber, T. 1996. Enumeration and characterization of microbial populations using optical and electrical zone-sensing particle counters, In *Automated microbial identification and quantitation*, edited by W. P. Olson. Buffalo Grove, IL: Interpharm Press.

Carter, J. 1996. Evaluation of recovery filters for use in bacterial retention testing of sterilizing-grade filters. *PDA J. Pharm. Sci. Technol.* 50:147–153.

Center for Drugs and Biologics. 1987. *Guideline on sterile drug products produced by aseptic processing.* Rockville, MD: Food and Drug Administration, Center for Drugs and Biologics, and Office of Regulatory Affairs.

Fifield, C. W., and T. J. Leahy. 1983. Sterilization filtration. In *Disinfection, sterilization and preservation*, edited by S. S. Block. Philadelphia: Lea & Febiger.

Fiore, J. V., W. P. Olson, and S. L. Holst. 1980. Depth filtration. In *Methods of plasma protein fractionation*, edited by J. M. Curling. London: Academic Press.

Greenberg, A. E., J. J. Connors, D. Jenkins, and M. A. H. Franson, eds. 1981. *Standard methods for the examination of water and wastewater,* 15th ed. Washington, DC: American Public Health Association.

Leahy, T. J. 1983. Dissertation, Graduate School of the University of Massachusetts at Amherst.

Liu, B. Y. H., and G. A. Kuhlmey. 1977. In *X-ray fluorescence analysis of environmental samples*, edited by T. G. Dzubay. Ann Arbor, MI: Ann Arbor Science Publishers.

Lou, Y., M. E. Klegerman, A. Muhammad, X.-P. Dai, and M. J. Groves. 1994. Initial characterization of an antineoplastic, polysaccharide-rich extract of *Mycobacterium bovis* BCG, Tice substrain. *Anticancer Res.* 14:1469–1476.

Meltzer, T. H. 1987. *Filtration in the pharmaceutical industry.* New York: Marcel Dekker.

Nema, S., and K. E. Avis. 1993. Loss of LDH activity during membrane filtration. *J. Parenter. Sci. Technol.* 47:16–21.

Nilsson-Thorell, C. B., N. Muscalu, A. H. G. Andren, P. T. T. Kjellstrand, and A. P. Wieslander. 1993. Heat sterilization of fluids for peritoneal dialysis gives rise to aldehydes. *Peritoneal Dial. Intl.* 13:208–213.

Olson, W. P. 1983. Process microfiltration of plasma proteins. *Process Biochem.* 18(5):29–33.

Olson, W. P. 1987. Sterilization of small-volume parenterals and therapeutic proteins by filtration. In *Aseptic pharmaceutical manufacturing: Technology for the 1990s,* edited by W. P. Olson and M. J. Groves. Buffalo Grove, IL: Interpharm Press.

Olson, W. P., R. O. Briggs, C. M. Garanchon, M. J. Ouellet, E. A. Graf, and D. G. Luckhurst. 1980. Aqueous filter extractables: detection and elution from process filters. *J. Parenteral Drug Assoc.* 34:254–267.

Parkin, J. E., M. B. Duffy, and C. H. Loo. 1992. The chemical degradation of phenylmercuric nitrate by disodium edetate during heat sterilization at pH values commonly endcountered in ophthalmic products. *J. Clin. Pharm. Ther.* 17:307–314.

Schmutz, C. H., and S. F. Muhlebach. 1991. Stability of succinylcholine chloride injection. *Am. J. Hosp. Pharm.* 48:501–506.

Seiberling, D. A. 1987. Clean-in-place and sterilize-in-place (CIP/SIP). In *Aseptic pharmaceutical manufacturing: Technology for the 1990s,* edited by W. P. Olson and M. J. Groves. Buffalo Grove, IL: Interpharm Press.

Simonetti, J. A., and H. G. Schroeder. 1986. Particle retention of submicrometer membranes. In *Fluid filtration, liquid, volume II,* edited by P. R. Johnston and H. G. Schroeder. Philadelphia, PA: Am. Soc. Testing Materials.

Waibel, P. J., M. Jornitz, and T. H. Meltzer. 1996. Diffusive airflow integrity testing. *PDA J. Pharm. Sci. Technol.* 50:311–316.

Waterhouse, S., and G. M. Hall. 1995. The validation of sterilizing grade microfiltration membranes with *Pseudomonas diminuta.* A review. *J. Membrane Sci.* 104:1–9.

# 8

# Air and Surface Sterilization with Chemicals

*Wayne P. Olson*

Oldevco
Beecher, IL

## THE REQUIREMENT FOR
## STERILE AIR AND SURFACES

The contact surfaces for injectable drugs usually are of stainless steel, glass, or elastomeric compounds that can be steam sterilized. Ideally, if possible, equipment surfaces that are not in product contact, but in proximity, should be made sterile. A number of surfaces are not amenable to steam sterilization but must be microbe free, or might preferably be sterilized by another means. A technique increasingly in favor is the use of hydrogen peroxide ($H_2O_2$) in the vapor phase. Examples include the sterilization of the interiors of centrifuges, the interiors of freeze dryers, the work surface in a sterility test hood, the air and surfaces in a sterility test isolator, and the air and surfaces in a filling barrier.

High efficiency particulate air (HEPA) is conceded to be sterile where the filter(s) is intact. However, HEPA entering a nonsterile chamber serves to reduce (by desiccation or displacement), not eliminate, the bioburden adsorbed to surfaces or airborne within that enclosure. What is needed to kill vegetative and spore-forming microbes are highly reactive chemicals of relatively low molecular weight that are microbicidal and sporicidal as vapors, aerosols, or liquids. Vapors are preferred as they can penetrate into niches inaccessible to liquids and are quickly eliminated from an enclosure by displacement, with the inefficiencies that result from mixing. Since plug flow seldom occurs, 4–5 volumes are required to effect the removal of a toxic vapor. The vapor systems presently in use are $H_2O_2$, with or without steam; peracetic acid (PAA); and chlorine dioxide ($ClO_2$). Sterilants used in different applications include ethylene oxide (EtO); formaldehyde (HCHO); and isopropyl alcohol (IPA) containing propylene oxide. Other toxic gases (beta-propiolactone, glutaraldehyde in the vapor phase, glycidaldehyde, methyl bromide, nitric oxide) are in varied use but are not in significant current use for pharmaceutical or device manufacturing. Ozone is increasingly of interest for the same reasons as $H_2O_2$ and PAA: It is a highly effective oxidant and leaves no toxic residue.

## Definitions

Sterilization is a process for the destruction of all life-forms, without exception. The life-forms usually taken to be most resistant to kill are bacterial spores (e.g., *Bacillus stearothermophilus* is highly resistant to moist heat [steam] sterilization). Aseptic practice (asepsis) excludes the introduction of organisms of any type into a system. An axenic culture contains one type of organism only (e.g., genetically modified *Escherichia coli*). A disinfectant kills most and perhaps all microbes on a given surface or within a given volume, but may not be sporicidal. An antiseptic is a disinfectant that is safe for human use. For example, phenol is a relatively satisfactory disinfectant that kills most microbes, but is unsatisfactory for human use. Ethyl alcohol is an antiseptic that kills most vegetative organisms and is safe for human use. A **biological indicator** (BI) is a well-characterized organism that may be included in every sterilization load to ensure that sterility has been achieved.

All of the agents considered in this chapter are nonspecific for the organisms killed and destroy organisms as the result of chemical or physical reactions. $H_2O_2$ is generally regarded as quite safe for human use (i.e., it is a common household, over-the-counter [OTC] antiseptic), but most of the others are not.

## EVALUATION OF A STERILIZING GAS

Most sterilizing gases are either oxidizers or alkylators and, as with sterilization by moist heat, evaluation is with resistant bacterial spores. Oxidizers tend to be corrosive to equipment; alkylators tend, at some concentration, to be carcinogenic (Brooks and Lawley 1964). Lest one automatically exclude alkylators from consideration, note that glutaraldehyde (-bis- or diglutaraldehyde) at 2 percent *in solution* is regarded as safe and is used very widely. HCHO, *in the vapor phase* in low-pressure steam, is used widely as a sterilant in Europe. With regard to mode of action, monofunctional alkylators, such as HCHO, should be more effective in generating point mutations (e.g., alkylation of guanine) than bifunctional alkylators (e.g., interstrand cross-linking of guanines). Cytotoxicity is greater with bifunctional alkylators.

The most heat-resistant and chemically resistant life-forms are bacterial spores. For example, the spore most resistant to heat and to $H_2O_2$ is *Bacillus stearothermophilus*; hence, that organism is used as a BI for the effectiveness of moist heat, or $H_2O_2$, or a combination of the two. Various standards organizations and the U.S. Food and Drug Administration (FDA) have designated the kill of at least $10^6$ of a resistant BI as proof of the appropriateness of a sterilization method. With heat, or with atmospheric steam plus $H_2O_2$, the BI is the aforementioned

*Bacillus stearothermophilus.* For radiation, the resistant species is *Bacillus pumilus*, and so forth.

There are three approaches to the evaluation of any sterilization method. One is to prepare packets (usually in porous Tyvek™) of the appropriate spores ($10^6$ or $10^7$) and, secreting the packets in many parts of the device or surface, exposing them to the gas and examining each packet for growth or nongrowth (the device may be a blow/fill/seal bottle, for example). The investigator then increases the gas concentration with each experiment until the minimal gas concentration that sterilizes reproducibly is found, then goes beyond that concentration so as to have a safety factor.

The problem with this approach is that the degree of safety is known approximately, but without good accuracy and precision. Accordingly, a BI appropriate to the sterilization conditions should be included in the load and, at cycle's end, the BI should be recovered, inoculated into sterile growth medium, and incubated for two weeks. No turbidity indicates no viable BIs. Small, lethal gas sterilizers (the cycles of which are not validated), such as those used in the practice of dentistry, make extensive use of BIs (Scoville 1994).

The second and preferred method is to utilize the spore packets and to *quantitate* the surviving spores at different times after the onset of gas exposure. This generates a *D* value, which is a measure of the time required for a one log kill under one set of conditions. If the *D* value then is multiplied by 6, one knows how long resistant spores must be exposed to one set of conditions to achieve a 6 log kill. Figure 8.1 (modified from Wang and Toledo 1986) is a semilogarithmic plot of percent surviving *Bacillus subtilis* spores as a function of time, when the spores are exposed to $H_2O_2$ in the vapor phase or in aqueous solution. The straight-line portions of these semi log plots provide the *D* values (*D* values are used in determining the lethality of chemicals as well as heat). These methods are referred to as the *overkill* approach.

The third approach is to sterilize under conditions that are less rigorous than those required for the kill of a highly resistant organism. This practice, which is declining, is applied almost exclusively to sterilization of somewhat heat-labile solutions or surfaces with moist heat (steam) and requires the rigorous

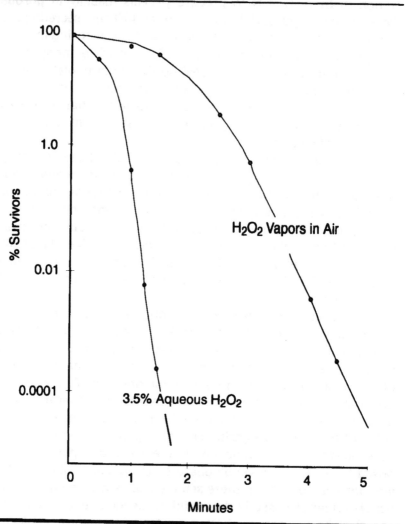

**Figure 8.1.** $D$ value determinations for $H_2O_2$ in the gas phase (top curve) and in an aqueous phase (bottom curve). *(From Wang and Toledo 1986)*

characterization of the microflora on the articles to be sterilized. BIs are run with each load.

It is important to emphasize that many conditions may alter a $D$ value (e.g., temperature variations within the sampled field, chemicals that may react with the gas [e.g., proteins], conditions under which the organism was raised or induced to sporulate, etc.). Hence a *D* value is valid for a given strain of organism under a defined set of conditions. The utility of *D* values is as a basis for comparison with kill in the conditions under test.

If one determines the *D* values with different concentrations of a lethal gas and multiplies each *D* value by 6, one can plot the series of conditions under which a 6 log kill can be made. A plot of time to a 6 log kill as a function of lethal gas concentration can be plotted as $y = 1/(f)x$ where $y$ = time to kill 6 logs of the BI and $x$ = lethal gas concentration (i.e., time to kill is proportional to the inverse of the gas concentration). A plot of this type follows the power law wherein $y = Ax^{-z}$ which is of the same form as $y = 1/(f)x$. As will be shown for $ClO_2$, EtO, and $H_2O_2$, published data fit this model well. The advantage of deriving the function is that interpolations and extrapolations can be done. Lysfjord et al. (1995) show a first-order plot (Figure 12.26, p. 265) of *D* value as a function of $H_2O_2$ concentration. Their calculated least-squares line is $y = 0.2732974 \times 10^{-0.000044x}$.

A log-log plot of minutes to 6 log kill versus gas concentration is a reasonably good linear approximation to $y = 1/(f)x$. This type of function is suitable for interpolation, but not necessarily for extrapolation. That is to say, at gas concentrations greater than the maximum used in derivation of the curve, kill may be significantly less than predicted by extrapolation of a log-log curve. Whether the log-linear or the log-log plot is the better approximation is not yet clear, but there is no question that this is not a simple linear relationship.

Departures from experimental conditions change *D* values. For example, the same organism that is steam autoclaved on a thermostable plastic may be killed at a different rate from the organism on a smooth 316 stainless steel surface. Similarly, it has been shown that organisms that are adsorbed to a surface are more resistant to $H_2O_2$ than the organisms in suspension (Yagi et al. 1995). Simply, when a stainless steel pipe is steam autoclaved, the steel heats up like a frying pan, in which case a biofilm is heated from both the vapor phase and the support. Plastics, as poor conductors of heat, provide a degree of protection to biofilms. Similarly, if the entire surface of an organism is exposed to a lethal chemical, the organism should receive a lethal injury more rapidly than where only a portion of the surface is available for chemical attack, as in biofilms.

## HYDROGEN PEROXIDE

At this writing (1997), $H_2O_2$ is the sterilant of choice because it is effective, leaves no residue (more about residues below), and poses a minimal risk to humans in the work environment. It is an oxidant. The mutagenic and toxic effects of hydroxyl radicals, and the protective microbial responses, have been described by Storz et al. (1990). Schirmann and Delavarenne (1979) provide an excellent review of the chemical reactivity of hydrogen peroxide.

$H_2O_2$ is produced in high purity and concentrations ranging up to 90 percent. Stabilizers are added to the commercial product. $H_2O_2$ retains antimicrobial activity over a range of conditions (Turner 1983). Concentrations used in the sterilization of equipment are 30–35 percent.

$H_2O_2$ appears to disinfect not with the superoxide radical $O_2^-$ but rather via the hydroxyl free radical, which forms when catalytic amounts of $Fe^{2+}$ or $Cu^{2+}$ are present. Iron within an organism is catalytic, and all bacteria and fungi contain some level of the metal in the cytochromes, if not elsewhere. In fact, iron and copper ions alone are microbicidal. The microbicidal synergy of $H_2O_2$ and metallic cations is enhanced by the addition of a metal chelator (Sagripanti 1992). Ultraviolet (UV) light at 254 nm and $H_2O_2$ concentrations greater than 1 percent, followed by heating to 85°C for 1 minute, achieved about a 5.5 log kill of *Bacillus subtilis* (Bayliss and Waites 1979). The UV light facilitated hydroxyl radical formation.

The organism most resistant to $H_2O_2$ appears to be *Bacillus stearothermophilus,* and the strain recommended for test and validation purposes is ATCC 12980 (Rickloff 1988), although other strains have been used successfully (e.g., FS 1518). *Bacillus subtilis* var. *niger* is also used on occasion. The current widespread use of $H_2O_2$ is due as much to its safety and breakdown to innocuous products (water and oxygen) as to microbicidal effectiveness. *D* values for 3 percent $H_2O_2$ versus various organisms range from too rapid to measure for *Neisseria gonorrhoeae* to 18.30 minutes for *Candida parapsilosis* (Turner et al. 1975).

Production of $H_2O_2$ vapor is by flash vaporization of a 31 percent (w/v) solution into warm dry air. The VHP 1000 Biodecontamination System™ (STERIS, Mentor, OH) is used widely to automate the sterilization of air volumes ranging from 10 to 1500 ft$^3$. The same vendor sells *Bacillus stearothermophilus* spores in Tyvek™ packages as a BI. The STERIS packages currently are under consideration by the U.S. Pharmacopeia (USP) as the recommended method for monitoring $H_2O_2$ gas sterilization. Although gas phase $H_2O_2$ is very effective, it is far more efficient as a sterilant in the aqueous phase (Figure 8.1). This is true of many gas phase sterilants.

The time to inactivation of $10^6$ *Bacillus subtilis* var. *globigii* spores is the inverse of the operating temperature. At 4°C, the inactivation time is 8 minutes, but at 27°C, the inactivation time is 32 minutes. The difference may be due to the more rapid decomposition of $H_2O_2$ at the higher temperature. However, the stability of $H_2O_2$ is such that at 100°C, in the absence of catalysts, there is about 2 percent decomposition in 24 hours; significant decomposition does not commence below 130°C. The atmospheric boiling point for pure $H_2O_2$ is 150.2°C. Decomposition at 100°C may be negligible. This author suggests that the inverse relationship of $H_2O_2$ toxicity to temperature at moderate temperatures merits additional investigation.

What is not clear is the effect of spore hydration. Hydration has a profound effect on spore kill, perhaps with regard to spore coat permeability, perhaps with regard to chemical reactivity. This author also suspects that, for spores hydrated as uniformly as possible, the inactivation time will drop again at temperatures significantly above ambient (i.e., there is a most-effective temperature between 0 and 140°C, at which temperature spores survive most briefly). However, this is speculative; it has not been shown.

Time to kill as a function of $H_2O_2$ concentration is shown in Figure 8.2, which has been calculated from the admittedly thin data (4 *D* values) of Pflug et al. (1995). However, additional data with other sporicidal gases confirm the model. Figure 8.3 is Figure 8.2 with the x-axis extended so that the nature of the function

**Figure 8.2.** Minutes to a 6 log kill of *Bacillus stearothermo-philus* spores at various concentrations of $H_2O_2$ in steam at 100°C. *(Calculated from the D values of Pflug et al. [1995])*

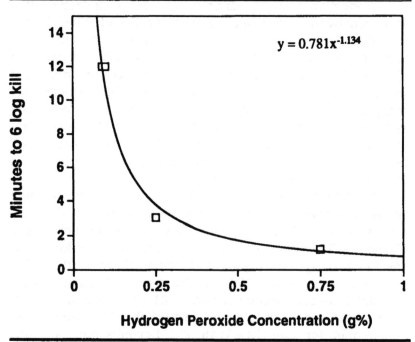

can be better appreciated. Clearly, beyond the inflection point, increasing concentrations of the vapor phase peroxide cause modest decreases in time to kill. However, in a manufacturing process (as distinct from, for example, the sterilization of an isolator used for sterility testing), reducing sterilization time from 1 minute to 15 seconds can have a dramatic effect on throughput.

Microbial kill is (in the main) a first-order, not square-wave, process, even when pains are taken to create square-wave conditions and reduce diffusion effects. There is, of course, the randomness of the chemical events when an oxidant or alkylator reacts with an enzyme critical to cell reproduction or a structural macromolecule that is less critical. Setlow and Setlow (1993) have shown that small, acid-soluble proteins (SASPs) in *Bacillus*

**Figure 8.3.** The function of Figure 8.2 with the x-axis extended, to emphasize the shape of the linear plot.

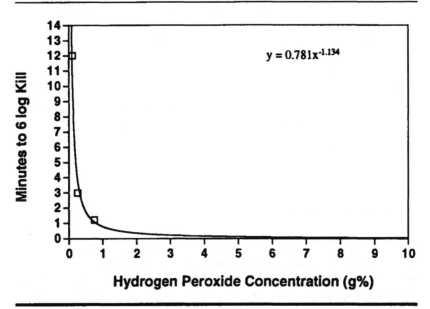

*subtilis* spores protect DNA (deoxyribonucleic acid) from damage by $H_2O_2$, and also protect spore DNA from heat damage (Setlow 1992). SASPs are an example of the panoply of variations in the internal biochemistry of organisms that may contribute to the randomness of death in microbial populations.

After vapor-phase $H_2O_2$ sterilization, some plastics [polystyrene, polyurethane (PU8), the Sil/PU6 blend of silicone and polyurethane, poly(methyl methacrylate) (PMMA), fluorosilicone acrylate, and poly(2-hydroxyethyl methacrylate (HEMA)] are cytotoxic unless aerated. Silicone and polyethylene sorb low levels of $H_2O_2$. Consequently, validation of the aeration of plastics following $H_2O_2$ sterilization is necessary (Ikarashi et al. 1995). This observation is especially applicable to devices. Since metals catalyze $H_2O_2$ degradation, stainless steel parts may pose less of a problem.

The primary applications of $H_2O_2$ sterilization that are widely applicable in most manufacturing and quality control areas relate to the sterilization of isolators used in sterility testing and the sterilization of the air and surfaces over/around barrier isolator filling lines (various authors in Wagner and Akers 1995). All of this is with peroxide in steam (referred to as "peroxide plus"). Fortsch et al. (1993) have developed an $H_2O_2$–based system for ophthalmic instrument sterilization at temperatures below 50°C. The authors claim that the corrosion of stainless steel that occurs with peroxide plus steam is obviated at the lower temperature.

## PERACETIC ACID

Peracetic acid (PAA) is produced in a mixture of acetic acid, sulfuric acid, water, and $H_2O_2$. The acetic acid is oxidized, and the resulting structure is $CH_3C(O)O:OH$ (Swern 1949). The characteristic of peracids is the -O:O- structure that is found in all other peroxides including $H_2O_2$. PAA is more active as a sterilant than $H_2O_2$ by several orders of magnitude (Eggensperger 1979), is more corrosive, and a more powerful oxidant, probably because the energy required for an organic peroxide to dissociate into free radicals at the O:O bond is significantly less than for $H_2O_2$. However, as with hydrogen peroxide, the breakdown products (oxygen, acetic acid, and water) are volatile and leave no toxic residue. Aqueous, aerosolized, and vapor forms of PAA are sporicidal (Baldry 1983; Block 1991). The aqueous and aerosolized forms are used far more than the vapor forms. PAA is sold as a 40 percent solution. Renalin™ (Renal Systems, Minneapolis, MN) is 4 percent PAA and 10 percent $H_2O_2$ in water and is used to clean and sanitize dialysis systems for reuse.

The optimum sporicidal activity of vapor-phase PAA is at a relative humidity of 80 percent. Almost certainly, as with the other gases, reaction with proteins, DNA, or RNA (ribonucleic acid) is greatly facilitated by moisture, since most of the macromolecules of the bacterial spore are sufficiently devoid of water so as to be in a semicrystalline state. Even at low temperatures, PAA is very active. Other advantages, relative to $H_2O_2$, are the liposolubility (hence rapid penetration into organisms) and freedom from inactivation by catalases and peroxidases.

Block (1991) presents tables (9-6 through 9-10) detailing the antimicrobial activity of PAA solutions and aerosols, and in tables 9-11 through 9-16 on antimicrobial activity of PAA aerosols. Whereas it is an exceptionally good sterilant, PAA degrades at a rate of 1–2 percent per month as contrasted with less than 1 percent per year in the cold for $H_2O_2$ (Block 1991).

Sterility test isolators at Upjohn are sterilized with PAA (Davenport 1989). Turner (1983) referenced the work of a number of authors in this application. A particular advantage, as with $H_2O_2$, is that PAA tends not to be adsorbed to surfaces.

One percent and higher concentrations of PAA in water promote tumor formation in mice, but lower concentrations do not, hence PAA is known as a weak carcinogen (Bock et al. 1975), as contrasted with $H_2O_2$ that does not promote tumor formation and is not carcinogenic. However, from the data reported by Block (1991), PAA appears to be more powerful a sporicide than $H_2O_2$ by at least 4 logs.

Benzoyl peroxide (Ph-C(O)O:OH) is seldom, if ever, used industrially, perhaps because it may explode when heated. However, it is used widely in the food industry as an oxidizing agent for the bleaching of oils because it has excellent liposolubility. It is formulated as an OTC preparation for dermal self-application by acne patients.

## CHLORINE DIOXIDE

Chlorine dioxide ($ClO_2$) is a very strong oxidant that smells like chlorine gas. The permissible gas phase exposure limit for an 8-hour work shift is 0.1 ppm (per the U.S. Occupational Safety and Health Administration [OSHA]) which also is the level at which the characteristic odor can be detected. It is used increasingly in water treatment because, unlike chlorine, $ClO_2$ does not form chloro-amines and trihalomethanes (Aieta and Berg 1986).

Gonzalez and Meyer (1990) sprayed $ClO_2$ into a sterility test isolator that stood unused for 3 hours and then was HEPA-swept for 2 hours. Over that 2 hours, the chlorine level dropped to 2 ppm. Validation of sterilization was with $10^6$ *Bacillus subtilis* var. *niger* spores (ATCC 9372) on glass. Sterilization times are inversely related to gas concentration and the relative humidity of the air volume to be sterilized (Jeng and Woodworth 1990). The Jeng and Woodworth $D$ values have been multiplied by 6 and plotted as a function of $ClO_2$ concentration (Figure 8.4). Although a function based on three points is suspect, note that it takes the same form as the same plot for $H_2O_2$ plus steam and the EtO plot (see below).

As with ethylene oxide, $ClO_2$ has been used to sterilize pallet loads of devices in their sterile barrier packages.

**Figure 8.4.** *Bacillus subtilis* var. *niger* spore kill as a function of chlorine dioxide concentration. *(Calculated from Jeng and Woodworth [1990])*

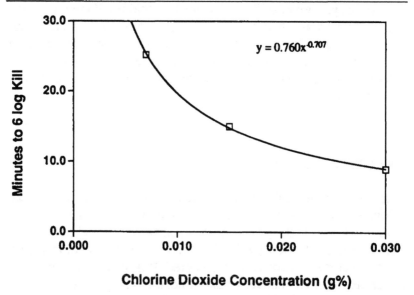

$$y = 0.760x^{-0.707}$$

## ETHYLENE OXIDE

EtO is an alkylator and is in disfavor with OSHA and the U.S. Environmental Protection Agency (EPA) because EtO is a suspected carcinogen, toxic, and mutagenic (Glaser 1977). The gas also tends to linger in plastics, and the degassing of EtO–sterilized devices has become rather important. The great advantage of EtO is penetration into all parts of a device or system, through cardboard, cellophane, poly(vinyl chloride), and, to a lesser degree, polyethylene. Russell (1982) has done an excellent review. The DNA damages of alkylators, and bacterial DNA damage-inducible responses that protect cells against the mutagenic and toxic effects of alkylators, have been described by Saget and Walker (1994).

Time to kill for 6 logs of spores as a function of EtO concentration (calculated from the data of Phillips 1949) is shown as raw data in Figure 8.5 (note the similarity to Figures 8.2–8.4) and as a log-log plot in Figure 8.6. The equation for the log-log line is

$$y = -0.804x + 2.634$$

and $r^2$ for the linear approximation is 0.983; hence, intermediate values can be interpolated without gross error. However, the log-log function appears to be curvilinear, and extrapolation beyond 884 mg EtO/L for an estimation of time to kill (at, for example, 1,000 mg EtO/L) may underestimate the time considerably.

The time to kill data are $D$ values that vary with temperature and moisture (developed below), and also with the support for the spores. Pinto et al. (1994) studied $D$ values for *Bacillus subtilis* var. *niger* (ATCC 9372) and found highly significant differences between paper, aluminum foil, and plastic (polyamide) as supports. $D$ values were highest on paper, intermediate on aluminum foil, and lowest for the polyamide. One might speculate on these rather thin data that the values for paper are highest because paper is porous, exposing much of the spore surface, and because the paper likely is poorly reactive with EtO. The values for plastic may be lowest because the plastic acts as a sink for EtO, such that EtO diffuses from spores into the support.

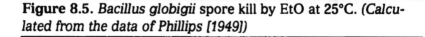

**Figure 8.5.** *Bacillus globigii* spore kill by EtO at 25°C. *(Calculated from the data of Phillips [1949])*

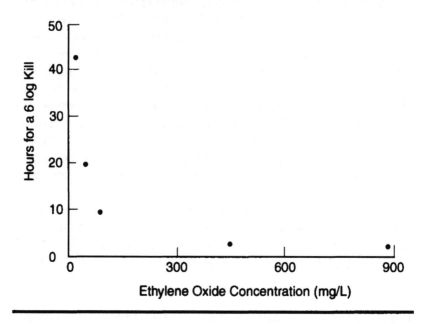

Aluminum foil may be intermediate in effect on *D* values because the foil shields a portion of the spore surface from EtO absorption, but does not provide sink conditions.

EtO kill increases with temperature (Figure 8.7). The relationship of kill time to temperature (calculated and plotted from the data of Liu et al. 1968) appears to be arithmetic but biphasic. Phillips (1949) claims a temperature coefficient for *Bacillus globigii* of 2.74 for each 10°C increase in temperature. This is approximately consistent with the Liu et al. (1968) data for *Bacillus subtilis* at temperatures below 58°C. At temperatures above 58°C, the temperature coefficient drops to about 1.8. This author suggests that the decrease in the coefficient may be attributable either to increased decomposition of EtO or to the extent of spore hydration. The data are reproducible (i.e., Phillips [1949] and Liu et al. [1968] had similar results).

**Figure 8.6.** Data from Figure 8.5 plotted as a log-log function.

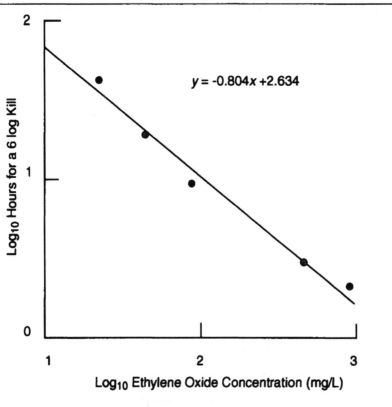

$y = -0.804x + 2.634$

Log$_{10}$ Hours for a 6 log Kill

Log$_{10}$ Ethylene Oxide Concentration (mg/L)

Bacterial spores are most susceptible to EtO when they possess an optimal amount of moisture. The "conditioning" of *Bacillus globigii* spores and exposure to EtO at relative humidities less than 33 percent results in incomplete kills. Spores equilibrated at relative humidities greater than 33 percent have longer times to kill, so 33 percent appears to be an optimum *with EtO*.

Dried spores, and beads of cross-linked hydrophilic macromolecules, hydrate and swell when moistened. With swelling, the permeability of the beads increases; this may also be the case with spore coats. Watt (1981) studied water-vapor sorption isotherms of *Bacillus stearothermophilus*; at a relative humidity of

**Figure 8.7.** *Bacillus subtilis* spore kill by EtO as a function of temperature. *(Calculated from the data of Liu et al. [1968])*

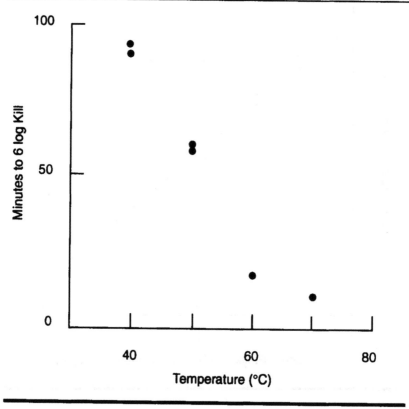

about 80 percent, the adsorption of water commenced to climb exponentially, and water content was consistently higher at 20°C than at 100°C.

   With regard to the effects of moisture, apart from simple physical effects, chemical reactivity certainly is impacted by the presence of bound and unbound water. The esterification of carboxyl groups, or the alkylation of amino, sulfhydryl, or hydroxyl groups, almost certainly is facilitated by unbound water. An excess of water dilutes EtO and gives rise to ethylene glycol ($HOCH_2CH_2OH$), which is far less reactive than EtO. Sublethal amounts of EtO do not prevent germination, but do prevent multiplication of *Bacillus subtilis* var. *niger* (*Bacillus globigii*), which often is used as a BI for EtO. Therefore, it is reasonable to assume that there exists a moisture optimum within a spore that promotes kill by sporicidal gases. This may also mean that a device that is exposed to a suboptimal EtO cycle may contain viable organisms that are damaged and do not grow out in a one-week sterility test.

EtO is mutagenic. It seems likely that the formation by EtO of the hydroxyethyl derivative of guanine at N7 and adenine at N3 (Brooks and Lawley 1964) interferes not only with translation and protein synthesis, but also with replication of the genome, hence interferes with multiplication. The hydroxyethylation of guanine and adenine must be facilitated by a modicum of unbound water.

The simplest method of determining EtO gas concentrations is the indirect method, whereby an EtO chamber is pressurized using gas cylinders containing certified mixtures of EtO (Gillis 1986). The direct method consists of recovering a gas sample from the chamber via a valve, with analysis by gas chromatography or infrared analysis. Moisture also can be analyzed by gas chromatography or infrared analysis. Methods for residual EtO have been developed at the U.S. FDA Office of Science and Technology (Lao et al. 1995).

In the presence of finely divided metallic oxides, or metallic hydroxides or anhydrochlorides, EtO can be explosive. More than 3 percent EtO is flammable in air; thus, commercial mixtures usually are 10 percent in EtO and 90 percent in $CO_2$. EtO is stable in $CO_2$, in water at neutral pH, and in freon (which, because it is an environmental contaminant, is no longer produced by DuPont, formerly the principal manufacturer). At this time (1997), EtO use is in slow decline in the sterilization of medical devices. For a recent review see Gillis (1986).

The principal use of EtO is in the vapor-phase sterilization of heat labile devices. Sterilization is by alkylation of carboxyl, amino, sulfhydryl and/or hydroxyl groups, thereby inactivating proteins and DNA. Sterilizing concentrations are 400–1000 mg/L, and relative humidity requirements for microbicidal effectiveness range from 30 to 60 percent. Relative humidities as high as

80–90 percent often are used with EtO, although the kill rate seems not to increase beyond 58 percent (Baird-Parker and Holbrook 1971). Bruch and Bruch (1970) regard relative humidity as a critical factor. One hundred to 200 mg EtO/L air is fatal to humans in a few seconds (Lueck 1980). Time for EtO sterilization is inversely related to EtO concentration and to temperature (Ernst and Shull 1962); typical temperatures for EtO sterilization range from 40°C to 70°C. The mode of action of EtO probably is the hydroxyethylation of purine and pyrimidine bases, and the esterification of phosphate groups, in DNA and RNA (Russell 1976).

One of the great advantages of EtO is that, since the gas penetrates cardboard quite well, entire pallets of the wrapped and packaged product can be bulk sterilized immediately prior to shipping. However, EtO sterilization immediately prior to warehousing and shipping also has the potential to invite inadequate outgassing. It would be of interest to know what proportion of suspect EtO–contaminated products had been processed on a shipping pallet.

EtO is too reactive to be effective as an aqueous solution, other than very short term. Ethylene glycol and chlorohydrin are formed (Bruch 1972), which are toxic. Since long degassing times are required to remove residues from plastics and because EtO may be carcinogenic, it is a target of the EPA, OSHA, and the FDA. EtO is decreasingly the gas phase sterilant of choice.

STERIS and other manufacturers produce EtO sterilizers.

## FORMALDEHYDE

HCHO is an alkylator and usually is formulated (40 percent in water) with methanol (10 percent) to prevent polymerization. As such, it is referred to as formalin. In air, sterilizing concentrations are 3–10 mg/L in a relative humidity of 75 percent.

Despite the low formula weight of 30, HCHO penetration into plastics is poor. Formaldehyde has excellent microbicidal activity at temperatures > 65°C, and leaves residues of less than 5 micrograms/cm$^2$ (residues are readily measurable on standard filter paper). Washer-disinfectors for surgical instruments and other equipment commonly use HCHO in Europe. More recently in Europe, low temperature steam and gaseous formaldehyde is in increasing use (Nystrom 1991). Because HCHO does not penetrate significantly into plastics, it generally is regarded as an excellent surface sterilant for clean rooms.

The killing rate is reported to be linearly related to HCHO concentration (Report 1958, cited in Russell 1982). If that is so, it is inconsistent with the data for $H_2O_2$, $ClO_2$, and ethylene oxide.

The BI of choice is *Bacillus stearothermophilus* (Nystrom 1991). However, it would appear that a *Salmonella* bacteriophage is also a good BI (Nilehn 1972), which poses the possibility that some viruses may be equally as, or more, resistant to thermal and other destructive methods as bacterial vegetative forms and spores. The inactivation factor considered appropriate for *Salmonella* phage is $5 \times 10^7$.

Handlos' (1977) method of residue analysis is simple and adequate for most surfaces. Exposure levels of HCHO to workers appear not to pose an issue with available equipment (Nystrom 1991).

HCHO gas can be generated in three ways:

1.  Evaporation of a formalin solution (40 percent HCHO, 10 percent methanol in water)

2.  Volatilization of paraformaldehyde

3.  The addition of formalin to potassium permangenate

Six logs of *Bacillus subtilis* spores can be killed in 90 minutes at an HCHO level of 0.3 mg/L at 20°C and 100 percent relative humidity (Acklund et al. 1980). The kill rate almost certainly can be increased with increasing temperature.

HCHO has a penetrating, unpleasant odor even in modest quantities. Since it is used in the embalming of corpses, the odor has unpleasant associations for most people. There are three interesting points about HCHO:

1.  The minimal amount that one can smell corresponds to impermissible concentrations in the air of the workplace.

2. The gas tends not to permeate into plastic; the implication is that one might do a short-term HCHO sterilization of the exterior of a bag containing parenterals, without putting the contents of the bag at risk.

3. HCHO works well with steam, hence a very short-term sterilization with HCHO in 100°C (atmospheric) steam should be feasible.

However, as with other gaseous sterilants, degassing is necessary.

## ISOPROPYL ALCOHOL WITH 3 PERCENT PROPYLENE OXIDE

IPA is lethal in minutes in water over a wide range of concentrations (10–91 percent w/v) to most vegetative organisms (Morton 1983). It is also virucidal, but less so than ethanol (Klein and Deforest 1963). It is ineffective against bacterial spores unless propylene oxide, an alkylating agent, is added. Propylene oxide has been used in the sterilization of foods (Alguire 1963) as contrasted with EtO, which is used primarily in the sterilization of medical devices. Incorporated into IPA at 3 percent, propylene oxide kills bacterial spores (Hart and Ng 1975) by DNA alkylation. Presumably, it is carried into the interior of a spore by IPA to react with the microbial chromosome. It is used widely as a fumigant and a soil sterilant and, like IPA, is flammable. Therefore, although aerosols of IPA–propylene oxide can sterilize air, they may be unsuitable for use with isolators. IPA leaves no significant residue, but propylene oxide leaves a residue of propylene glycol, which is nontoxic and accepted as safe for foods.

As a rule of thumb, any system that kills bacterial spores kills viruses and vegetative organisms. The IPA–propylene oxide mixture likely is suitable as a liquid sterilant, but requires evaluation for use in isolators, because IPA accumulation may damage gloves, half-suits, or polymeric walls. Because of the potential for the generation of occasional sparks by metal equipment, and since both IPA and propylene oxide are flammable, this mixture is not recommended for the wipe down of conveyors and unscrambling tables in filling rooms. Primary applications include spills involving organisms in research or production areas.

## SYSTEMS FOR THE DELIVERY OF CHEMICAL STERILANTS

The system in most widespread use for the automated delivery of $H_2O_2$ and other sterilants is the VHP 1000 Biodecontamination System™ (STERIS, Mentor, OH), which is shown in Figure 8.8. With appropriate services, the system can control the entire cycle (dehumidification, conditioning, sterilization, aeration to remove the sterilant).

In addition to delivery into the isolator, room, or system, some means must exist to remove the sterilant from air exiting the system during aeration. Since $H_2O_2$ degrades to oxygen and water, residues are not a consideration. However, for PAA, HCHO, and other gas-phase sterilants, it is preferable to remove or inactivate the chemical prior to venting to atmosphere.

This author's inclination is to column sorbents because they have high capacity and reliability and usually are regenerable. For example, beaded silica is readily available and relatively inexpensive. Water-moistened Sephadex™ is a very expensive sorbent, but will react with both oxidizers and alkylating agents that form a covalent adduct with hydroxyl groups. The best gas sorbents for many purposes likely would be very fine glass fibers, which are available for packing into columns or as pleated cartridges. Any such sorbents require validation.

## VALIDATION AND THE CHOICE OF BIOLOGICAL INDICATORS

The means of sterilizing critical surfaces and volumes must be shown to be effective with a known bioburden of an organism resistant to the method of kill, and shown to be reproducible, else the system is not properly controlled.

**Figure 8.8.** VHP 1000 Biodecontamination system™ (STERIS).

The dependence on validation with resistant spores is appropriate. A dependence on one strain throughout the industry seems inappropriate where well-characterized strains are available in-house and the resistance to a given killing agent is well known. However, *Bacillus stearothermophilus* remains the indicator organism of choice for $H_2O_2$ and for peroxide plus steam. *Bacillus subtilis* var. *niger* is the the indicator of choice for most other lethal gases.

The general assumption is that spores are the organisms most resistant to chemical inactivation. Some vegetative bacteria are more resistant to EtO than are *Bacillus stearothermophilus* spores. This author is inclined to suspect that some viruses may prove to be more resistant to vapor-phase chemicals than certain vegetative bacteria, simply because there are fewer critical structures that can be targeted, but this remains to be shown.

## When Things Go Wrong in the Gas Chamber

Some comments are appropriate about nonsterility of articles in loads. The best analyses of what goes wrong relate to the small gas sterilization units used by dentists, but the information is applicable to industrial-scale systems. The following list is modified from Molinari et al. (1994):

- Use of sterilization wraps not intended for chemical vapor sterilization

- Sealed containers, foil, or cloth as wraps

- Wet equipment

- Packs are not spaced for rapid gas dispersion

- Worn door gaskets or seals

- Inadequate mixing of gases in the chamber

Scoville (1994) makes a particular point of the stratification of gases within a chamber in which a vacuum is not pulled, with the result that the distribution of the lethal gas is far from uniform. The points are that workers require thorough indoctrination and equipment maintenance is crucial. In these days of corporate downsizing, reductions in either can have a tremendously negative effect on the bottom line. Often, this must be explained to CEOs and CFOs.

## INTERPRETATIONS OF THE DATA

The chapters in this text are intended to be practical. However, some insights into mechanisms might induce the reader to discover novel applications and methods. For example, one wonders why the effect of at least several of the lethal gases is nonlinear with concentration. Two of the simpler possibilities include the following:

1.  *Location and Orientation of the Organisms.* The lethal gas must react on or through the available surface of the organism. A bacterium that is exposed alone on a flat surface very likely is more susceptible than one that is masked by other organisms, or is partially embedded in a surface such as a soft plastic (the amount of gas absorbed must be a function of the surface area of the microbe that is exposed). The most exposed organisms should react first. This seems logical, but the writer is unaware of a detailed proof. One of the obvious sequellae is that one takes pains to clean a piece of equipment and reduce the burden of removable micrometer-sized particles (including microbes) before seeking to sterilize.

2.  *Distribution of Moisture Among Organisms.* As there are distributions of properties within organisms, it seems likely that where moisture is limited in a closed system (like a vial or ampoule), the moisture is distributed among organisms in a manner that may or may not be Gaussian. In freeze-dried material (e.g.,

proteins), the moisture likely is resident in the formed elements, like the bacteria and viruses. Moist organisms should be most susceptible to chemicals since an aqueous environment is best for optimal reactivity of the chemical(s) with the DNA or RNA of the organism. What this likely means is that microbial kill by lethal gases is optimal at some level of moisture at or above 1 percent, in which case the optimum moisture level probably is higher than the "bone dry" level currently targeted for dried products by the pharmaceutical industry. Research on this issue would be highly useful and merits funding.

Similarly, it would be helpful to have a clearer view of the role of temperature in the lethality of these gases. The inactivation of $H_2O_2$, $ClO_2$, and the other gases mentioned in the chapter is proportional to temperature. Presumably, where a gas reacts with itself, inactivation also is proportional to concentration. Studies on gas inactivation *and* microbial kill over a wide range of temperatures do not appear in the literature that this author has accessed. The kill rate of EtO as a function of temperature appears to be linear between 40°C and 60°C and between 60°C and 80°C (Liu et al. 1968), but the kill rate at the higher temperatures is about half the rate at the lower temperatures. EtO *may* degrade at a dramatically greater rate above 60°C, which would account for the change in kill kinetics. Over linear segments of the thermochemical destruction curve, extrapolation of $D$ values follows the Stumbo (1965) relationship:

$$\log D_2 - \log D_1 = 1/z \times (T_1 - T_2)$$

where $D_2$ corresponds to the $D$ value at temperature $T_2$, and $D_1$ corresponds to the $D$ value at temperature $T_1$.

The applied side of microbial kill with lethal gases is moderately well documented. The scientific basis, or understanding, of how these events occur is inadequately documented.

## ACKNOWLEDGMENT

Judy Button, Baxter Healthcare Renal Division, McGaw Park, Illinois, rendered Figures 8.2–8.4 on a Macintosh system from data supplied by the author.

## REFERENCES

Acklund, N. R., M. R. Hinton, and K. R. Denmeade. 1980. Controlled formaldehyde fumigation system. *Appl. Environ. Microbiol.* 39:480–487.

Aieta, E. M., and J. D. Berg. 1986. A review of chlorine dioxide in drinking water treatment. *J. Am. Water Works Assoc.* 78:62–72.

Alguire, D. E. 1963. Effective sterilization with 100% ethylene oxide. *Bull. Parenteral Drug Assoc.* 17:1–8.

Baird-Parker, A. C., and R. Holbrook. 1971. The inhibition and destruction of cocci. In *Inhibition and destruction of the microbial cell*, edited by W. B. Hugo. London and New York: Academic Press.

Baldry, M. G. C. 1983. The bacteriocidal, fungicidal, and sporicidal properties of hydrogen peroxide and peracetic acid. *Appl. Microbiol.* 54:417–423.

Bayliss, C. E., and W. M. Waites. 1979. The combined effect of hydrogen peroxide and ultraviolet irradiation on bacterial spores. *J. Appl. Bacteriol.* 47:263–269.

Block, S. S. 1991. Peroxygen compounds. In *Disinfection, sterilization, and preservation*, 4th ed., edited by S. S. Block, Philadelphia: Lea and Febiger.

Bock, F. G., H. K. Meyers, and H. W. Fox. 1975. Cocarcinogenic activity of peroxy compounds. *J. Natl Cancer Inst.* 55:1359–1361.

Brooks, P., and P. D. Lawley. 1964. Alkylating agents. *Br. Med. Bull.* 20:91–95.

Bruch, C. W. 1972. Toxicity of ethylene oxide residues. In *Industrial sterilization*, edited by G. B. Phillips and W. S. Miller. Durham, NC: Duke University Press.

Bruch, C. W., and M. K. Bruch. 1970. Gaseous disinfection. In *Disinfection*, edited by M. A. Bernarde. New York: Dekker.

Davenport, S. M. 1989. Design and use of a novel peracetgic acid sterilizer for absolute barrier sterility testing chambers. *J. Parenteral Sci. Technol.* 43:158–166.

Eggensperger, H. 1979. Disinfektionsmittel auf der basis persaure abspaltender verbindungen. *Zentralbl. Bakteriol. Hyg. I. Abt. Orig.* B 168: 517–524.

Ernst, R. R., and J. J. Shull. 1962. Ethylene oxide gaseous sterilization. *Appl. Microbiol.* 10:337–344.

Fortsch, M., J. W. Pruter, J. Draeger, F. Helm, A. Sammann, H. Seibt, and H. Ahlborn. 1993. $H_2O_2$ low temperature plasma sterilization. New possibilities for use with eye surgery instruments. *Ophthalmologe* 90:754–764. Abstract from Medline Express.

Gillis, J. R. 1986. Ethylene oxide sterilization and validation for practical pharmaceutical aseptic production. In *Validation of aseptic pharmaceutical processes*, edited by F. J. Carleton and J. P. Agalloco. Buffalo Grove, IL: Interpharm Press.

Glaser, Z. R. 1977. Special occupational hazard review with control recommendations for the use of ethylene oxide as a sterilant in medical facilities. DHEW (NIOSH) No. 77-2200. Rockville, MD: National Institute of Occupational Safety and Health.

Gonzalez, J. P., and D. Meyer. 1990. Advanced aseptic processing: Barrier system technology as applied to production filling lines. In *Proceedings of the PDA/PMA Sterilization Conference.* Washington, DC: Parenteral Drug Association and Pharmaceutical Manufacturers Association.

Handlos, V. 1977. Formaldehyde sterilization. I. Determination of formaldehyde residuals in autoclave-sterilized materials. *Arch. Pharm. Chem. Sci. Ed.* 5:163–169.

Hart, A., and S. N. Ng. 1975. Effect of temperature on the sterilization of isopropyl alcohol by liquid propylene oxide. *Appl. Microbiol.* 30: 483–484.

Ikarashi, Y., T. Tsuchiya, and A. Nakamura. 1995. Cytotoxicity of medical materials sterilized with vapour-phase hydrogen peroxide. *Biomaterials* 16:177–183.

Jeng, D. K., and A. G. Woodworth. 1990. Chlorine dioxide gas sterilization under square-wave conditions. *Appl. Environ. Microbiol.* 56: 514–519.

Klein, M., and A. Deforest. 1963. Antiviral action of germicides. *Soap Chem. Spec.* 39:70–72, 95–97.

Lao, N. T., H. T. Lu, A. Rego, R. H. Kosakowski, and R. D. Hume. 1995. Interlaboratory comparison of analytical methods for residual ethylene oxide at low concentration levels in medical device materials. *J. Pharm. Sci.* 84:647–655.

Liu, T., G. L. Howard, and C. R. Stumbo. 1968. Dichlorodifluoromethane-ethylene oxide mixture as a sterilant at elevated temperatures. *Food Technol.* 22:86–89.

Lueck, E. 1980. *Antimicrobial food additives.* Berlin: Springer-Verlag.

Lysfjord, J. P., P. J. Haas, H. L. Melgaard, and I. J. Pflug. 1995. Barrier isolation technology: a systems approach. In *Isolator technology, applications in the pharmaceutical and biotechnology industries.* Buffalo Grove, IL: Interpharm Press.

Molinari, J. A., M. J. Gleason, and V. A. Merchant. 1994. Sixteen years of experience with sterilization monitoring. *Compendium* 15:1422–1432.

Morton, H. E. 1983. Alcohols. In *Disinfection, sterilization and preservation*, 3rd ed., edited by S. S. Block. Philadelphia: Lea and Febiger.

Nilehn, B. 1972. A method for the quantitative microbiological check of heat decontaminators. *Scand. J. Infect. Dis.* 4:245–253.

Nystrom, B. 1991. New technology for sterilization and disinfection. *Am. J. Med.* 91(Suppl. 3B):264S–266S.

Pflug, I. M., S. Schaffer, H. Melgaard, J. Lysfjord, and P. Haas. 1995. Sterilizing the inside of the barrier isolator using atmospheric steam plus hydrogen peroxide. In *Advanced barrier technology*, joint PDA/ISPE conference, 17–18 January, in Atlanta, GA.

Phillips, C. R. 1949. The sterilizing action of gaseous ethylene oxide. II. Sterilizations of contaminated objects with ethylene oxide and related compounds: Time, concentration, and temperature relationships. *Am. J. Hyg.* 50:280–288.

Pinto, T. J. A., T. Saito, and M. Iossif. 1994. Ethylene oxide sterilization: III. Influence of carrier nature in a biological monitor performance. *J. Pharm. Sci. Technol.* 48:155–158.

Report. 1958. Disinfection of fabrics with gaseous formaldehyde. Report of the committee on formaldehyde disinfection. *J. Hyg.* 56:488–515.

Rickloff, J. R. 1994. Hydrogen peroxide gas and its use in sterilizing barrier isolators. In *Proceedings of the barrier isolation technology seminar.* Philadelphia: International Society for Pharmaceutical Engineering.

Russell, A. D. 1976. Inactivation of non-sporing bacteria by gases. In *The inactivation of vegetative microbes,* edited by F. A. Skinner and W. B. Hugo. London: Academic Press.

Russell, A. D. 1982. *The destruction of bacterial spores.* London: Academic Press.

Saget, B. M., and G. C. Walker. 1994. The Ada protein acts as both a positive and negative modulator of *Escherichia coli's* response to methylating agents. *Proc. Natl Acad. Sci. USA* 91:9730–9734.

Sagripanti, J. L. 1992. Metal-based formulations with high microbicidal activity. *Appl. Environ. Microbiol.* 58:3157–3162.

Schirmann, J.-P., and S. Y. Delavarenne. 1979. *Hydrogen peroxide in organic chemistry.* Paris:Edition et Documentation Industrielle.

Scoville, J. R., Jr. 1994. False negative interpretations of biologic indicators or the importance of using two biologic indicator test strips. *Compendium* 15:1472–1478.

Setlow, P. 1992. I will survive: Protecting and repairing spore DNA. *J. Bacteriol.* 174:2737–2741.

Setlow, B., and P. Setlow. 1993. Binding of small, acid-soluble spore proteins to DNA plays a significant role in the resistance of *Bacillus subtilis* spores to hydrogen peroxide. *Appl. Environ. Microbiol.* 59:3418–3423.

Storz, G., L. A. Tartaglia, and B. N. Ames. 1990. Transcriptional regulator of oxidative stress-inducible genes: Direct activation by oxidation. *Science* 248:189–194.

Stumbo, C. R. 1965. *Thermobacteriology in food processing.* New York, NY: Academic Press.

Swern, D. 1949. Organic peracids. *Chem. Rev.* 45:1–52.

Turner, F. J. 1983. Hydrogen peroxide and other oxidant disinfectants. In *Disinfection, sterilization and preservation,* 3rd ed., edited by S. S. Block. Philadelphia: Lea & Febiger.

Turner, F. J., J. F. Martin, C. Wallis, and E. G. H. Spaulding. 1975. Hydrogen peroxide disinfection of contact lenses. In *Resistance of microorganisms to disinfectants: Second Intl Symposium,* edited by W. B. Kedzia. Warsaw, Poland: Polish Acad. Sci.

Wagner, C. M., and J. E. Akers, eds. 1995. Isolator technology: Applications in the pharmaceutical and biotechnology industries. Buffalo Grove, IL: Interpharm Press.

Wang, J., and R. T. Toledo. 1986. Sporicidal properties of mixtures of hydrogen peroxide vapor and hot air. Food Technol. 40 (12):60–67.

Watt, I. C. 1981. Water vapor adsorption by Bacillus stearothermophilus endospores. In Sporulation and germination, edited by H. S. Levinson, A. L. Sonenshein and D. J. Tipper. Washington, DC: American Society for Microbiology.

Yagi, Y., K. Takino, and Y. Uchitomi. 1995. Study on the sterilization of adhering bacteria in ultrapure water manufacturing system. *12th Semiconductor Pure Water Chem. Conf.* (1993):63–74; *Chem. Abst.* 123: 152419.

# 9

# Integrated Systems for the Aseptic Filling of Sterile Product

*Hans L. Melgaard*

*Paul J. Haas*

Despatch Industries
Minneapolis, MN

Pharmaceutical manufacturers are rapidly adopting integrated barrier isolators to enclose systems for the aseptic filling of products. To better understand the motivation for this trend, a comparison of traditional aseptic filling suites with integrated barrier isolation systems is necessary. There are both quantitative and qualitative differences between these two approaches.

From a qualitative perspective, the major difference is the separation of personnel from the environment surrounding the open vial. Operators account for up to 70 percent of the particulate found in a clean room. Studies have consistently shown significant reductions in airborne particulate when people are removed from the area. The removal of personnel also allows for the use of atmospheres other than air (nitrogen, etc.), without provision for respirators. Operation at temperatures outside the normal ambient range are also more easily accommodated when people are not present within the confines of the filling equipment environment. Continuously monitored parameters within a barrier isolation enclosure can be used as an engineering measurement of changes. Particulate occurring within the system is suitable as a measure, as there are no people in the enclosure generating randomly high quantities of potentially viable particulate. This provides for trend line analysis and prediction, which is not possible to the same degree with people present.

This difference between the conventional aseptic filling suite and an integrated barrier isolation system has long been recognized.

> *Clean rooms are ventilated with microbial free air and positively pressurized against ingress of contaminated air from adjacent areas. Airborne contamination can, therefore, only arise from within the room and, except for unusual circumstances, the source of microorganisms is the occupant of the room.* (Whyte and Bailey 1985)

Integrated barrier isolator technology has the added benefit of keeping the product away from people. Although particularly toxic compounds will require a special core, even in an integrated barrier isolator, this is much easier to accommodate with the barrier in place.

The inclusion of the entire filling line within a single mechanical enclosure also allows for relatively easy transport of the entire line to remote locations, with reduced concern for the building interface into which the equipment will be placed. An additional benefit resulting from the encapsulation of the small environment directly above the active open vial area is in the ability to perform external maintenance on the system without necessarily breaking the sterility of the barrier environment. Additionally, barrier isolators can be cleaned-in-place (CIP) and sterilized-in-place (SIP) between product runs, reducing the risk of product carryover.

The only perceived negative for the application of integrated barrier isolation systems is the instantaneous accessibility of any exposed portion of the filler line. In the case of integrated barrier isolation systems, intervention is accomplished through gloved ports and/or mechanical manipulators operating within the enclosure itself.

There are also distinct quantitative economic differences between the traditional filling suite and integrated barrier isolation systems. Table 9.1 identifies the relative cost differentials that result from the use of integrated barrier isolation systems when compared to the traditional filling suite approach. The exact quantification of each of these items is heavily dependent on the quantity and type of product being filled, the number of lines being operated, and the specific classification into which the integrated barrier isolator is placed. It appears, at present, that the operation of barrier isolator filling lines in Class 100,000, or in controlled but unclassified areas, will be allowed within the GMP guidelines of the FDA (Muhvich 1995).

## HARDWARE REQUIREMENTS FOR AN INTEGRATED ASEPTIC FILLING BARRIER ENCLOSURE AND ACCESS TO EQUIPMENT

The barrier enclosure must be of robust construction, ergonomically friendly, and easily opened for changeover. In addition, it

Table 9.1. Economic Comparative Matrix for Filling Line Equipment

| Cost Factor | Conventional Filling Line | Barrier Isolator Filling Line |
|---|---|---|
| Equipment | Lower | Higher |
| Facility | Higher | Lower |
| Utility | Higher | Lower |
| Gowning | Higher | Lower |
| Personnel | Higher | Lower |
| Relocation | Higher | Lower |

must maintain the interior environmental integrity for extended periods of time. The degree of robustness depends, in part, on the process being contained. Cytotoxic materials and other potent compounds require additional considerations for operator protection. When used to contain potent compounds, the barrier isolator must be substantial enough to prevent inadvertent sterility failures. In general, this requires a hard-wall barrier isolator enclosure as opposed to a flexible or soft-wall enclosure. Examples of hard- and soft-wall enclosures are shown in Figures 9.1 and 9.2. The importance of hard-wall enclosures becomes more evident when inlet and outlet openings are required for product or commodity passage, as any deflection of the outside envelope of the isolator will result in a corresponding pressure change across the exit opening. Another way of looking at this would be to state that the barrier isolator has a known volume. If you deflect the outer surface to decrease the interior known volume by 5 percent, and then allow it to return rapidly to its original profile, over

**Figure 9.1.** Soft-wall barrier isolator enclosure.

**Figure 9.2.** Hard-wall barrier isolator enclosure.

that period of time, 5 percent of the interior volume will have to be made up from some other source. This can cause a momentary disruption in the otherwise positively pressurized barrier isolator, which may draw an external environment into the barrier and, therefore, violate the sterile integrity of the chamber.

The materials of construction must not be detrimentally affected by the CIP and SIP cycles required to render the inside of the barrier isolator clean and sterile. In general, this means a minimal amount of elastomer products should be used. Preferred materials, in most cases, are 316 stainless steel and glass, with any elastomers used as sealants or glove surfaces to be as compatible as possible with the CIP and SIP cycles. This generally precludes the use of materials such as neoprene.

Barrier isolators must accommodate external human activity. The workings inside the barrier isolator should be visible and, should intervention be required during a process cycle, have means provided, either mechanically or through gloved access, to reach the components involved. Glove box manufacturers have developed guidelines for the angle of glass within an isolator enclosure (American Glovebox Society 1994). These are consistent with the ergonomic requirements to interface with a barrier isolator. In addition, various ergonomic design software is available to allow positioning of mannequin operators of any gender or physical size. A stick figure shown in operating position at a barrier is shown in Figure 9.3. In general, the narrower the barrier isolator, the more easily accessible the components inside become.

In committing to major construction programs involving barrier isolators, it is prudent to construct a simple paperboard model, such as that in Figure 9.4. This provides for real-time human interaction with the mocked up components within a barrier isolator and may reveal difficulties that can be addressed prior to the actual fabrication. The long-term goal is to eliminate human intervention in a barrier isolator system, but the current generation of barrier isolators has not yet demonstrated the unattended reliability required to eliminate the potential for intervention.

As most barrier isolator systems will involve equipment that will, from time to time, have to be modified for alternative products or processes, it is well to have easily opened doorways or external removable panels to provide full access to the mechanisms within. Doors, such as those illustrated in Figure 9.5, provide easy access to all of the components within the barrier isolator.

To be compatible with the CIP cycle, whether it is a manual or automatic cycle, as many of the interior surfaces as possible should be sloped to drain locations. The recommended slope is 1/2″ per foot, or 40 mm per meter. In addition, any areas that may trap liquid should be designed out of the equipment and avoided in the construction of the barrier isolator. It is also appropriate to have any sealing material mating glass to framework designed with a slope from the glass to the framework to avoid liquid pooling areas. Figure 9.6 shows the preferred glass to framework seal for drainage.

Once a barrier isolator is clean and sterile, maintaining the sterile conditions within the barrier isolator requires knowledge

**Figure 9.3.** Ergonomic model of barrier isolator operator.

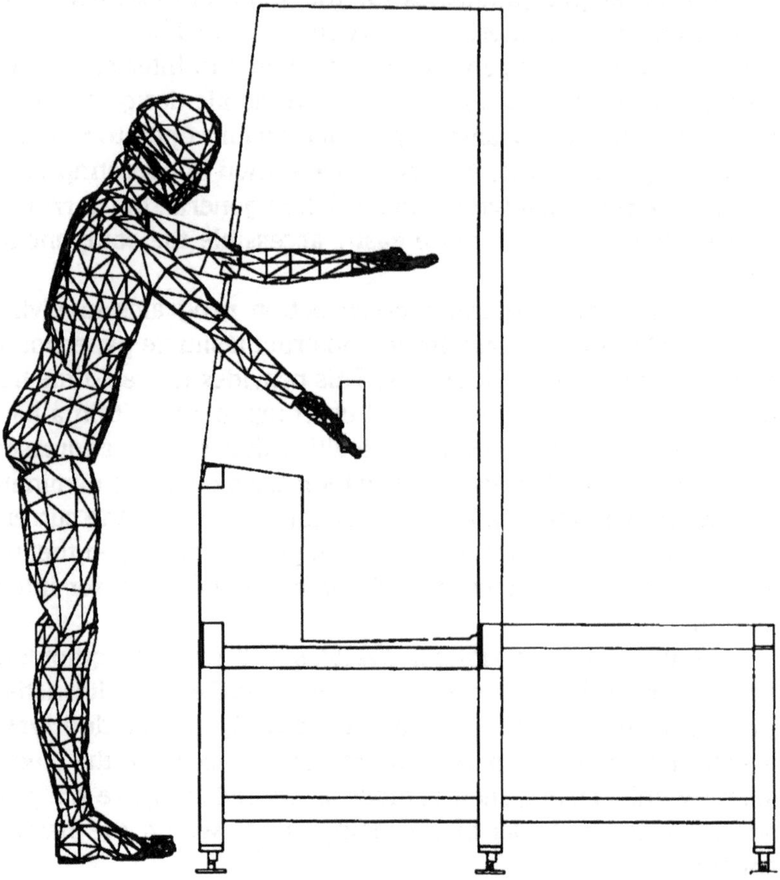

of the pressure balance inside the isolator and the sterility assurance levels (SALs) of all commodity transfer systems feeding into the barrier isolator. In general, sterility maintenance can be achieved through the use of cascaded interior pressures fed from HEPA– (high efficiency particulate air) and ULPA– (ultralow particulate air) filtered air sources. Once positive pressure differentials are proven, they may be monitored through the course of operation to ensure that the sterility is not contravened by

**Figure 9.4.** Barrier isolator.

**Figure 9.5.** Barrier doors.

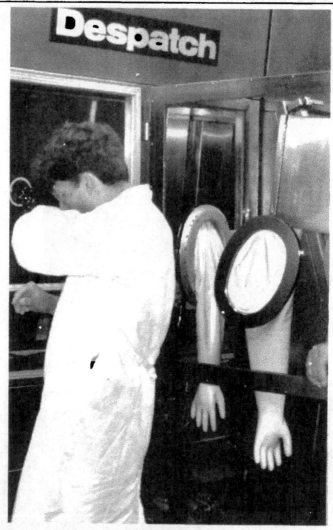

pressure upsets. Long-term operation in a sterile mode has been demonstrated in the work done by Michael Porter and Les Edwards of Merck (1995), and has also been confirmed through operation over an extended period of time of the LUMS (Lilly, Upjohn, Merck) prototype, as reported in May 1995 to the U.S.

**Figure 9.6.** Window seal method.

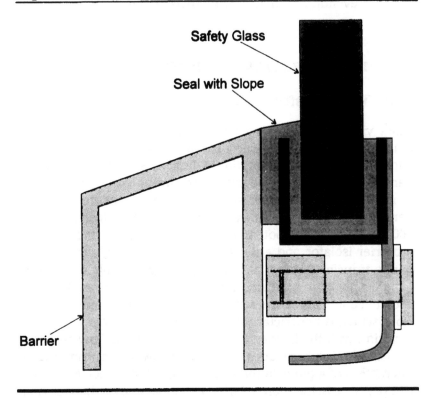

Food and Drug Administration (FDA 1995). The testing at Merck was particularly significant since it involved at least one inadvertent pressure decay within the barrier caused by a power failure that, over a period of 17 hours, did not cause detection of a nonsterile condition by exposure plates within the barrier. Open mouseholes existed at both ends of the equipment, and it was operating in an unclassed and uncontrolled environment.

Maintenance of sterility within the barrier, once the sterilization cycle has been validated, is primarily assured by using continuous measurements of the pressure differential from within the barrier to the outside environment. Differentials of

0.05 inches water column (w.c.) or less have been shown effective at preventing particle ingress (FDA 1995).

## COMMODITY PROCESS EQUIPMENT

### Glass Vial Transfer

Glass vial transfer is typically the largest volume transfer required to move from outside a barrier isolator into a barrier isolator during a filling operation. Glass vials with no appendant hardware can typically be processed through a depyrogenation tunnel that renders them sterile and endotoxin free. Depyrogenation temperatures in excess of 280°C result in more than a 100 log reduction in potential bioburden. The interface between a barrier isolator and a depyrogenation tunnel typically occurs with a transfer through an orifice plate area, while maintaining a higher pressure gradient in the barrier isolator than the tunnel cooling zone. This allows for a certain amount of continuing outflow, which is beneficial also from a potential glove volume interaction with the barrier. Having significant outflow occurring on a continuing basis goes a long ways toward ensuring that no interaction of a glove in the wall will result in a neutral or negative pressure being momentarily present inside the barrier isolator with respect to its outside environment.

Tests have been conducted on a number of different barrier isolators, where the internal pressure was monitored for fluctuations with respect to the external environment. Operating with even relatively low overpressure, and low airflow through the system, a doubled glove removal was not capable of driving a normal system to a negative condition. Figure 9.7 graphically shows the pressure profile using a two-glove rapid removal as a test. The apparatus used to derive this profile is shown in Figure 9.8.

Another method of transfer into a barrier isolator involves the use of presterilized containers, or containers placed in overwrapped packages that are transferred through a surface sterilizing system. Ultraviolet high-intensity lamps may be used

**Figure 9.7.** Dual glove withdrawal pressure graph.

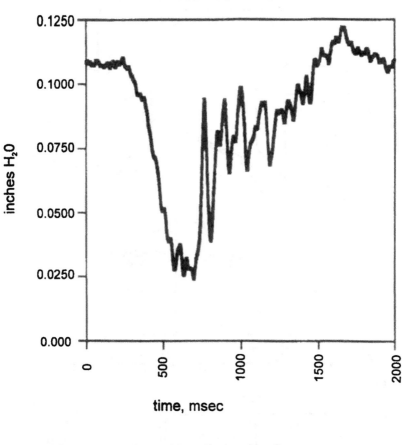

Glove Volume = 404 in³
Chamber Volume = 102,214 in³

to surface sterilize this type of overwrapped container entering a barrier isolator (Davenport and Melgaard 1995). Other methods, such as the application of vapor hydrogen peroxide to surface sterilize containers entering a barrier isolator, have also been reported (Walker 1995).

An alternate method of transfer involves the use of rapid transfer ports. Rapid transfer ports, used with a container closed

**Figure 9.8.** Chamber pressure test setup.

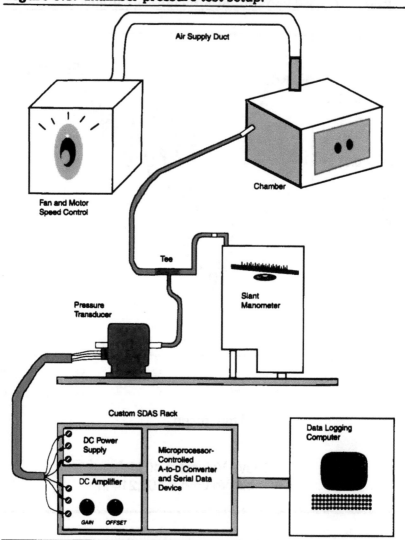

in a sterile area with components inside, remain sterile when docked with a barrier isolator wall. Rapid transfer ports, such as the one shown in Figure 9.9, have been outfitted with automatic opening equipment requiring only the removal of the interior contents of the port once docked to the barrier isolator. Concern

**Figure 9.9. Rapid transfer port.**

has been raised about the point of interface where the seals of the captured external closure and the external door portion of the rapid transfer port meet. A method using short-term elevated temperature thermal sterilization of this ring of concern has been demonstrated by Central Research Laboratories. This is illustrated in Figure 9.10.

**Figure 9.10.** Rapid transfer port cycle.

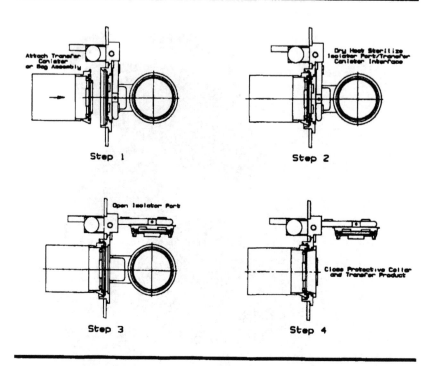

Step 1

Step 2

Step 3

Step 4

## Blow/Fill/Seal

Blow/fill/seal processes may also be conducted within barrier isolators. A potential exists to enclose fully a blow/fill/seal machine, allowing it to be within a presterilized environment.

## Stopper Transfer

Stoppers are typically the second largest volume component transferred into a barrier isolator. Several methods are currently in use for transferring into a barrier isolator. One process utilizes a self-contained sterilizing transfer vessel that couples directly to the top of a barrier isolator and allows stoppers to be transferred or metered into hoppers within the barrier isolator. The size and complexity of this form of stopper processing requires a multi-story facility to accommodate the stopper connection to the barrier.

A rapid transfer port scheme has also been demonstrated for transferring stoppers into a barrier. This method involves a moving flange and an external bag containing the commodity stoppers in a presterile state, where the sterilization has been accomplished either externally with a steam autoclave or through the use of gamma irradiation. A system using this methodology is shown in Figure 9.11.

Ultraviolet pass-through systems have also proved successful in transferring components into barrier isolators. The validation of an ultraviolet pass-through system for this application was described by Stewart Davenport in a paper presented to the PDA/ISPE symposium on Advanced Barrier Technology in Atlanta in February 1995 (Davenport and Melgaard 1995).

**Figure 9.11.** Rapid transfer port sterile stopper transfer system.

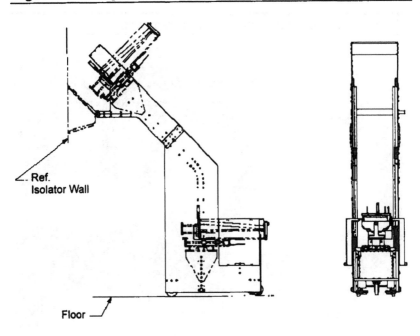

Ref.
Isolator Wall

Floor

This method leaves no residual on the surface, and the decontamination cycle is generally less than one minute in length. This transfer method is illustrated in Figure 9.12. Continuous versions of this equipment are given in Figure 9.13 (Melgaard 1995).

Pass-through steam autoclaves can also be used abutting the wall of the barrier isolator. The time duration and the potential drying requirements generally make this a less desirable mode of transfer.

## AIR PROCESSING AND PRESSURE BALANCE EQUIPMENT

The processing of air, or other atmosphere within a barrier, is ultimately very important to the overall performance of the system. This includes considerations relative to normal operation,

**Figure 9.12.** Glove port stopper bag removal.

**Figure 9.13.** Continuous ultraviolet pass-through.

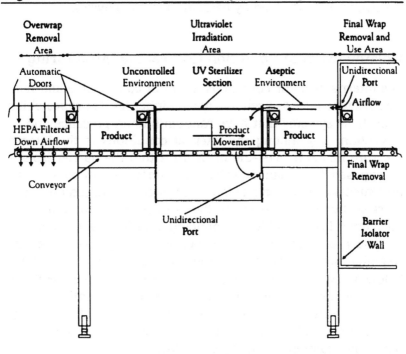

and also the SIP and CIP phases of the equipment. There are essentially two major classifications.

The first involves a closed system where no open flow occurs during normal operation. This type of a system is applicable to batch processes, where all of the components for a manufacturing lot can be combined within the barrier at one time. Systems of this type require a simple overpressure control and do not have multiple dynamic considerations.

The second, and more complex form, is a barrier isolator that requires the constant introduction of product to be filled as well as container enclosures. This type of barrier isolator generally incorporates openings or mouseholes, or depyrogenation tunnel access, which allows the commodity product to enter and leave the barrier isolator. To operate in a sterile mode, flow must continually exit these openings. Flow considerations relative to the opening are defined by basic air and gas physics.

## Isolator Internal Condition Control and Monitoring

The internal isolator conditions are both controllable and measurable, which lends itself to a much higher level of repeatability than those systems that have random human presence and intervention. Key among the internal conditions is the internal pressure maintenance—the pressure within the barrier isolator as opposed to its surrounding conditions. Incoming flow requirements of makeup gas into the system are based on both a cross-sectional open area and the pressure differential between the barrier inside and its surrounding area. Typical pressure differentials, as used in the past, between filling suites and surrounding areas have been on the order of 0.05 inches w.c. The following calculations show the flow through an open orifice of large size due to this magnitude of pressure differential.

The calculation of the velocity required to maintain a specified pressure differential across an opening starts with the Bernoulli equation for steady state flow:

$$p = \left( \frac{V^2}{2g} \right)\left( \frac{\delta}{5.192} \right) = \frac{V^2}{1096.7(\delta)}$$

$p$ is the pressure in inches of w.c., $V$ is the velocity in ft/min, and $\delta$ is the density of air in lb/ft.

$$V = 1096.77\left( \frac{P}{\delta} \right)^{0.5}$$

Air at standard temperature and pressure is 0.075 lb/ft$^3$; therefore,

$$V = 4 \times 10^3 \, (p)^{0.5}$$

$$Q = V \times A$$

Example: If the pressure differential across the wall of the sterile enclosure is 0.05 inches w.c. and A is 0.25 ft$^2$, then

$$V = (0.05)^{0.5} \times 4 \times 10^3 = 8.94 \times 10^2 \text{ ft/min}$$

$$Q = 8.94 \times 10^2 \text{ ft/min} \times 0.25 \text{ ft}^2 = 223 \text{ ft}^3/\text{min}$$

This flow ($Q$) is with containers present and approximates the gross airflow requirement to maintain the pressure differential between the sterile enclosure and the ambient environment.

A measurement was made close to the exit hole of the enclosure and shows that at no time, even with rapid removal, was there a flow reversal in the inside of the enclosure from positive to negative.

In the case of potent compounds and other materials that must be isolated from operators, it is possible to build bidirectional flow tunnels, where the pressure inside the initial enclosure is positive with respect to the outside environment and yet flows counter to a flow from the outside environment through a higher level negative located in the middle of the entry and exit tunnel. This is shown in Figure 9.14. Double-walled systems with a higher external pressure of HEPA–filtered air can be used to insure both containment and sterile surround.

Particulate control is also critical in isolator design. The most important consideration is the filtration of outside air that is used as makeup air for the barrier isolator. All filter banks should have challenge and checking ports available for test. Since viable particles may be present, the general technique used in barrier isolators is to recirculate the chamber air through a HEPA filter, and add filtered makeup air required to keep the positive pressure inside the barrier isolator. Generally, these are done in a fashion similar to Figure 9.15.

In addition, the makeup air system is usually provided with one or two roughing filters to extend the life of the HEPA or ULPA filter used as the final filter on the makeup air. The probability of a particle passing through from the makeup air system into the isolator enclosure, and finally through the recirculation filter on the isolator enclosure to the product environment below, is shown in Figure 9.16. This is not to say that this is the probability of having a particle of any type at the level of the product. This is the probability of a particle making its way through the multiple levels of filters to reach the sterilized, inside

**Figure 9.14.** Bidirectional flow entry.

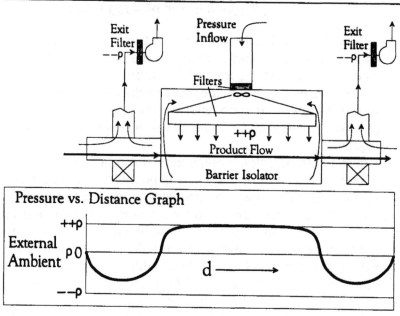

critical area. The probability of any particle reaching the critical area must be factored to include the probability of a viable particle reaching that area. Using the known references, the probability of a viable particle as opposed to an inert particle is on the order of $10^{-5}$. So, the probabilities shown in Figure 9.16 must be reduced by that probability in order to correctly define the probability of a particle of a viable nature reaching the inside of the barrier from its external environment.

Flows inside the chamber should be accomplished to avoid turbulence and extract particulate at its probable source. Pressure balance within the enclosure can be maintained through the use of pressure transducers tied to variable frequency fan-drive systems providing the pressurized in-flow for the barrier system. These respond rapidly to upsets and are very stable under normal control conditions. An example of this type of pressure

**Figure 9.15.** Airflow inside and into isolator.

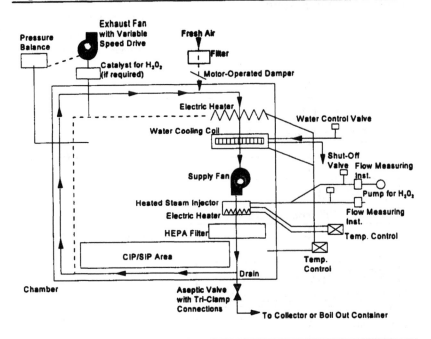

balance scheme is shown in Figure 9.17. Whether the air-handling equipment is an integral portion of the barrier or re-motely located is a matter of individual preference and the geometry of the surrounding area. Both types of systems have been successfully operated in a nonclassified environment.

## FILLING EQUIPMENT

Filling equipment is critical to the proper operation of a barrier isolator filling system. Historically, filling equipment has in-volved the use of relatively broad, flat tables on which the vari-ous apparatus have been installed. This methodology does not lend itself well to operation within a barrier, nor to the CIP and

**Figure 9.16.** Particulate levels at different points in a barrier isolation airflow system.

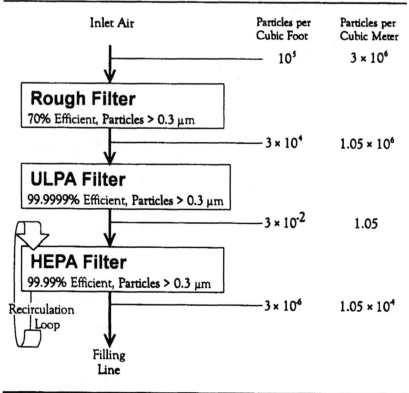

| Inlet Air | | Particles per Cubic Foot | Particles per Cubic Meter |
|---|---|---|---|
| | | $10^5$ | $3 \times 10^6$ |
| **Rough Filter** 70% Efficient, Particles > 0.3 μm | | $3 \times 10^4$ | $1.05 \times 10^6$ |
| **ULPA Filter** 99.9999% Efficient, Particles > 0.3 μm | | $3 \times 10^{-2}$ | $1.05$ |
| **HEPA Filter** 99.99% Efficient, Particles > 0.3 μm | | $3 \times 10^{-6}$ | $1.05 \times 10^4$ |

Recirculation Loop

Filling Line

SIP cycles to which the barrier enclosed equipment would be subjected. Improvements to a partial barrier design have been accomplished, such as that shown in Figure 9.18. More recent techniques have been developed where the entire filling system is installed and oriented so as to occupy a minimal front-to-back footprint, as shown in Figure 9.19. This equipment, first developed by TL Systems, has been incorporated into a number of filling lines. The geometry of this filler is such that the front-to-back reach is easily accomplished with gloves from the operator side of the barrier isolator.

**Figure 9.17.** Pressure balance control system.

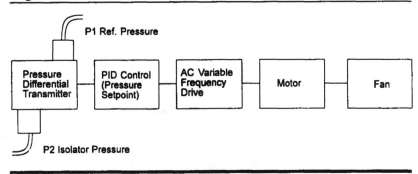

All filling apparatus within a barrier isolator must be capable of maintaining its physical integrity throughout the CIP and SIP cycles, in addition to being operable with minimal adjustment during running conditions that require human intervention

**Figure 9.18.** Partial barrier design improvements.

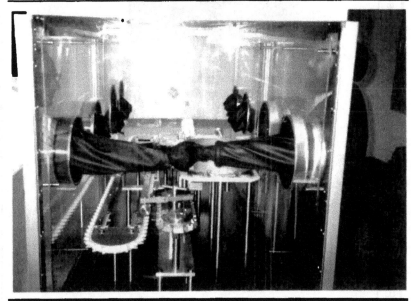

**Figure 9.19.** Minimal front-to-back filling system footprint.

through glove ports or manipulators. Considerations need to accommodate SIP of the wetted path, changeout of filler needles, the incorporation of checkweigh stations, the connection of the wetted path in a sterile manner to the commodity feed container, and the physical access for a major line changeover. The synchronization of the commodity components with the actual filling requires major integration effort between all the in-feed and transfer components of the system. Even though the filling equipment is at the heart of the functional operation, without the complete coordination of the commodity items fed into the barrier isolator, smooth production cannot be attained. Figures 9.20 through 9.22 show the evolution of filler footprint from the original large table design to a system compatible with easy glove reach to all critical areas within a barrier isolator.

## SIP AND CIP EQUIPMENT

### CIP/SIP of Isolators

Any barrier must also be occasionally opened. Following the opening of a barrier isolator, a CIP and SIP process must be completed to render the inside of the barrier clean and sterile for filling operations. The CIP process can be operated with several different approaches. A wand can be provided within the enclosure that could be manipulated through glove ports. This is the simplest, and perhaps the least repeatable, of the methods of cleaning in place. Spray balls may be used with pattern coverage. This is probably the most repeatable coverage. A spray ball operated in a barrier isolator is shown operating in Figure 9.23. In either case, if fillers are directly above the equipment, a filter screen is required to prevent wetting of the HEPA filters. An example of a filter protection screen is illustrated in Figure 9.24. Rounded corners in all cleanable areas and proper drain locations assist in completing the removal of foreign material from within the barrier isolator.

Sterilization of the barrier can be accomplished through a number of different methods. Ethylene oxide and peracetic acid

**Figure 9.20.** Section view of conventional vial filling/weighing/stoppering equipment with barrier isolator installed.

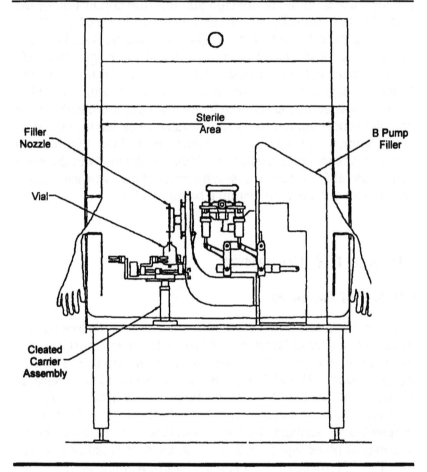

have been used, but are not favored because of toxicity and chemical residue concerns. The method currently used on test isolators has largely been a cold vapor hydrogen peroxide sterilant as provided by the AMSCO VHP™ generator (Steris Corporation). Despatch and TL Systems recently developed a combination atmospheric steam/hydrogen peroxide system (called Peroxide Plus™) for use in barrier isolators. The sterilant

**Figure 9.21.** Phase II: section view of improved vial filling, checkweighing, and stoppering equipment.

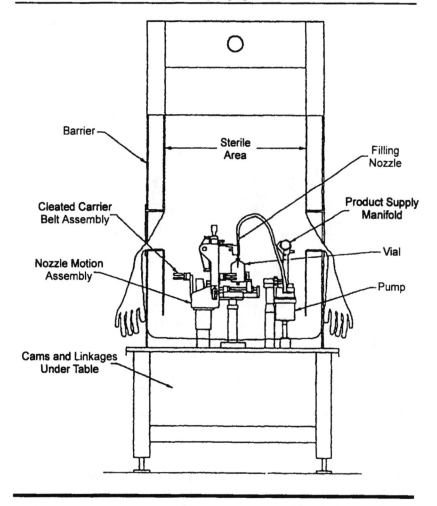

concentration of hydrogen peroxide, whether it is vapor form, or peroxide and steam, appears to be very effective against resistant spores such as *Bacillus stearothermophillus* and *Bacillus subtilis* when used in concentrations of approximately 2,500–10,000 ppm, or one-fourth to one percent. Table 9.2 shows the

**Figure 9.22.** Phase III: section view of redesigned vial filling, checkweighing, and stoppering equipment.

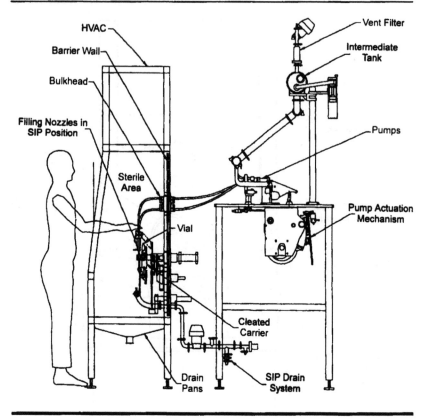

range of D values for 7 resistant spores in 11 separate experiments using the Peroxide Plus™ system at a concentration of 2,500 ppm $H_2O_2$. The apparatus used for the D value determination is shown in Figure 9.25. Some concern has been raised relative to the effect of condensed $H_2O_2$ on the D value. Experiments conducted during December 1993, where spores were deposited in the center of an indented disc (see Figure 9.26) where the steam/hydrogen peroxide mixture was allowed to condense, showed minimal increases in the effective D value. The results

**Figure 9.23.** Spray ball piped into barrier isolator.

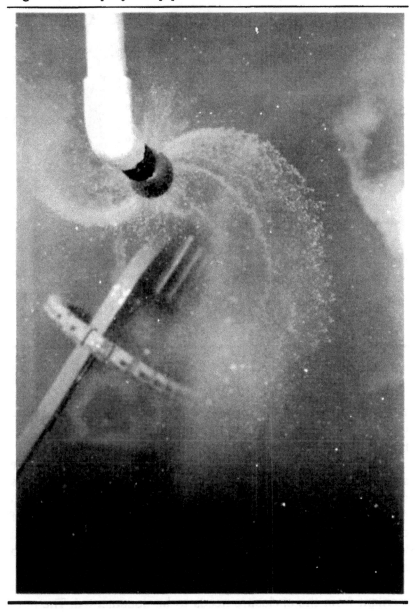

**Figure 9.24. Offset perforations in diffuser panels.**

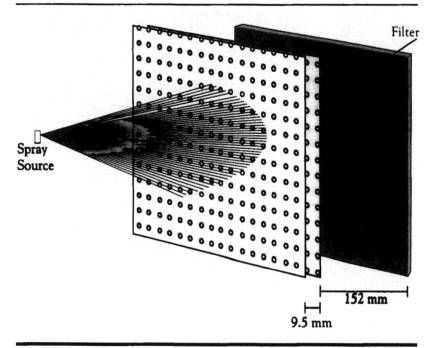

are shown in Table 9.3 (Pflug 1993). Other methods, such as ozone or atmospheric steam, have also been used for the sterilization of barrier isolators.

It is critical as part of the sterilization cycle to verify the sterilant concentration. One method provided to do that is shown in Figure 9.27 (patent pending). This method works with the Peroxide Plus™ sterilant to give a repeatable indication of concentration. It is also necessary to ascertain sterilant residual levels; several types of apparatus are commercially available to accomplish this.

Following the completed sterilization cycle, removal ·of all sterilant from the enclosure needs to be documented. The normal level to be reached before initiation of filling, using $H_2O_2$, would be less than 1 ppm residue.

Table 9.2. D Value Results for 7 Spores Tested at Hole 3 ($H_2O_2$ = 2500 ppm) (Data Gathered in 11 Separate Experiments)

| Hole 3, 1B2B Expt. Number | B. stearo. (1) PB30T | B. stearo. (2) PD01T (S) | B. stearo. (3) QF01T (A) | B. coagulans BG45T | B. subtilis (1) SA22 | B. subtilis (2) E3BBT | B. macerans YBAT |
|---|---|---|---|---|---|---|---|
| IP3090 | 0.35* | | | | | | |
| IP3098 | 0.57* | | | 0.28 | | 0.17 | |
| IP3105 | 0.80* | | | 0.49 | | 0.14 | |
| IP3112 | 0.50* | | | | | 0.21 | 0.53 |
| IP3350 | 0.44* | | | | | | |
| IP3355 | 0.47* | | | | | | |
| IP3362 | 0.32* | | | | | | |
| IP4117 | 0.46* | | | | | | |
| IP4137 | 0.45* | 0.48 | 0.37 | 0.20 | 0.20 | | 0.33 |
| IP4144 | 0.54* | 0.56 | 0.36 | 0.23 | 0.24 | 0.16 | 0.39 |
| IP4151 | 0.48* | 0.44 | | | | | |

*Indicates D values normalized to 2500 ppm $H_2O_2$. All values are in minutes.

**Figure 9.25.** Atmosphere steam hydrogen peroxide test **apparatus.**

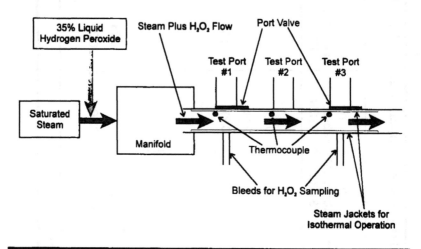

## Materials Selection and Compatibility

Filler and isolator materials must be compatible with the CIP/SIP processes. With respect to the use of hydrogen peroxide as the sterilant, either in vapor form or in combination with steam, there are a number of materials that are fully compatible with the use of the sterilant. These materials are also, in most cases, impervious to the penetration of peroxide. This is particularly true of metals. Plastics and elastomers would normally be held to minimal content to avoid any absorption and desorption questions with respect to the sterilant. In the case of the steam/hydrogen peroxide system, the materials must also be designed to repeatedly withstand 100°C temperatures during the sterilant cycle. Note also that raising the components to this temperature also has a disinfection property for nonsurface contact areas. This is a major benefit when used with complex equipment.

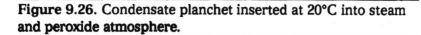

**Figure 9.26.** Condensate planchet inserted at 20°C into steam and peroxide atmosphere.

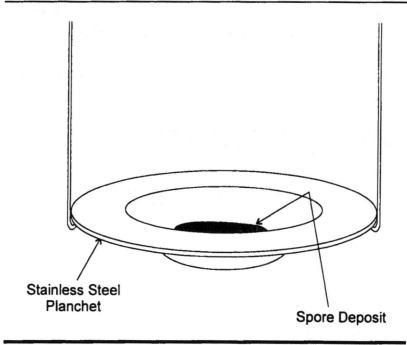

Stainless Steel
Planchet

Spore Deposit

# INTEGRATED SYSTEM PHYSICAL REQUIREMENTS AND OPERATIONAL RATIONALE

## Physical Configuration Alternatives

Physical configuration alternatives are no longer limited by the need to incorporate portions of the line within a predesignated sterile area. This allows for a certain amount of freedom with respect to the orientation and footprint of the equipment. Some of the potential physical arrangements for a barrier isolator filling system are shown in Figures 9.28, 9.29, and 9.30. These illustrate an in-line system, an L-shaped system, and a U-shaped system. The best physical configuration for a specific application will depend not only on the physical space available, but also on what method of transfer works best between the different segments of

**Table 9.3. D Value Results from Peroxide Plus™ Condensate Study**

| $H_2O_2$ Level | Condition | D Value |
|---|---|---|
| 7,600 ppm | D value with condensate on disc (planchet) | 0.29 min |
| 8,300 ppm | D value with vertical planchet | 0.21 min |

**Figure 9.27.** $H_2O_2$ concentration measuring system.

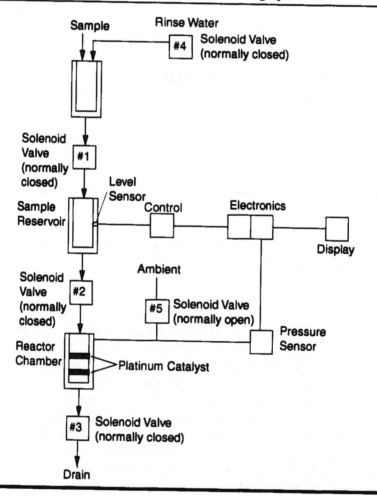

**Figure 9.28.** Integrated filling line in barrier isolator, in-line system.

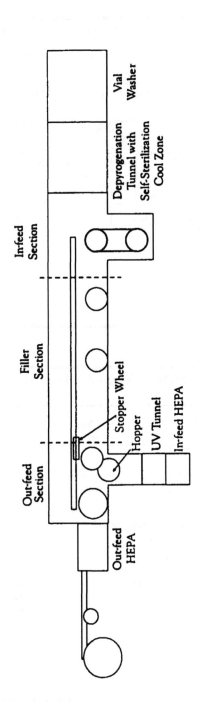

**Figure 9.29.** Integrated filling line in barrier isolator, L-shaped system.

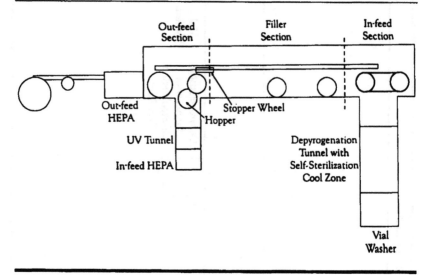

**Figure 9.30.** Integrated filling line in barrier isolator, U-shaped system.

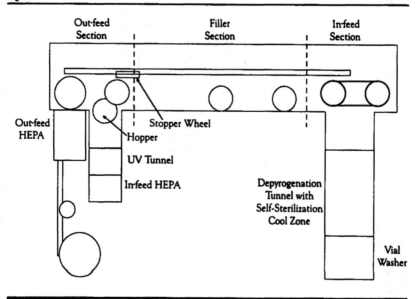

the barrier isolator filling line. In general, the major components of a filling line for vials are the washer, the depyrogenation tunnel and filling/stoppering system, and the exit conveyor. Combined with the ancillary commodity transfer ports, these make up the majority of the line configuration. If the out-feed destination is a lyophilization process, where unsealed product is transferred to an additional unit, the out-feed configuration becomes somewhat more complex, and mechanical transfers or human-aided transfers become a requirement.

If fully stoppered vials are exiting from the barrier isolation filling system, an external capper may be a final packaging component on the line. This may require a HEPA–filtered transfer area following the discharge mousehole where the stoppered but uncapped vials transit to the capper location external to the barrier isolator.

## Facility Requirements

Although barrier isolator filling systems are relatively self-contained, they are dependent on certain basic facility requirements, which generally include the following: clean compressed air, an ULPA–filtered fresh air inlet, makeup air to the system, water for cooling or cooling loop to provide the cool-down following depyrogenation of glassware, and conditioning equipment for the atmosphere within the barrier isolator. This conditioning equipment could involve refrigeration for operation in the filling area at below ambient conditions, or it could include capability to operate within a nitrogen or other controlled atmosphere requirement. As the operators are no longer inside the barrier isolator when the filling process is in progress, a number of atmosphere alternatives that have not, in the past, been easily adaptable to a filling line are possible.

Additional facility requirements include appropriate CIP solutions, the SIP source and control equipment, and electrical power. Depending on individual requirements, there may also be need for a data port connection to a facility host system monitoring the operation and alarms for the barrier isolator filling system.

## Operational Cycle

The operational cycle for preparation and CIP/SIP for a barrier isolator will depend heavily on the method chosen for CIP and SIP of the equipment. Typical documented operational cycles are shown below for the AMSCO VHP™ process and the Despatch Peroxide Plus™ process. Both of these systems have shown successful operation while sterilizing the interior of large-scale barrier isolator filling systems with in-feed and out-feed connections for the vial flow path (see Figures 9.31 and 9.32).

---

**Figure 9.31.** Steam/hydrogen peroxide process.

---

### STEPS

1. Air Heat      4. Steam Purge
2. Steam Soak    5. Air Dry
3. Sterilization    6. Air Cool
   (Steam + H₂O₂)

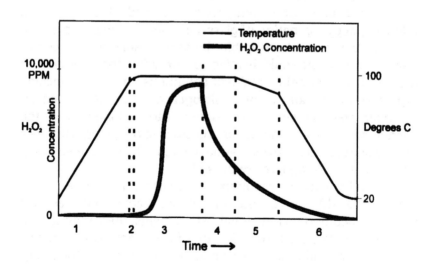

---

**Figure 9.32.** Hydrogen peroxide gas process.

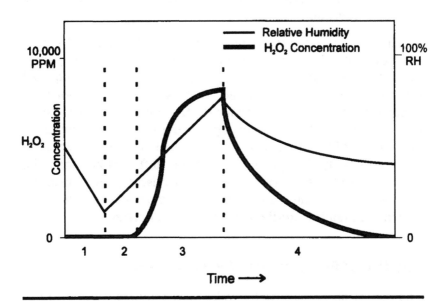

## STEPS

| | | | |
|---|---|---|---|
| 1. | Dehumidify | 3. | Sterilization |
| 2. | Condition | 4. | Aeration |

## Validation Approaches

Since the equipment is more or less freestanding, a certain amount of equipment prevalidation can be accomplished at the manufacturer's or integrator's location. This might include the following:

- Operational qualification of the sterilization cycle for the interior

- Operational qualification of the sterilization cycle for the wetted path

- Operational qualification of the various measurement equipment used in conjunction with monitoring of the barrier isolator

- Operational qualification of the pressure differential maintenance and outflow measurements under actual operating conditions, as well as documentation of all of the various sensors and alarms utilized for system operation

Much discussion has taken place relative to the use and limitations of media fills for validation. If the claimed SAL is greater than $10^{-3}$ (i.e., $10^{-4}$ or $10^{-5}$, etc.), the number of vials required for validation as the claimed sterility level rises goes up exponentially; it rapidly approaches the point where validation of increased sterility level through media fills is not a practical alternative. Monitoring of the conditions inside a barrier isolator, particularly for pressure gradient differential, once the sterilization of the interior of the barrier isolator has been accomplished and validated, is probably a more useful, long-term tool in maintaining sterile operation of the barrier.

## The Use of Physical Data for Validation

The use of physical data for validation requires monitoring of several aspects of the operation within a barrier isolator. Most important of these is the pressure differential to the ambient or nonsterile area. Additional measurements that are useful in tracking the performance of a barrier isolator are particle count measurements within the active enclosure. Operating history would dictate when yellow and red levels are achieved with respect to any of the measured parameters. A large amount of work has gone into the determination of what pressure differential will maintain sealed transfer into a barrier isolator (Porter and Edwards 1995).

## Duration of Sterility Maintenance

The production filling time for a barrier isolator system that has been sterilized and operating is dependent on a number of factors, but perhaps foremost of these is the length of time that the product in-feed sterilizing liquid filters can be maintained sterile. Current plans for validation validity range from 24 hours to 7 days, depending on the manufacturer's experience and data (FDA 1995). Most of today's aseptic filling operations are conducted in conventional Class 100 pressure-separated rooms with gowned operators, but this will likely change.

## SUMMARY

Extensive development and testing have shown the capability of barrier isolators to replace the conventional aseptic filling suite. The potential for increased quality combined with economic benefits will continue to move this technology to the forefront. The possibility of having an unattended automated line is a future potential.

## REFERENCES

Akers, J. E., and C. Wagner. 1994. Barrier technology. Third International PDA Congress and Workshops, 14–18 February, in Basel, Switzerland.

Akers, J. E., and J. P. Agalloco. 1993. Aseptic processing—current perspective. In *Sterilization technology: A practical guide for manufacturers and users of health care products.* New York: Van Norstrand Reinhold.

American Glovebox Society. 1994. *Guidelines for glove boxes.* AGS-G001-1994.

Bradley, A., S. P. Probert, C. S. Sinclair, and A. Tallentire. 1991. Airborne microbial challenges of blow/fill/seal equipment: A case study. *Journal of Parenteral Science and Technology* 45 (4).

Casamassina, F. J., J. W. Hulse, and R. P. Tomaselli. 1993. Controlling medical product contamination. In *Sterilization technology: A practical guide for manufacturers and users of health care products.* New York: Van Norstrand Reinhold.

Davenport, S. M., and H. L. Melgaard. 1995. Ultraviolet pass-through as a transfer technology in barrier and isolator systems. ISPE/PDA Joint Conference on Advanced Barrier Technology, 17–18 January, in Atlanta, GA.

Farquharson, C. 1994. Aseptic filling in a network of rigid isolators. The 3rd PDA International Congress, 14–16 February, in Basel, Switzerland.

FDA. 1995. Barrier Users Group Symposium (BUGS), unpublished report to the FDA, 31 May.

Federal Standard 209D. 1988. Clean room and work station requirements.

Frieben, W. R. 1993. Presentation at FDA Open Conference on Sterile Drug Manufacturing, 12 October, in Bethesda, MD.

Haas, P. J., H. L. Melgaard, J. P. Lysfjord, and I. J. Pflug. 1993. Validation concerns for parenteral filling lines incorporating barrier isolation techniques and CIP/SIP systems. Paper presented at PDA Second International Congress, 24 February, in Basel, Switzerland.

Hoffman, G. 1992. Presentation at Barrier Isolation Technology Conference, 24–27 August, in Minneapolis, MN.

Killick, P. F. 1992. Facility design—isolation technology. Fourth International Congress of Pharmaceutical Engineering, 8–10 September, in Vienna, Austria.

Loy, L. H., and J. F. Melahn. 1993. Current stopper processing methods and handling techniques. Barrier Isolation Technology Conference, 19–20 August, in Minneapolis, MN.

Lysfjord, J. P., P. J. Haas, H. L. Melgaard, and I. J. Pflug. 1993. The potential for use of steam at atmospheric pressure to decontaminate or sterilize parenteral filling lines incorporating barrier isolation technology. Paper presented at PDA Spring Meeting, 10 March, in Philadelphia, PA.

Melgaard, H. L. 1994. Barrier isolation issues. *Pharmaceutical Engineering* (November/December): 24–31.

Melgaard, H. L. 1995. The use of ultraviolet energy to disinfect/sterilize surfaces in a continuous mode. Barrier Isolation Technology Conference, 4–6 December, in Rockville, MD.

Melgaard, H. L. and I. J. Pflug. 1993. Nature and quality of the air leaving the filters at the top of a barrier isolator. Unpublished, 7 December.

Muhvich, K. 1995. Remarks from Barrier Isolation Technology Conference (BITCON), 4–6 December, in Rockville, MD.

Pflug, I. J. 1990. *Textbook for an introductory course in microbiology and engineering of sterilization processes.* 7th Ed.

Pflug, I. J. 1992. Microbiological testing program of the TL/Despatch barrier system. Paper presented at Barrier Isolation Technology Conference, 26–27 August, in Minneapolis, MN.

Pflug, I. 1993. Report of steam-hydrogen peroxide experiments carried out and analyzed during October, November, and December, 1993. Environmental Sterilization Laboratory, University of Minnesota.

Pflug, I. J., H. L. Melgaard, C. A. Meadows, J. P. Lysfjord, and P. J. Haas. 1993. Rigid isolation barriers for aseptic filling lines decontamination with saturated steam at atmospheric pressure of sterilization with steam plus hydrogen peroxide. Paper presented at the Kilmer Memorial Conference on the Sterilization of Medical Products, 13–15 June, in Brussels, Belgium.

Pflug, I. J., H. L. Melgaard, S. M. Schaffer, and J. P. Lysfjord. 1994. The microbial kill characteristics of saturated steam at atmospheric pressure with 7,500 and 2,500 ppm hydrogen peroxide. Paper presented at PDA Spring Meeting, 10 March, in Chicago, IL.

Porter, M. E., and L. M. Edwards. 1995. Microbiological and physical limits testing of a locally controlled environment (LCE) prototype filling system. Joint PDA/ISPE Conference, 17–18 January, in Atlanta, GA.

Walker, N. 1995. Continuous production isolator—syringe filling. ISPE Barrier Isolation Technology Conference, 4–6 December, in Rockville, MD.

Whyte, W. 1984. The influence of clean room design on product contamination. *Journal of Parenteral Science and Technology* 39 (3):103–108.

Whyte, W., and P. V. Bailey. 1985. Reduction of microbial dispersion by clothing. *Journal of Parenteral Science and Technology* (Jan.–Feb.): 5.

# 10

## Gamma Radiation Sterilization of Pharmaceuticals

*Brian D. Reid, Ph.D.*

GamRay CONSULTING, INC.
Merrickville, Ontario, Canada

*Barry P. Fairand, Ph.D.*

SteriGenics International, Inc.
Hayward, California

Radiation processing technology is perhaps the least understood by the pharmaceutical industry compared to other technologies such as steam, dry heat, or sterilant gas. However, the advantages of this technology over other currently used methods make it prudent to learn more about radiation processing technology and how to use it successfully. Gamma radiation processing is the exposure of a product to ionizing radiation from an isotopic source in a controlled manner. The source of ionizing radiation may be Cobalt 60 ($^{60}$Co) or Cesium 137 ($^{137}$Cs).

This controlled application of ionizing energy ensures that the dose delivered to the product is sufficient to reduce the bioburden to the desired level. It also ensures that any changes to the properties of the material being treated are minimized.

How this process may best be applied to your pharmaceutical raw materials and/or finished products is the subject of this chapter. Radiation sterilization can be characterized as follows:

- It is a "cold" process. The temperature of the product typically increases by a few degrees above its starting point during sterilization.

- It is a safe and clean process. No residues are left on, or in, the product, and no toxins are emitted into the environment.

- It is an easily monitored and controlled process. The single operating variable for gamma radiation is time—the length of time the product is exposed to the radiation source.

- It is a dosimetric release process. The medical disposable industry has achieved such a high degree of process assurance through the development/validation of dosimetry, that dosimetric product release is permitted. That is, the product is released for use/distribution on proof that the required dose was received—no biological indicators (BIs) are necessary or even recommended (ANSI/AAMI/ISO 1994).

- It is a penetrating process. Gamma rays, as opposed to electron beams, steam or gaseous sterilants, are able to penetrate through large containers of product. This permits sterilization of the end product in its final shipping format.

- It is an effective process. Radiation's effect on microbial populations is well understood and documented. There is no doubt about its lethal effect, and the doses required to achieve specified degrees of sterility assurance are available in table format when ANSI/AAMI/ISO methodology is used.

- It is an economical process. Extensive (30 years) practical use and development by the medical disposables industry has resulted in radiation sterilization costs being competitive with ethylene oxide costs and, in larger-volume scenarios, significantly less costly (Brinston 1991).

With the advent of new biologically derived pharmaceuticals, an additional need has come to light: ensuring that the final product is viral inactivated as well as being sterile. Gamma radiation processing offers a simple, safe, and effective means of inactivating all viruses (Ginoza 1968). It can accomplish this without the use of potentially carcinogenic sensitizers that are needed for other methods. It can eliminate the need for expensive viral filters and can replace or supplement other methodologies that are capable of inactivating only certain types of viruses.

This chapter will present a concise outline of how to develop both a product qualification program and a viral inactivation protocol for pharmaceuticals. We will discuss the use of radical scavengers, the effects of media, and the treatment of frozen products. Data on viral inactivation as it exists in the literature will be related to modern irradiation technology. The use of gamma radiation processing presents an opportunity to lower production costs as well as provide enhanced safety to pharmaceuticals of biological origin. We will begin with the basics. Why use gamma radiation processing anyway? How does it work, and what needs to be known about the equipment and process conditions?

The ability of gamma radiation to inactivate microorganisms has been well documented (Polley 1962). New documentation relating to viruses or new strains/reclassifications of microorganisms is continually being added (Reid 1996). The major benefit of using irradiation sterilization as the terminal step in the manufacturing process as opposed to autoclaving, or dry heat methods, is the lack of, or reduction in, product degradation with this technology (Reid 1995).

The process has been in use in the medical device industry for over 30 years. Ample evidence as to its efficacy exists in the literature. Materials and processes have been developed to reduce the impact of radiation on the product. It is the intent of this chapter to present some of the process developments that will facilitate the use of this technology for the terminal sterilization of pharmaceutical products. It will also assist those wishing to improve the microbial quality of raw materials entering the manufacturing process. Clean materials reduce the bioburden impact on a cleanroom facility.

## HISTORY OF GAMMA RADIATION PROCESSING

Research carried out in the late 1960s and early 1970s focused on the treatment of pharmaceuticals with high doses of radiation (Dzieglielewski et al. 1973). While there were many materials that could be demonstrated to withstand high doses of radiation, this was not always the case. Often color and viscosity changes, as well as undesirable chemical changes, occurred (Bor 1981; Pope et al. 1978). These early results tended to discourage the consideration of radiation sterilization technology in the pharmaceutical area, where it was believed that a terminal sterilization dose of 25 kGy, which was excessively high, would be required.

However, with the advances made in aseptic processing, we now have products and materials that are much cleaner, from a microbial point of view, and thus, are likely to require much lower radiation doses to achieve a $10^{-6}$ sterility assurance level (SAL). This change in the manufacturing environment provides an opportunity to terminally sterilize, or at least enhance the SAL, of a much larger number and range of drugs. It may be the only way to sterilize many biologicals, or biologically derived products, because of their sensitivity to heat or excessive viscosity, thus, making filtration impossible. The ability to use gamma radiation processing to treat these products has become especially important with the need to provide assurance of viral inactivation for human-, animal-, and tissue culture-derived products.

The pharmaceutical industry has historically relied on steam, dry heat, EtO, filtration and chemical processes to meet sterilization or microbial load reduction requirements. Now, with

EtO sterilization technology coming under intensive regulatory and public health scrutiny, and with increased regulatory pressure on the industry to improve on the microbial bioburden or SAL of aseptic processes, "new" technologies, in particular gamma radiation processing, are being reexamined. This renewed interest has come at a most appropriate time. The increased use of radiation processing for other industrial purposes, such as the sterilization of medical devices and the sanitization of cosmetic raw materials, has led to the development of more efficient and economical irradiation equipment and processes. It has also generated additional scientific data on how to apply and validate the technology effectively. In addition, the regulatory climate is favorably disposed to the use of this technology when it is properly documented.

## WHY USE GAMMA RADIATION?

### Terminal Sterilization

Terminal sterilization is used once the product is sealed in its final container. It provides the assurance that the product is sterile and viral inactivation has been achieved. There are certain cases where a raw material may need to be supplied as a sterile product (e.g., an antibiotic powder that is to be supplied to a contract packager). In such instances, the material can be double- or triple-bagged before being sterilized. This will facilitate bringing the material into the cleanroom manufacturing area.

### Reliable, Reproducible Viral Inactivation

Gamma radiation inactivates all types of viruses (DNA [deoxyribonucleic acid], RNA [ribonucleic acid], enveloped, and nonenveloped). Mycoplasmas can also be inactivated without destroying the culture medium by using the appropriate radiation conditions (Bender et al. 1989; Gergov et al. 1988; Wyatt et al. 1993a,b).

## Complete Product Penetration

The unique penetrating ability of gamma radiation ensures that all of a product is treated to the minimum dose required. The ability of gamma radiation to penetrate any existing packaging material allows the selection of the most appropriate packaging to protect the raw material or product. There is a very large selection of radiation-resistant packaging materials on the market today.

## Ease of Validation

Gamma radiation is the simplest sterilization process to validate. Once the dose required by the product is set, only the time must be determined. There is no complicated transportation mechanism, no vacuum system, no gas/steam mixture to monitor, and no power variables (i.e., voltage, current) to regulate.

## No Posttreatment Quarantine

Product release is immediate. It is called *dosimetric release.* No BIs are required. The dosimetry system used means that the results are available within minutes of the end of treatment. As this indicates the dose delivered to the product, there is no need for sterility testing. The product can be shipped directly to the customer. There are also no "residuals" with gamma radiation processing. Gaseous treatment methods require a lengthy post-exposure time to remove residual gas. Frequently, by-products of the reaction of the gas with water are left behind. These by-products at worst can be toxic (Dorman-Smith 1991) and can sometimes cause sensitivity reactions in patients. These residues have been detected for up to 7 days (Reid 1987, personal communication) and sometimes even longer.

## Novel Product Design Potential

Dual chamber syringes and vials, transdermal patches, and liposome carriers, for example, offer unique ways to deliver/protect the drug and still allow the final product to be terminally sterilized. Because there is no need to consider the diffusion of a gas into or out of the product, multiple packaging layers can be safely employed.

## Reduction in Endotoxin Level

Two papers have recently been published demonstrating the ability of *gamma, but not* electron beam radiation, to reduce endotoxins (Guyomard et al. 1987, 1988). This is an added feature that will further encourage the use of gamma radiation processing.

## Environmentally Friendly

Because Cobalt 60, which is the principal isotope that is used in radiation processing applications, is a man-made isotope and decays to inactive Nickel 60, it presents no long-term storage problems. Its half-life of 5.26 years means that a pencil containing 10,000 Curies (Ci)[1] is essentially inactive after 150 years. It is *not* produced from spent fuel rods. Most sources can be used effectively for up to 25 years before they need to be replaced. The biological shield around a facility ensures that the radiation level outside is no greater than the normal background radiation for that area. No toxic gases are used that require expensive recovery systems to prevent their release to the environment. No expensive maintenance is required as with autoclaves. The working environment is also quieter and cooler than with the latter systems.

## Cost-Effectiveness

Cost-effectiveness is without a doubt the hardest area to address. Is radiation processing more expensive than other methods? It depends on how the cost analysis is done. For example, if raw material rejection, lost production time due to raw material

---

1. The authors choose to use the older terminology here as the SI units are too large and become meaningless.

rejection, product rework, product recall, lost market, and image or delayed entry expenses are included, then radiation processing is certainly *cost-effective* because the assurance received from using the process means that these other activities are unlikely to occur. Reduced microbial loads in raw materials means lower bioburdens in the manufacturing area, which means a lower probability of product contamination or cross-contamination. This in turn means reduced disinfectant costs and could mean greater shelf life, product protection, or reduced requirement for product preservatives. All of these factors must be weighed when assessing the cost-effectiveness of the process. The ability to ship to customers or to receive raw materials immediately after treatment means reduced inventory and reduced downtime in the production area. In addition, it could mean a faster response time to market demands (Brinston 1995).

# GENERAL ASPECTS OF GAMMA RADIATION PROCESSING

## Physical Characteristics of Ionizing Radiation

An understanding of how the gamma sterilization process works requires an understanding of the concept of radiation, its sources, and a basic appreciation of the physics and chemistry of radiation interactions in matter. Radiation interaction processes are similar to thermal or chemical processes in that the changes brought about in a material are caused solely by the deposition of energy (McLauglin and Holm 1978). Gamma radiation is pure electromagnetic energy that propagates through space. Gamma rays make up only one small part of the total electromagnetic spectrum that embraces many other forms of radiation, including thermal sources and radio waves. Gamma rays, also referred to as photons, essentially function as a precursor in the sterilization process (Laughlin and Genna 1996). Energy is transferred to electrons that in turn transfer energy to the material to produce the beneficial effects. From this viewpoint, photons can be identified with electron beams that are also used in sterilization applications. The difference is that in the gamma process electrons are produced homogeneously throughout the material, while the source of electrons is external to the material in the electron beam process. The energy of the electrons that are produced in both processes exceeds the binding energies of molecules (i.e., < 12 eV [McLauglin and Holm 1973])[2]. The result is that virtually any chemical bond may be broken and any potential chemical reaction may take place. The key word is "may." There are definite probabilities based on bond energies.

## Gamma Radiation—Definition and Sources

As shown in Figure 10.1, gamma rays are found at the high-energy and high-frequency end of the electromagnetic spectrum. Radio waves are found at the low-frequency end of the spectrum; other familiar sources of radiation, including microwaves, infrared radiation (i.e., heat), light, ultraviolet radiation, and X rays, fill in the spectrum. All of these sources of radiation belong to the same electromagnetic spectrum and differ only in the type of emitter and frequency.

Historically, radioisotopes have served as the principal source of gamma rays for sterilization applications. A radioisotope is an element that contains an unstable nucleus that spontaneously decays at a predictable rate to form another element. The decay process is often accompanied by the emission of gamma rays. Cobalt 60 is an example of a radioisotope that has found widespread application as a source of gamma rays for the sterilization of various materials and for radiation therapy.

## Production of Cobalt 60

Cobalt is a metal that is found in nature as an ore. In its stable state, the Cobalt nucleus contains 27 protons and 32 neutrons. It has an atomic number of 59, which equals the total number of protons and neutrons in the nucleus. Cobalt 60 is formed by exposure of the stable isotope Cobalt 59 to neutrons in a nuclear reactor. A neutron is absorbed by the cobalt nucleus to create $^{60}$Co (Figure 10.2a, b). The resultant radioisotope, which has a

---

2. 1 eV is equivalent to $1.6 \times 10^{-12}$ ergs or $0.38 \times 10^{-19}$ cal.

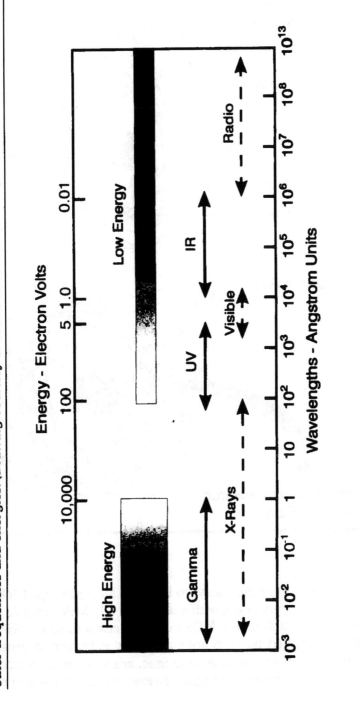

**Figure 10.1.** Electromagnetic spectrum showing the range of energies for gamma radiation in relation to other frequencies and energies. *(Drawing courtesy of SteriGenics International)*

**Figure 10.2 a & b.** Creating $^{60}$Co from $^{59}$Co and the disintegration of $^{60}$Co. *(Drawing courtesy of MDS Nordion)*

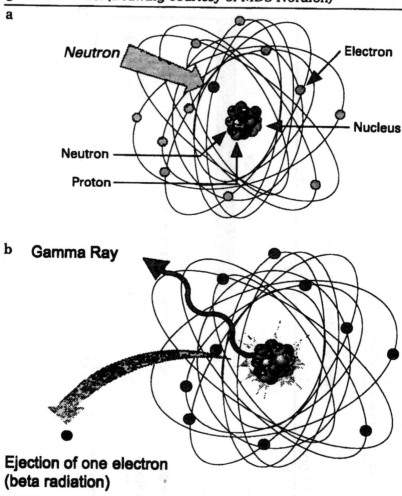

a

Neutron

Electron

Nucleus

Neutron

Proton

b   Gamma Ray

Ejection of one electron
(beta radiation)

half-life of 5.26 years, decays to form the stable isotope Nickel 60 ($^{60}$Ni). Two gamma rays, with energies of 1.17 MeV and 1.33 MeV, are emitted each time a $^{60}$Co nucleus decays to form $^{60}$Ni. The photons emitted by $^{60}$Co are sufficiently energetic to penetrate deeply into most materials, and thus, are an effective source for bulk sterilization applications; however, the photon energies are below the threshold energy that is needed to create other radioisotopes (McLaughlin et al. 1989). Therefore, irradiation with $^{60}$Co cannot make materials radioactive.

The source shape for industrial applications is that of a thin rod or "pencil" about 18" in length. Refined cobalt is a metal powder that is sintered into a cylinder roughly 1/4" in diameter by 1" in length. This is known as a "slug." The slugs are nickel plated to help retain any other impurities that might be present in the cobalt. Six to eight of these are placed in a zircaloy tube. The tubes are placed in a special holder and sent to a nuclear power reactor. There they absorb neutrons and become radioactive. It takes approximately 18 months to convert 7 percent of the Cobalt 59 to Cobalt 60. The zircaloy tubes are returned in special steel-covered lead containers to companies such as MDS Nordion where they are then encased a third time in stainless steel (316L) rods. Each rod holds two zircaloy tubes and houses approximately 10,000 Ci of $^{60}$Co.

These sealed pencils are the source of gamma rays that are used to sterilize pharmaceuticals as well as many other products. The number of photons emitted per second by a radioisotopic source is a function of the number of curies in the source. Each curie of $^{60}$Co emits $7.4 \times 10^{10}$ photons per second. Since it is common practice to stock an irradiator with more than a million curies of isotope, products that are processed in an irradiator are literally bathed by an intense flux of photons. This allows large volumes of medical device products to be treated in relatively short periods of time (i.e., in several hours or less). This may not necessarily be best for pharmaceuticals where photon flux (i.e., dose rate and therefore, local heat generation) can play a significant role (Sterways 1993).

## A Typical Gamma Radiation Facility

Figure 10.3 shows a typical gamma irradiation facility. It consists of a biological shield, a source shield, a mechanism to raise the source, a mechanism to transport the product past the source, and deionizers to keep the water free of chlorine that would attack the stainless steel of the pencils. In addition, there is an air

**Figure 10.3.** A typical gamma irradiation facility. Automatic pallet irradiator. *(Drawing courtesy of MDS Nordion)*

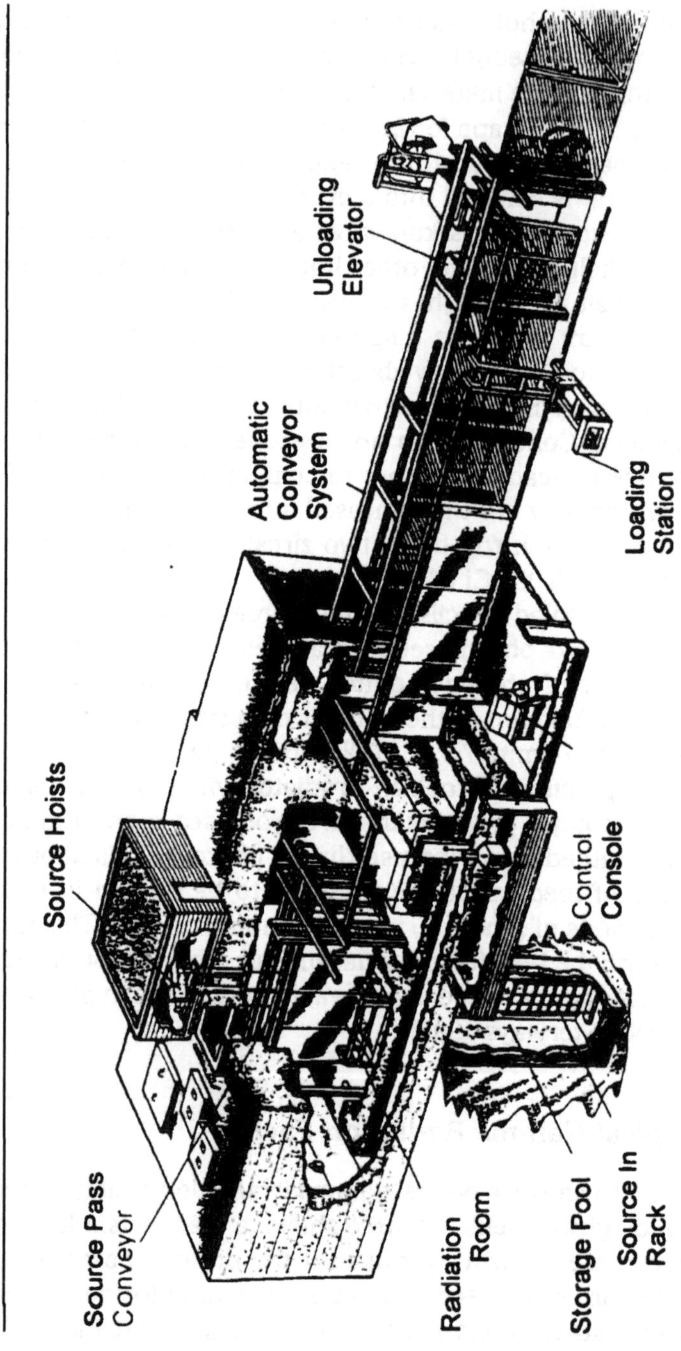

Source Hoists

Automatic Conveyor System

Unloading Elevator

Loading Station

Source Pass Conveyor

Radiation Room

Control Console

Storage Pool

Source In Rack

exchanger to remove the ozone that is generated when the source is in the raised or operating position.

In addition to this large commercial facility, other types of facilities are in use that are somewhat smaller in size. A smaller unit that still uses the biological shield, but does not need the pool of water to shield the source, instead utilizes a specially modified shipping container (Figure 10.4). The container sits in a pit in the floor of the room. Product is placed on rotating and movable turntables. This feature allows the simultaneous treatment of a wide variety of goods at different dose rates and to different total doses. In addition, the small volume of product on the turntable ensures that the maximum to minimum dose ratio ($D_{max}/D_{min}$ ratio) is very close to one.

A third type of unit is one where throughput has been sacrificed in order to achieve very uniform dosing of the product. That is, the $D_{max}/D_{min}$ ratio is very close to one. In this type of irradiator, individual boxes of product can be conveyed through the irradiation cell rather than large containers that are typically filled with many cubic feet of product. A standard carrier holding a single box of product is shown in Figure 10.5. Irradiators of this type are used in product validation studies and for the processing of small lots of high unit value product.

A fourth type is the research irradiator (Figures 10.6 and 10.7). This is a self-contained unit using either $^{137}$Cs or $^{60}$Co as the radionuclide source. The advantage of $^{137}$Cs is that it has a very long half-life (~30 years), so it seldom needs to be replaced. However, the dose rate and energy from a $^{137}$Cs source does not match that found in a commercial facility using $^{60}$Co. Therefore, it is more usual to see a $^{60}$Co source used. The major application of the $^{137}$Cs units would appear to be for the treatment of blood to kill white blood cells, thus eliminating graft versus host disease in transfused patients. These smaller irradiators have a small irradiation chamber, which is a limitation, but they serve the purpose well.

However, because cesium exists as a salt, it cannot be safely used in larger irradiators that utilize a pool of water in which to store the radioactive source. Attempts have been made to devise dry-storage units for the use of this isotope, but up to now, there has been limited commercial success. Recently, a new company,

**Figure 10.4.** A special gamma irradiation facility for pharmaceuticals. (*Drawing courtesy of MDS Nordion*)

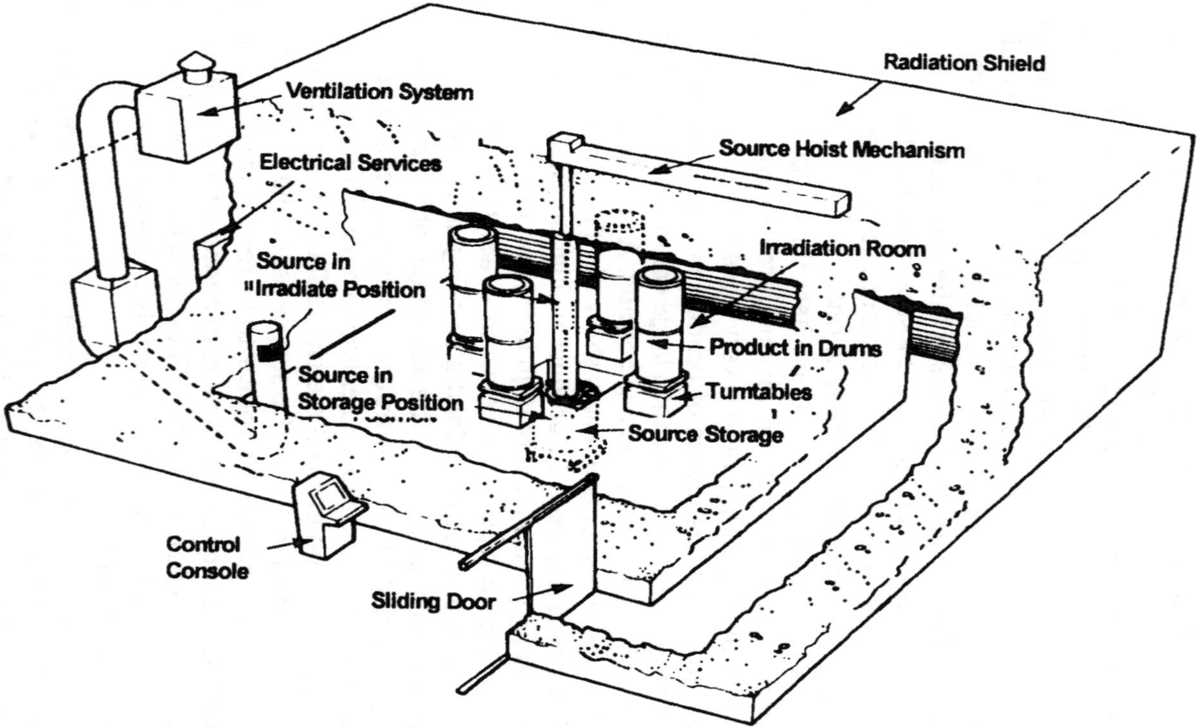

**Figure 10.5.** A typical product carrier used in the ExCell gamma irradiation facility. *(Drawing courtesy of SteriGenics International)*

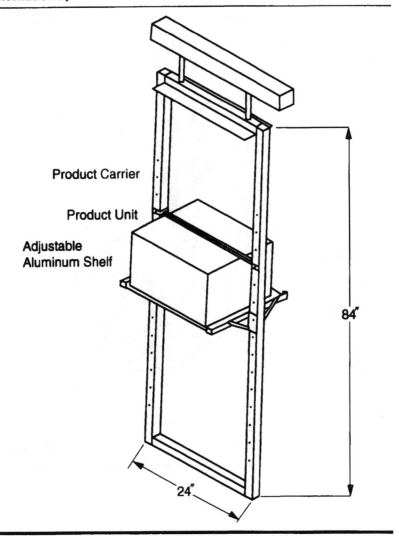

**Figure 10.6**. A small research irradiator: the Gammacell 220. *(Drawing courtesy of MDS Nordion)*

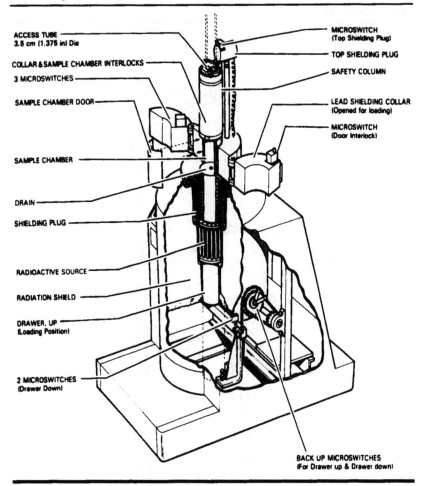

**Figure 10.7.** A blood irradiator using a $^{137}$Cs source. Typically used to treat blood for the prevention of graft vs. host disease. *(Drawing courtesy of MDS Nordion)*

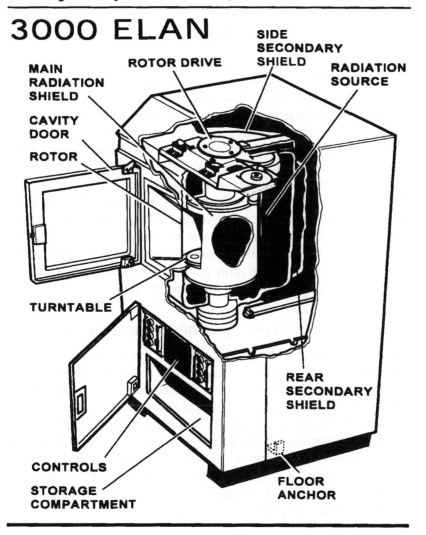

# 3000 ELAN

SIDE SECONDARY SHIELD

ROTOR DRIVE

RADIATION SOURCE

MAIN RADIATION SHIELD

CAVITY DOOR

ROTOR

TURNTABLE

REAR SECONDARY SHIELD

CONTROLS

STORAGE COMPARTMENT

FLOOR ANCHOR

Gray\*Star, located in New Jersey, has devised a rather novel approach that may be more successful. However, a commercial unit of this type has not been built.

## Dosimeters and Dosimetry

A critical part of the radiation processing cycle involves measurement of the radiation environments to which pharmaceuticals are exposed. This environment must be measured using quantitative techniques. Dosimetry is the measurement tool for accomplishing this objective. In this section, fundamental terms that are relevant to the metrological aspects of dosimetry are defined. Classes of dosimeters, their hierarchical position in the measurement process, and methods of analysis are discussed. Standards for the use of dosimetry systems, calibration techniques, and traceability to recognized standards laboratories are reviewed. The section ends with a discussion of selection criteria and the use of routine dosimetry systems that are used for the measurement of radiation environments in both validation and production phases of the processing cycle.

### Fundamental Terms and Types of Dosimetry Systems

Gamma energy must be absorbed by the material of interest (i.e., pharmaceuticals) to effect change in its properties and destroy microorganisms. *Absorbed dose* is defined as the mean energy imparted to an incremental unit of matter, divided by the mass of that matter. The international or SI unit of absorbed dose is the joule per kilogram, which is termed the gray (abbreviation Gy). The term kGy is now used to define the radiation dose delivered to a product and has replaced the older term *rad* (or radiation absorbed dose). The relationship of the gray to the rad is shown in the following equation:

$$25 \text{ kGy} \equiv 2.5 \text{ Mrad}$$

A *dosimeter*, which is used to measure absorbed dose, is defined as a device that, when irradiated, exhibits a quantifiable and reproducible change in some property of the device that can be related to absorbed dose in a given material using appropriate analytical techniques. Many materials show change in some property when irradiated; however, dosimeters must exhibit a quantifiable and reproducible change. These properties separate dosimeters from most other materials. Dosimeters are a subset of systems that include analytical instrumentation and techniques for measurement of change in properties. A dosimetry system is a system that is used for determining absorbed dose, consisting of dosimeters, measurement instruments, and their associated reference standards and procedures for the systems use.

Reference standard dosimeters represent one of several classes of dosimeters. The hierarchical order begins at the top with primary standard dosimeters that are of the highest metrological quality, recognized as national standards and maintained by national standards laboratories. Calorimeters and ionization chambers are two commonly used primary standard dosimeters (McLaughlin et al. 1989; ICRU Report 35; ICRU Report 14). Primary standards are retained by the standards laboratories to calibrate radiation fields against which other classes of dosimeters are calibrated. Reference standard dosimeters are the highest metrological quality devices that are available for use in irradiation facilities. These dosimeters are used to calibrate radiation fields and routine dosimeters. Several reference standard dosimeters are available for use in high-dose applications. Examples include dichromate aqueous solutions, alanine and cericcerous sulfate dosimetry systems (ASTM E1401; ASTM E1607; ASTM E1205). The stable and reasonably rugged properties of these reference standards also allow them to be used as transfer standards (i.e., an intermediary system) for the calibration of radiation fields and routine dosimeters. Routine dosimeters are used for on-site monitoring of the radiation process and are calibrated against primary or reference standard dosimeters. Examples of routine dosimeters include poly(methyl methacrylate), radiochromic film, and cellulose acetate (ASTM E1276; ASTM E1275; ASTM E1650).

Dosimeters are divided into two types dependent on their method of analysis: physical systems or chemical systems (McLaughlin et al. 1989). Physical systems offer basic methods for measuring absorbed dose and are usually associated with primary standard dosimeters such as calorimeters and ionization chambers. The principle of operation of the calorimeter is simple. Energy is deposited as heat in an adiabatic chamber and recorded as a temperature rise in the thermally isolated mass. In ionization chambers, energy is deposited in a medium that produces electron-ion pairs whose energetics are well known. A voltage is applied to the medium, and current flow is measured with an electrometer.

Chemical systems can be either liquids or solids. Dichromate and ceric-cerous are examples of liquid dosimeters, where absorbed energy changes the ion concentration in the aqueous solutions that can be measured with spectrophotometers as a change in absorbance of the medium. Ceric-cerous offers an alternate method of analysis using potentiometric techniques (ASTM E1205).

Solid chemical systems, which include poly(methyl methacrylate), radiochromic thin films, and cellulose films, all change color from absorbed dose. Spectrophotometers are typically used to measure change in absorbance of the dosimeters. In alanine, radical formation is measured using electron paramagnetic resonance techniques (ASTM E1607).

### Standards, Calibration Methods, and Traceability

An ANSI/AAMI/ISO standard was developed by ISO Technical Committee 198 to serve as an international standard for radiation sterilization of healthcare products (ANSI/AAMI/ISO 1994). In recognition of the importance of dosimetry to the radiation sterilization process, a section of the standard is dedicated to a discussion of dosimeters, dosimetry, and associated equipment. The standard provides information for selecting and using dosimetry systems to measure absorbed dose in irradiators and delineates types of dosimetry systems that may be employed on a routine basis as a means of quality assurance in radiation processing of healthcare products.

This standard is complimented by several standards for selection, calibration, and use of dosimetry systems that are published by the American Society for Testing and Materials

(ASTM E1401; ASTM E1650; ASTM E1261). Standard practices for the use of individual dosimetry systems guide the user in application of the dosimetry system to obtain quantitative measurements of absorbed dose. Methods of dosimetry system calibration are discussed in ASTM E1261. Calibration consists of irradiating dosimeters to several known absorbed doses over the range of use, analysis of dosimeter response using calibrated analytical instrumentation, and generation of a calibration curve.

There are three methods of calibration of routine dosimeters, each offering its own advantages. These include the following:

- Irradiation of the routine dosimeters in an irradiation calibration facility (e.g., located at a national standards laboratory)

- Irradiation in an in-house calibration facility that has an absorbed dose rate measured by reference or transfer standard dosimeters

- Irradiation of the routine dosimeters together with reference or transfer standard dosimeters in the production irradiator

Regardless of the calibration procedure, traceability to national standards must be demonstrated. For example, in the United States, this requires traceability to the National Institute of Standards and Technology (NIST); in the United Kingdom, it is the National Physical Laboratory (NPL). Due to the international use of irradiation sterilization, which includes extensive trade in irradiated products and the limited number of standard laboratories that provide calibration services, the International Atomic Energy Agency (IAEA) also has set up a high dose intercomparison program that promotes dosimetric accuracy in products that are processed in irradiator facilities (Nam 1983). Other recognized laboratories that meet stringent requirements can also be used for calibration purposes (ASTM E1400).

## Routine Dosimetry

The selection of an appropriate, routine dosimetry system for use in a specific application is guided by several factors, including type of irradiator, frequency of use, economics, environmental conditions, and time allotted for analysis and certification of absorbed dose. No ideal routine dosimetry system exists; therefore, it is important in the selection of a routine dosimetry system to match its performance with specific application criteria (ASTM E1261). Some of these criteria include stability and reproducibility of the system, which includes pre- and postirradiation stability; dependence on environmental factors (e.g., temperature, humidity, light, dose rate); ability to correct for systematic errors; and practical considerations which include ease of use and analysis, ruggedness, availability, and cost. Advantages and disadvantages of individual dosimetry systems are listed in ANSI/AAMI/ISO (1994).

Departure of routine dosimeters from the ideal system could invoke concern over the accuracy of measurement; however, through proper use of these systems and attention to standards on use of the chosen system, overall uncertainty in the measurement, which includes systematic errors as well as random errors, should be within about 6 percent at a 95 percent confidence level (ASTM E1276; ASTM E1275; ASTM E1650; ASTM E1707). Dosimetry, properly employed, may constitute one of the more accurate measurements in the radiation processing cycle of pharmaceuticals.

## THE SCIENCE BEHIND GAMMA RADIATION PROCESSING

### Interaction of Gamma Radiation with Matter

Photons that are emitted by $^{60}$Co transfer their energy to matter principally through scattering events with orbital electrons of atoms that make up the target material. This scattering process is named after Arthur Compton, the scientist who first described the energetics of the process (McLauglin and Holm 1973). The process is physically described in Figure 10.8. As seen from Figure 10.8, the incident photon scatters off an orbital electron, transfers part of its energy to the electron, and is deflected in a new direction at a reduced energy. In the Compton scattering

**Figure 10.8.** Compton Scattering. ($h\nu$ = incident energy of photon; $h\nu'$ = scattered energy of photon; $T$ = kinetic energy of electron; $m = m_0/(1 - v^2/c^2)^{1/2}$; $m_0$ = rest mass of electron; $v$ = velocity of electron; $c$ = velocity of light)

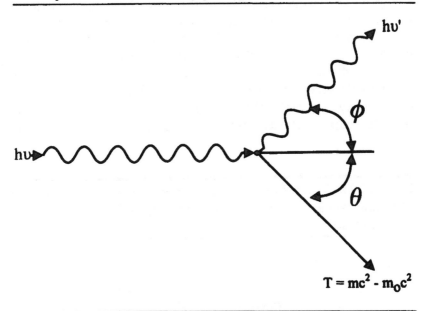

process, sufficient energy is transferred from the photon to the orbital electron to cause the electron to be ejected from its bound state around the atom (i.e., the atom is ionized). The high energy electron, termed the *primary electron*, rapidly dissipates its energy through a large number of low-energy scattering events with other atomic electrons in the material. These secondary electrons cause further ionization and excitation of atoms, excitation and dissociation of molecules, and possible heating of the material. It is these low-energy events that lead to the resultant physical and chemical changes in a material as well as assist in the destruction of pathogenic microorganisms. Thus, as we have already noted, the gamma rays only act as a precursor to the ultimate absorptive events that lead to the sterilization of products and the alteration of material properties.

Each photon emitted by the [60]Co radioisotopic source typically undergoes several Compton scattering events within the target material before its energy is effectively dissipated. In a unit density material such as water, the mean distance between Compton scattering events is on the order of several centimeters, whereas the range of the primary electrons is limited to a few millimeters. In lower density materials, these distances are correspondingly longer. The multiple scattering events associated with the interaction of photons in a material are shown in Figure 10.9. The large mean free path of [60]Co photons, even in unit density materials, allows large volumes of material to be treated to approximately the same absorbed dose. Due to the

**Figure 10.9.** Scattering of photons from [60]Co as they strike and penetrate an object. *(Drawing courtesy of SteriGenics International)*

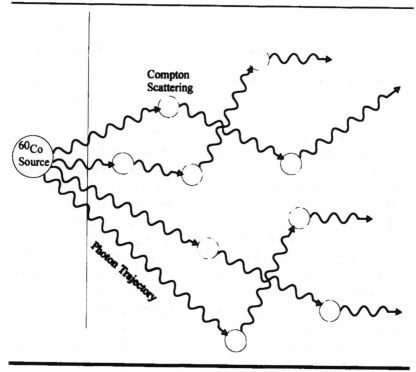

extremely large number of photons that interact with a unit volume of material, the interaction zones associated with the primary electrons overlap, thus ensuring homogeneous treatment of the target material.

However, when we deal in the realm of pharmaceuticals, the subsequent complex interactions, which produce a cascade of low-energy tertiary radiation, become more important. Eventually, these low-energy photons and electrons cause atomic and molecular excitations, displacements, chemical reaction transfers, and thermal processes (McLauglin and Holm 1973). Most interactions affect molecules only for short times; some are more or less permanently affected.

Tertiary radiation (delta rays, or very low-energy electron cascades) can be the source of extensive local effects. The ratio of ionization to excitation of atoms and molecules decreases as the various photons and electrons decrease in energy. Eventually, ionization is completely replaced by excitation, fluorescence, nonradiative chemical reactions, bond polarizations, heat dissipation, and atomic and molecular displacements. These last events are of primary importance in radiation biology and also for biologically based pharmaceuticals (Ley 1973).

## Practical Consequences

As we have seen, changes induced by the Compton scattering of photons are concentrated around the track of the primary electron, which has a range of a few millimeters or less in unit density materials. At the atomic and molecular level, these changes appear almost instantaneously (i.e., on the order of $10^{-10}$ second or less) in the form of ionized species, dissociated and excited molecules, and thermal processes (Antoni 1973). Recombination processes also begin to occur in the same time frame; however, long-lived species (i.e., certain free radicals) may be chemically active for hours or days. For practical reasons, and as a first approximation, it is generally assumed that one is dealing with a monoenergetic spectrum, defined geometries, and constant environmental conditions (McLauglin et al. 1989). However, the wary user of irradiation facilities will be careful to note that unexplained variations in results may be due to changes in the "assumed" constants. There really are only two methods for determining the dose received/delivered and its adequacy. The first is physical and involves the use of dosimeters; the second is biological and involves determining the capability of an organism to grow.

The energetics of these processes are quantifiable. However, the dynamics of the process, particularly the different pathways for recombination to occur, which affects the ultimate physical and chemical properties of the material, make full quantification difficult. The influence of environmental factors from thermal, chemical, and electrical sources further complicates the problem. In spite of these shortcomings, much is known about the effect of radiation on pharmaceuticals. A good share of the knowledge comes from observations made at the macroscopic level, which are dealt with in subsequent sections of this chapter.

Knowledge about preferential pathways is being gained through the use of pulse irradiation and by trying to more accurately define the radiation milieu. This permits the optimization of the formulation in order to sensitize the microorganism while sparing the pharmaceutical. Eventually though, a balance must be struck between organism inactivation and product loss or the generation of toxic breakdown products. This is standard, straightforward chemistry. No new or unusual compounds have ever been created unintentionally by radiation. Gamma radiation sterilization results simply from the deposition of energy into a system. It is not unlike using a microwave oven—the more rapidly the energy is applied, the less edible is the food. Micro areas of localized heating can denature proteins as well as other molecules. This is one reason why the use of electron beam sources for the sterilization of pharmaceuticals is unlikely to be very great except for a few raw materials or final products.

## The Effects of Ionizing Radiation on Bacteria, Fungi, and Viruses

In radiation sterilization, the focus is on specific, predictable, biological, and material effects. The principle that an understanding of all of the events discussed above will permit us to predict results in more complex molecules, while true, is not practicably attainable. There is no substitute for actual experimental testing. For example, in the radiation energy region used for sterilization, there is strong energy and rate dependence of the absorption cross-section of highly conjugated molecular sites (McLauglin et al. 1989). These measurements do not adequately define what is happening to the locally irradiated cell components. This is particularly true in those regions responsible for cell division, at pyrimidine bases, or at active sites in enzyme molecules. The sites of reaction are too small and the changes too rapid for accurate measurements to be made (McLauglin et al. 1989).

The probability of local damage by an ionization or excitation event is complicated by in vivo protective mechanisms, such as energy storage by delocalization of $\pi$-orbitals, bond polarizations and relaxations, and heat conduction. In addition, the presence of protective agents, such as cysteine or cysteamine, also hinder the measurments. As McLaughlin and Holm (1973) state,

> The radiation resistance of a biologically active molecule can, in fact, be measured to some degree by determining the degree of conjugation and the concentration of other scavengers of DNA and enzyme-attacking agents (e.g., scavengers of solvated electrons and OH radicals, etc.).

However, more information is needed to characterize the process completely. It is the space-time relationship, charge distribution, and spatial relationships of electron density that are critical factors influencing radiobiological effects. At present, it is still difficult to know these things in sufficient detail to allow full prediction of the outcome.

## Mechanism of Inactivation

As we have seen previously, gamma rays ionize atoms and molecules as they pass through matter. These ionization events kill or inactivate microorganisms (bacteria, fungi, viruses, or parasites) by altering the molecular structure or spatial configuration of *biologically active* macromolecules. There are two key groups of molecules. The first is the nucleic acid. Because this is the source of all information required to replicate the organism, any unrepaired/unrepairable alteration will cause the cell to die. Proteins and enzymes within the cell constitute the second group of molecules, of which there are many copies. Damage to these molecules is, therefore, less critical as they can be replaced if the nucleic acid is undamaged.

The occurrence of these ionizations within or near a macromolecule produces a "lesion". A lesion is referred to as a "hit", and the macromolecule is referred to as the "target" (Dessauer 1922). The most sensitive target is usually the nucleic acid. A lesion in a nucleic acid results in either a single-strand break, a double-strand break (for double-stranded nucleic acids), or the loss or alteration of one or more bases. This concept, "target theory", is used to explain the events surrounding irradiation damage to living cells (Ginoza 1968). It provides the confidence needed to interpret inactivation curves and predict the degree of inactivation for a given radiation dose. The degree of inactivation of a microorganism is determined by its ability to reproduce within a cell. The ability to replicate (its "infectivity") is the final event in a series of events beginning with exposure to a source of radiation. These events are depicted in Figure 10.10. Investigators have been able to use this sequence to determine the dose required to inactivate a given process in the cycle. This is illustrated in Figure 10.11.

## Types of Radiation Effects

The lesions produced in DNA or RNA can arise from the direct or indirect effects of radiation. Direct effects are the damage done when an ionization event occurs in the mass of the nucleic acid. Indirect effects are the damage done by diffusible free radicals produced in the irradiated medium and subsequently transferred to a structure or cellular function (e.g., enzyme system). In practice, it is almost impossible to separate the effects of direct and indirect action from each other. The action of diffusible agents cannot be quantitated. Their mean life, diffusion constants, and reaction probabilities in collision with the target material are difficult to measure. Bacteria and viruses have both

**Figure 10.10.** The sequence of events for viral replication to occur. Interruption of the sequence at any point would register as viral inactivation in an assay. It demonstrates how difficult it is to assign a radiation effect to a particular point in the cycle. *(After Ley 1973)*

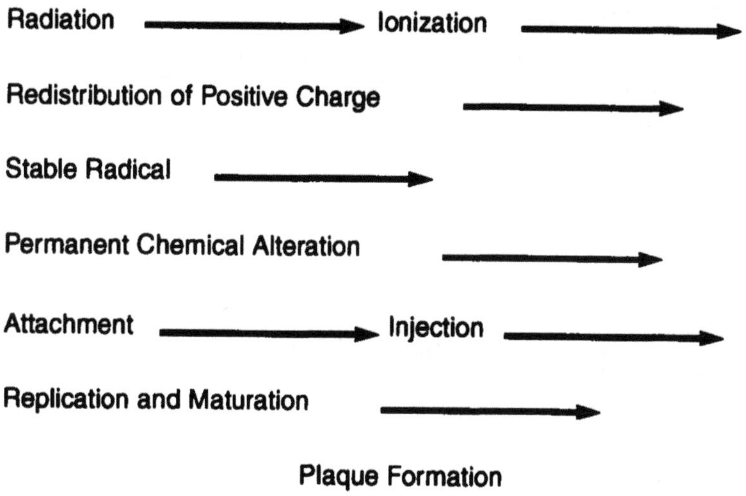

Radiation ⟶ Ionization ⟶

Redistribution of Positive Charge ⟶

Stable Radical ⟶

Permanent Chemical Alteration ⟶

Attachment ⟶ Injection ⟶

Replication and Maturation ⟶

Plaque Formation

been shown to provide self-protection against radicals, even if irradiated wet. No good explanation of this phenomenon has been proposed. It may be related to a similar phenomenon where bacteria and viruses require a "critical mass" in order to replicate. Otherwise, they languish in a resting state or actually die out.

The same two effects (direct/indirect) also apply to other molecules encountered in the organism. That is, energy can be deposited directly in a bond of the macromolecule that can cause a rearrangement of its structure, altering or destroying its normal function. The second is to generate free radicals, primarily from water contained within the cytoplasm. The free radicals thus generated react with the macromolecules to subvert their normal function. In either case, the result is the alteration of normal cellular metabolism, which leads to the loss of the reproductive capability of the microorganism (Ley 1973).

**Figure 10.11.** The dose response of various viral replication steps to relationship between total radiation dose and the inactivation of viruses. *(Reproduced from Ley 1973)*

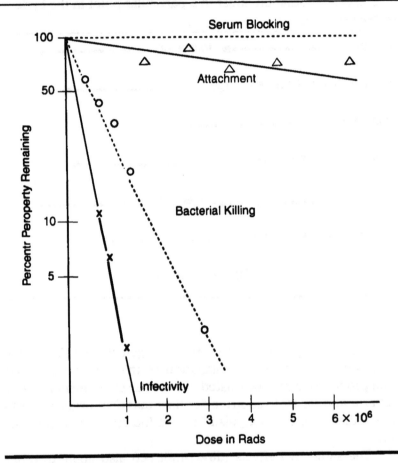

## Measurement of Inactivation

The number of organisms inactivated by a given radiation dose is a statistical phenomenon. It depends on the sensitivity of the biologically active macromolecule(s) to alteration (denaturation), the number of alterations elicited within the cell, and the ability of the cell to repair these alterations. Different organisms have different capabilities to withstand or repair such alterations. The size of the organism, its state of hydration, and the presence or absence of radical scavengers all affect the outcome of exposure to ionizing radiation. It has been demonstrated that bacteria, molds, fungi, and viruses are inactivated by a first-order or single-hit process (Ley 1973).

A dose/survival curve illustrates the relationship between the number of organisms that survived and the radiation dose delivered to them. In practice, this is done by exposing known populations of organisms to a range of radiation doses and determining the number of survivors in some form of nutrient media. This number of survivors is then expressed as a fraction of the original number. This fractional number is then plotted on a semilog scale. The response of various species of organisms gives rise to the family of curves shown in Figure 10.12. Thus, the choice of a sterilization dose for a particular product must be based on the knowledge of the types of organisms found in/on that particular product and with the knowledge that there is some probability of the existence of survivors.

This pioneering work was done by Gunter and Kohn (1956) and Antoni (1973) who distinguished the four different shaped curves and by Alper (1961) who designated them A, B, C1, and C2 for convenience. The simple exponential curve represented by type A, with its constant slope over the whole dose range, was demonstrated for bacteria as early as 1912 (Ley 1973). The semilogarithmic plot makes it easier to understand how a sterilization process must be based on the probability of the existence of survivors. The most common response to inactivation by radiation is represented by dose/survival curves of either type A or C1, with a shoulder at the beginning of the curve.

Figure 10.13 (Ley 1973) demonstrates this with data obtained for *Pseudomonas* species and *Salmonella typhimurium*, type A; the curve for the others being type C1. The curve with the largest shoulder represents that obtained for *Micrococcus radiodurans*. The data are from Krabbenhoft et al. (1967).

## The $D_{10}$ Value

If one makes the assumption that the exponential relationship is valid, then the probability of having a surviving organism can be

**Figure 10.12.** The different forms of survival curves of bacteria in response to increasing dose of gamma irradiation. *(Reproduced from Ley 1973)*

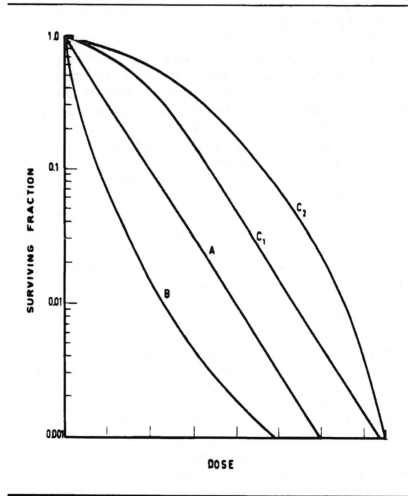

**Figure 10.13.** Demonstrating the different radiation resistances of bacteria to gamma radiation and the shape(s) of the survival curves. Key: 1 = *Pseudomonas* sp.; 2 = *S. typhimurium;* 3. *Strep. faecium* A21; 4. = *B. pumilus;* 5. *Cl. botulinum* type A; 6 = *M. Radiodurans. (Reproduced from Ley 1973)*

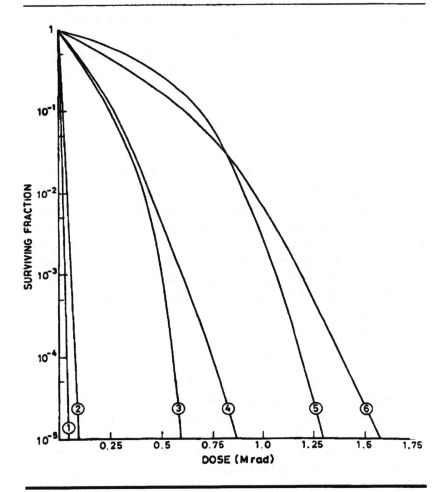

calculated from these survival curves. This point is especially critical. It is critical to the radiation sterilization process to characterize the curve so that the statistical accuracy of the slope provides the assurance that extrapolation to higher doses is accurate.

The $D_{10}$ value, sometimes referred to as the decimal reduction dose, is obtained from such a curve. This value is defined as the dose that will reduce a given population by a factor of 10. It is determined from the linear part of the curve by calculating the dose required to reduce the surviving fraction by one log. The linear plot is expressed mathematically as

$$Log_{10} \frac{N}{N_0} = -kD$$

where $N$ is the number of cells surviving a given dose $D$, $N_0$ is the initial number of viable cells, and $k$ is a constant equal to the slope of the curve. For example, the $D_{10}$ value for spores of *Bacillus pumilus* irradiated in air is 1.7 kGy. Therefore, a dose of 25 kGy would give an inactivation factor of $10^{14.7}$. In practice, this means that if articles were contaminated each with 100 such spores and then treated under similar conditions with 25 kGy, an examination of $10^{12.7}$ articles should reveal that one organism has survived.

## Difference in Resistance Between Species

From all of the above, it can easily be seen that the greater the degree of contamination, the higher the radiation dose that will be required to achieve the desired level of inactivation. Thus, the cleaner the product, the lower the radiation dose and the less radiation effect on the product. Further, the lower the initial bioburden, the greater the likelihood that the variety of organisms will be limited to one or two species, especially when products are manufactured using good manufacturing practices (GMPs).

Through a series of papers published in 1956 and quoted by Ley (1973), insight was gained as to how to apply these principles to the commercial use of gamma radiation processing. While these papers were not based on dose/survival curve experiments, they provide information on the comparative resistance of a wide range of bacterial spores, vegetative organisms, yeast, and molds. Darmady et al. (1961) did similar experiments in the United Kingdom, and Whitby et al. (ANSI/AAMI/ISO 1994), did similar work for the tables used by the Association for the Advancement of Medical Instrumentation (AAMI). These papers established the validity of bioburden-based dose selection and set the stage for the adoption of the process worldwide. It can now be seen why it is so important to determine not only the number of initial organisms, but also the species of organism present.

While we have come to expect that the process will deal effectively with even the most resistant species present, defining the sterilization process in terms of the numbers and type of contamination can be complicated by other factors. These include the physiological state of the organisms, the environmental conditions in which the organisms are irradiated, and the particular strain of a species. More recently, the influence of other processes that may have been used to reduce the bioburden of an ingredient have been recognized as having an influence on the selection of the correct radiation dose. The use of suboptimal sterilization methods can reduce the population of a sensitive organism while allowing more resistant ones to multiply. This latter effect will skew the distribution of organisms on which the radiation dose tables are based and can result in the use of too low a radiation dose.

## Factors Influencing Resistance

**Radical Scavengers.** Many compounds, particularly the sulfhydryls, when present in the medium during irradiation, protect against radiation damage even under conditions of direct action. Cysteine, cysteamine, and reduced glutathione are examples of protective agents. The protective process is known as *intermolecular energy transfer*. The inactivation curve will remain exponential, but its slope will decrease severalfold over that obtained without the reagent (Figure 10.14). Tryptone broth, nutrient broth, gelatin, and concentrated bacterial lysates have been shown to be good protective agents against diffusible

**Figure 10.14.** The effect of phosphate ion and sulfhydryl groups on the survival of viruses after different doses of gamma radiation. *(Reproduced from Ley 1973)*

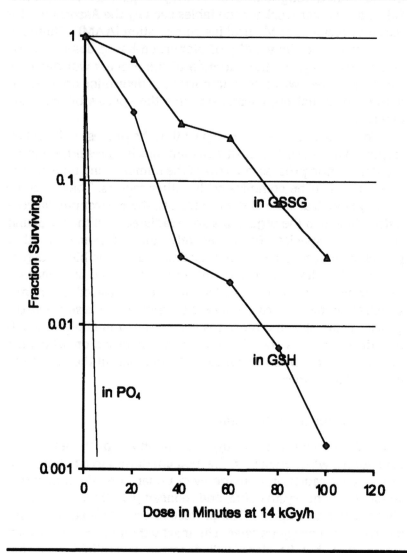

radicals (indirect effects). However, there is no clear distinction between the classes of compounds that, by intermolecular energy transfer, do or do not protect against the direct effects of radiation. Phosphate compounds, on the other hand, tend to act as a sensitizer, making the organism more sensitive to radiation damage (Ginoza 1968).

**Irradiation Conditions.** *Presence of Oxygen.* Targets irradiated in air generally show greater sensitivity than without air. An "oxygen lesion" is not the only type of damage biomacromolecules suffer by direct action. Anoxic conditions do not confer full protection; they show only that inactivation is less efficient than under aerobic conditions. Oxygen present during radiation strongly sensitizes fungi. For *Rhizopus stolonifer*, the dose permitting 0.1 percent survival of sporangiospores is about 3 kGy, compared to about 4 kGy if irradiated in anoxia (Sommer 1973). Substrates present during irradiation may contain protective substances and thereby increase the required dose. If fungus spores germinate and initiate colonies in such a substrate, the population of fungus structures that can serve as propagules may be dramatically increased.

*Heat Sensitization.* Heating followed by irradiation results in greater inactivation than the additive effects of the component treatments. This increased sensitivity to subsequent irradiation is seen as a reduction in the width of the shoulder of the dose-survival curve. In some cases, a change in the slope of the exponential portion may also be observed. If, instead of heating before irradiation, one heats after irradiation, the inactivation is usually much less than the former sequence (Sommer 1973).

A delay between heating and irradiation usually causes all or a part of the sensitization to disappear. Baldy et al. (1970) studied the recovery of conidia of *Penicillium expansum* after heating. He demonstrated that a holding period between heating and irradiation of about three days eliminated the heat-induced sensitivity of the spores. Since the added sensitivity was not lost if the spores were held at 0°C in anoxia, or were treated with respiratory inhibitors, Baldy et al. concluded that desensitization was the consequence of metabolic recovery.

**Effect of Ploidy on Resistance.** Latarjet (1964) demonstrated that haploid yeast cells were less radiation resistant than the diploid form. The diploid form is apparently more resistant because of the second set of chromosomes, thus providing further protection against recessive lethal mutations. Additional duplication of chromosomes, such as that occurring in polyploids, does not appear to provide added radiation resistance (Mortimer 1954). The spores of most common fungi are haploid. Diploidization appears to occur spontaneously and can be induced by irradiation. Stable diploid strains are evidently a rarity in nature (Sommer 1973).

**Repair Mechanisms.** A few fungus spores are known to have the capability to repair potential lethal radiation injuries. One species, *Rhizopus stolonifer* sporangiospores, after a dose of 3 kGy can be reduced in number to 0.5 percent if plated immediately. If plating is delayed by 48 hours, a dose of 5 kGy was required to obtain the same effect. The key factor for this recovery to occur seemed to be a temperature at which the fungus could survive without the initiation of germination. This points to the need of controlling all of the factors surrounding a product that has been irradiated. In other words, irradiation is no substitute for GMPs.

**Host Cell Repair.** Frequently, the question of the repair of radiation damage to the viral nucleic acid by a repair mechanism of the host bacteria has been asked. Latarjet (1964), Winkler (1964), and Sauerbier (1962) have not found such a mechanism. Fluke and Pollard (1955) did report finding a mechanism in T1 phage. However, according to Sauerbier and Latarjet, the repair of ionizing radiation damage, unlike the repair of ultraviolet dimerization, has not been demonstrated.

## Factors Unique to Various Types of Microorganisms

**Bacteria.** Bacteria are perhaps the next most sensitive organisms after mammalian cells and fungi to radiation damage. Viruses are more resistant because of their simpler construction. The work of Ley (1973), Gunter and Kohn (1956), and Alper (1961) demonstrate the great variation in the susceptibility of various species of bacteria to ionizing radiation. Of even greater interest, from a practical point of view, are the changes in this response based on the media in which the bacteria are found and their physical state at the time of irradiation. An appreciation of this can be determined by examining the compilation of values done by G.P. Jacobs, for the PDA Subcommittee on Irradiation Sterilization of Parenterals (Jacobs 1994). Not only is there an obvious difference in inherent resistance between microbial species but also even between strains of the same species. $D_{10}$ values derived from survival curves for 21 strains of *Lactobacillus* ranged from 5 to 14 krad (0.05 to 0.14 kGy) (Dupuy and Tremeau 1961). Using whole egg as medium Comer et al. (1963) examined 18 serotypes of *Salmonella*, and the $D_{10}$ values ranged from 0.5 to 0.7 kGy.

**Viruses.** Viruses appear to have a *theoretical* ability to repair radiation damage. For this to happen, two virus particles must infect the same cell and in some way work together to share genetic information and thus regenerate an intact virus particle (Pollard 1973). This is at best a rare occurrence and may be more theoretical in nature. Viruses have been shown to have the following properties: attachment, injection, protein/enzyme production, cell lysis, interference with other viruses, and eliciting antibodies in an immune reaction (Ginoza 1968). Each of these properties reacts differently to the insult of ionizing radiation; the response is always informative, but not necessarily predictable from virus to virus. It is predictable only within the virus strain.

Every virus conforms to "single hit" kinetics. There are no exceptions to this rule. Variations in response have been observed, but the response reflects an altered viral environment. This can occur if there is a mixture of sensitive and insensitive strains of virus present, or if there is a "clump" of virus particles (Pollard 1978). In these cases, the exponential relationship is not seen until, on the average, the clump has been reduced to one single infectious unit. Another factor may be the presence of protectors in the medium.

*For Single-Stranded RNA or DNA.* Virus preparations that are homogeneous with respect to radiation sensitivity and inactivation do not require cumulative damage. The survival curve slopes show no dependence on the dose rate. Frozen samples (-180°C and -60°C) show the same radiation sensitivity as dried samples irradiated at 30–40°C. However, this is not true if one compares values for viruses irradiated in the liquid state as opposed to frozen or dried viruses. Significant differences have been demonstrated under these conditions. Radiation hits on the capsid or the presence of the capsid do not affect the sensitivity of the nucleic acids it encapsulates. Sulfhydryl compounds and phosphate ions appear to modify the direct action of radiation (Figure 10.14).

*For Double-Stranded DNA.* The target size does not increase in direct proportion to an increase in the size of the virus (Lea 1961). Only double-strand scission or simultaneous base-pair damage can kill, and the efficiency for either of these possibilities is one-twentieth that of single-stranded nucleic acids. The host cell can repair the X-ray lesions occurring in double-stranded DNA, but not those occurring in single-stranded DNA, which is the case for ultraviolet-damaged DNA. The target is a single-stranded segment of the phage genome that specifies the expression of the "early" phenotypic functions required for the initiation of phage vegetative development in the host cell (Benzer 1952; Symonds and McCloy 1958; Barricelli 1956; Epstein 1958; Krieg 1959). Base destruction alone can inactivate viruses. Besides direct damage to the DNA, radiochemical alterations in the protein structures required for host attachment and for injecting DNA into the host cell may be responsible for part of the viral inactivation by ionizing radiation.

The most sensitive property of a virus is its infectivity, because this activity requires all of the attributes of the virus to be functional for the cycle of infection to occur. The lethal effect for viruses with a very simple structure, such as a capsid surrounding a short stretch of RNA or single-stranded DNA, is damage to the nucleic acid. Viruses containing double-stranded DNA have two copies of all of the molecular information needed. Thus, damage to only one strand may only lengthen the time required for the virus to mature. For radiation to be effective, it must either damage both strands or one segment of the DNA that is critical to the operation of the virus. An example would be the segment coding for the special DNA polymerase that is used by the virus. Double-stranded viruses are 10–20 times less sensitive to the effects of radiation than single-stranded viruses.

The problem of viral inactivation does not usually lie in the normal response of a virus to radiation, but may more likely be a result of clumping of the virus in groups in the original preparation. If the viruses occur in clumps, so that 200 or 300 could be thought of as a single unit, then it will be necessary to eliminate every one in each unit before the preparation is considered sterile. This results in a "multiple hit" type of inactivation. It is not a serious problem, because the final slope for inactivation is the same as would be the case if the viruses were distributed singly. It does mean that the "shoulder" effect takes on greater signficance and that a larger dose is required to reach the linear portion of the exponential curve. Measurements must be taken over a wider dose range in order to be sure that the infectivity is being lost.

Radiation also acts on the capsid and other aspects of the virus. The capsid and other structures are usually made of an assembly of many protein subunits, and damage to one or two of these may not be expressed. Even if the damage is confined entirely to the part concerned with attachment, then the virus will be damaged. However, in general, the effects on the capsid are far less damaging to the virus than the effects on the nucleic acid. A good example can be found in the work of Pollard (1973). He studied the ability of influenza virus to agglutinate red cells and found that the line showed no slope. To see an effect, doses that are 20–100 times higher must be used. That is why using gamma radiation for the production of vaccines is so successful. The virus is inactivated without losing its antigenicity.

For radiation sterilization or viral inactivation to be effective, it is essential to have full penetration of the material. This is not a problem with gamma radiation. It is the most uniformly penetrating method of sterilization. It is also important that all of the virus in the product be in the same physical environment. There must be no pockets of untouched virus or partially wet material. Lewis (1961) demonstrated that as water was removed from a bacteriophage preparation, the sensitivity lessened. The fully wet virus was the most sensitive; as dehydration proceeded, sensitivity decreased. This phenomenon may not be universal for all viruses. Sometimes, as in the case for heat, the fully dry virus is more sensitive. What is important is the realization that a preparation that contains a mixture of dry and damp virus may not be uniform with regard to radiation sensitivity. It may appear that there is a resistant population when it is actually due to environmental conditions.

**Fungi.** The inactivation of fungi is similar to bacteria, and those conditions that affect bacterial response to radiation affect fungi in the same way. Any peculiarities are primarily due to differences in growth habit and the more complex cytology, morphology, and life cycles of fungi (Sommer 1973).

The radiation sensitivity of fungus spores is influenced not only by genetic factors but also by the number of cells in a spore, by the number of nuclei per cell, and by ploidy. According to Sommer (1973), germination is not a reliable indication of spore survival in fungi because many irradiated spores die only after considerable germ tube development. The radiation resistance of fungus spores is usually much less, however, than the resistance of important spore-forming bacteria. Usually, a radiation dose that will inactivate bacterial contamination will also inactivate fungal contamination.

The most striking difference in the shape of the survival curves between relatively sensitive and resistant fungi is the width of the shoulder rather than contrasting slopes of the exponential portion. As with viruses or bacteria, a false resistance or false shoulder can be observed if spores have clumped together before they are plated. This can usually be avoided by using a surface-tension reducing agent.

*Fungus Structures Other Than Spores.* Determining the sensitivity of structures other than spores is fraught with problems of technique and interpretation. For example, the nature of mycelial growth precludes the use of bacteriological plating techniques, while the determination of cell numbers is difficult and frequently impossible in mycelial colonies. Young mycelia appear to be the most resistant, followed by conidia and sclerotia. Mature mycelia are the most sensitive (Sommer 1973). Inactivation of 50 percent of a population of young mycelia can be achieved with a radiation dose of 6 kGy. A dose of only 4.7 kGy inactivated the same level of conidia. Sclerotia were only slightly more resistant. In contrast, 50 percent of populations of mature mycelia could be inactivated with a dose as low as 3.5 kGy.

# VALIDATION OF GAMMA RADIATION PROCESSING

## The Validation Program

Prior to beginning a process validation program for radiation sanitization or sterilization, the following should be in place:

- The desired end product should be defined (specifications) in terms of its physical, chemical, microbial, and pharmacological properties.

- Specifications should be established for all raw materials/packaging.

- Assays used to determine end product and raw material purity should be validated.

- The sensitivity of the analytical test methods should permit accurate detection and quantification of the product of interest, as well as impurities, prior to the initiation of radiation processing, so that the effects of radiation on the product or impurities can be determined.

The validation program has three key steps: product qualification, equipment qualification, and process qualification (Hoxey 1991). These three steps will be reviewed in detail below.

## Validation Approaches

The validation of a radiation sterilization process may be performed using one of the three approaches outlined below. The most appropriate approach should be selected, and this selection justified/defended in any submission to a regulatory authority. This chapter addresses only the prospective validation approach. The other approaches must contain similar information.

## Prospective Validation

Prospective validation should be applied when new products or new formulations of existing products are being developed, or when a change of sterilization process (to radiation sterilization) is intended. The validation is conducted, documented, and evaluated, and the validation process and end product are "certified" (by in-house or regulatory authorities as required) *prior to* starting toxicological or clinical trials (if these are required).

## Concurrent Validation

When existing products and/or raw materials and/or supplies (e.g., bottles, septa) are routinely produced using a radiation sterilization process, but no accurate and complete materials, process, and test data exist, then concurrent validation should be applied. This approach should be considered only in situations where there exists a clear history of consistently high-quality product/material/supply production. It would be a wasted effort to attempt to validate an inconsistent "situation."

## Retrospective Validation

Justification for retrospective validation applies when existing products and/or raw materials and/or supplies (e.g., bottles, septa) are routinely produced using a radiation sterilization process for which accurate and complete materials, process, and test data exist. In this case, the validation could be based entirely on historic data. However, note the following:

- It must be established that the product/material/ supplies were consistently treated by the same process (i.e., with the same dose perhaps) and that the radiation sterilization equipment was operated according to relevant GMPs.

- The historic data should be capable of demonstrating that the irradiated product/material/supplies were clinically effective.

- Data related to periods in which different sterilization processes were employed should be used to compare the effects of these different processes on the product/material/supplies. Differences in the product/material/supplies attributable to the different processes used during the data collection periods should be evaluated and reported.

## Documentation

The documentation required for radiation treatment encompasses, but is not limited to, the following:

- Product qualification protocol

- Equipment qualification protocol

- Process qualification protocol

- Reference documents (e.g., standard test methodologies, instrument calibration data, measurement devices used and procedures)

- Personnel documentation (staff qualifications)

- Data gathered as part of the various validation activities (e.g., tests, studies, reference searches)

- A process qualification summary with activity reports, evaluations, overall findings, and conclusions

The ionizing radiation used for sterilization may be generated by an isotopic source, such as $^{60}$Co, electron beams, or X rays generated from electron beams striking a suitable target. There are significant differences among the three source types that affect the validation of the process. For instance, gamma radiation from $^{60}$Co is delivered slowly over a period of minutes to hours. An electron beam machine usually delivers the same dose in a fraction of a second. It is imperative that a product be independently validated for the specific type of radiation that will be used.

## Product Qualification

### *Effect of Gamma Radiation on the Product*

The most important results of product qualification are the *maximum tolerated dose* and the *maximum process dose*. The former is that dose at which the product undergoes unwanted or excessive degradation (i.e., beyond pharmacopeial or product specification limits). The later is arbitrarily set just below the maximum tolerated dose. This ensures that during routine processing, the maximum tolerated dose will not be exceeded. It takes into account the error in dose measurement by the dosimeter.

The *minimum process dose* is determined by the achievable $D_{max}/D_{min}$ ratio. This ratio is determined by the characteristics of the particular irradiator, the product loading pattern, and the product density. Thus, the SAL achieved is determined by the minimum process dose and the sensitivity of the bioburden.

This approach is quite different from that for medical devices, where a minimum dose is established and the maximum dose is set by the characteristics of the plastic. For many plastics, this sterilization window is quite wide. This is not true for drugs.

While the approach to establishing the irradiation dose is the inverse of that used for medical devices, nevertheless, the validation principles are the same. The process is validated to not exceed the maximum process dose, and the SAL achieved is determined by the bioburden and the minimum process dose delivered. This allows the use of the ANSI/AAMI/ISO 11137 tables to determine the SAL delivered.

In these qualification studies, we would encourage the initial use of widely separated radiation doses. The purpose is to answer two questions: Will the product/container withstand any radiation at all? What effects/changes should be sought? A very low dose, such as 5 kGy, will answer the first question. Doses such as 15 or 20 kGy will determine what happens near the "standard" dose ranges. A high dose, such as 30–50 kGy, should provide breakdown products in sufficient quantity to allow any adjustment to analytical detection techniques or analyses.

Since some parenteral products can be included in medical device "kits" or surgical packs, they will be subjected to a second round of sterilization. The reason for this is that medical packs or kits are not usually assembled under cleanroom conditions. Therefore, they need to be sterilized to treat the exterior surfaces of the presterilized components in the kit as well as any nonsterile components.

Therefore, while the initial sterilizing dose may be quite low, the product may need to be sterilized at a much higher dose. However, gamma radiation's unique benefit of penetration means that the contents will also receive that sterilization dose. Some drugs will not withstand this treatment; therefore, kits containing these drugs will need to be sterilized by another technology. This technology could be a sterilant gas or a plasma.

New techniques for irradiating drugs are also emerging. These require, for example, that the drug be treated in the frozen state. It is well known that water, which comprises the bulk of a parenteral or oral drug product, is the major source of free radicals. If the product can be safely frozen and thawed, the potential exists to provide a much greater range of terminally sterilized pharmaceuticals. This procedure takes advantage of the fact that any free radicals formed, are trapped in the ice crystal structure. Since they have less freedom to move about, they are more apt to recombine with each other than to react with other molecules.

*NOTE: Include control samples that travel the same route and are handled or exposed to the same conditions as the irradiated samples. Only in this way will the effects of transportation and storage (such as excessive heat, cold, or vacuum) be separable from those due to radiation.*

Products should be qualified for the different types of ionizing radiation technologies where appropriate or needed. It is unnecessary to qualify a product for different facilities using the same technology, provided the dose delivered can be kept within the specified $D_{max}/D_{min}$ limits. Within technologies, dose rate considerations are unlikely to be a factor; however, if different results are obtained between routine processing and the initial experimental work, this should be reinvestigated.[3]

By carefully selecting the maximum process radiation dose ($D_{max}$), it is possible to limit the amount of product degradation that will occur. It may be possible to select a dose at which no radiation-induced changes in the analytical profile can be detected. The minimum radiation dose ($D_{min}$) will be determined to a large extent by the total package configuration and the limitations of the irradiator. Nevertheless, the desired goal is to obtain a $D_{max}/D_{min}$ ratio as close to 1 as possible. This will maximize the SAL while decreasing product degradation. The radiation sensitivity of the bioburden, as contained in the product, should be confirmed first. This must be done to detect any radioprotective effect that the product might have on the bioburden. It may also show heightened sensitivity, which could be used to reduce the radiation dose or increase the achievable SAL. Once the radiation sensitivity of the bioburden has been confirmed, then the total radiation process can be validated.

## Effects of Microbial Limits

Radiation processing is especially suited to reducing the bioburden on incoming raw materials. This reduces the bioburden on the manufacturing process and reduces the radiation dose that needs to be applied to the finished product. The microbial limits established for many raw materials require that they be treated by some technology to reduce the natural bioburden to the levels required by the specification. By specifying radiation processing (or any other technology) for every lot of raw materials, consistent performance properties are assured. It would not be unusual for different lots of a raw material from the same or different suppliers to be treated by different methods to achieve a microbial specification. Each method can affect a different property of the raw material, leading to a need to adjust the formulation to meet the product specifications. The technologies available are radiation, heat, steam, gaseous sterilant, or pesticide.

---

3. See Sterways Patent for examples of dose rate effects using gamma radiation.

Each technology has the potential to alter the physical and/or chemical properties of the raw material. This in turn affects not only the behavior of the material in the product formulation but also the properties of the finished product. For example, a disinfectant residue or pesticide could result in allergic-type reactions in certain patients. If a consistent process were used to treat the raw materials, much narrower formulation limits could be used and a more consistent product would be produced. A consistent raw material and a consistent product mean a more cost-effective process (less manipulation, no reworks) and hence a more competitive product.

## History of Raw Materials

From the above, it can be seen why it is important to know what treatments have been applied to the raw materials before subjecting them to radiation processing. If a microbial limit specification exists, there is a very high probability that heat, steam, radiation, or gas has been used to treat the material to achieve the specified microbial limit. Each process will affect the material in a different way. Heat or steam may create degradation products that are more sensitive to radiation than is the original material. EtO will leave residues that are activated by radiation. Prior radiation exposure may cause the product to exceed the maximum dose limit if it is retreated as a final product. Synergistic effects are avoided if raw materials and final product are treated using the same process.

## Knowledge of Bioburden(s)

The requirement for information regarding the bioburden of the product or raw material is another poorly understood area of gamma radiation processing. If the process fails for any reason, it is likely due to an underestimation of the true product bioburden. Often in the rush for bioburden data, sufficient growth time is not allowed. This means that slow growers, or more fastidious organisms, do not show up on a 24–48 hour incubation test. They may not even show up on a 7-day test. A minimum of 14 days is the time of choice; often 21 days may be required. Test procedures are available from AAMI. Any underestimation of the true bioburden means that insufficient radiation will be applied to the product to effect the degree of microbial kill required. Therefore, it is vitally important that good bioburden data be gathered, not only by number but also by species. (Subspecies typing is not usually required.) These data allow a more direct comparison of the radiation resistance in the bioburden of this product with that used in the establishement of the standard ANSI/AAMI/ISO guidelines. Good bioburden data are the foundation on which successful radiation processing is based.

### Formulate/Design for Radiation Processing

Selecting an appropriate dose form (e.g., lyophilized product, microencapsulated product, frozen product, oil-based emulsion, or including radical scavengers or buffers to compensate for pH changes) will improve the ability of the product to withstand radiation processing. The elimination of water (other than that of crystallization) greatly improves the stability of the material to irradiation.

### Test the Product/Product Container

Hand in glove with formulation is product testing. It is important to understand how the product reacts to radiation. Once the basic response of the drug is understood, it should be further tested in its final packaging to detect any potential product-packaging material interactions. There are changes known to occur, for example, in the lubricants used for plastic packaging materials, which may affect the properties or stability of the irradiated product. Knowledge of the product breakdown products will assist in arriving at the best formulation. It must be strongly emphasized that no unique radiolytic products have been found in irradiated drugs. The breakdown products found after irradiation are identical to those found during manufacture of the drug, or are metabolites of the drug.

### Analytical Methods

The analytical techniques must be established and validated. Ideally they should be the same as those currently used to assess the efficacy/safety of the product.

## *Accelerated Aging Studies*

Accelerated aging studies are an accepted means of determining the shelf life of a product. However, it is the belief of the authors that it is too early in the development of this application for such studies to be undertaken. Accelerated aging studies, when initiated, should be done simultaneously with real-time studies. The results could then be used as an indicator of future trends. Frequent early sampling is encouraged to establish the trend. If there is to be an effect, what has been observed is an initial drop in the level of the active ingredient. This is followed by a paralleling of the normal degradation curve.

It should also be mentioned that no unusual compounds have been discovered after products have been irradiated. Degradation products, if found, are typically the initial starting materials, expected breakdown products (as determined from shelf-life studies), and (in some cases) metabolites of the product. Radiation introduces only energy, not new science. Therefore, the only reactions that can take place are those that are in the known pathways.

## Equipment Qualification

The second step is the qualification of the irradiation equipment available for use. Normally this will be done or provided by the operator of the irradiation equipment or by the manufacturer of the equipment. In the event that the irradiation equipment is to be operated by the pharmaceutical manufacturer or raw material supplier, this documentation should already be available. It need only be reviewed to ensure that a particular irradiator/ facility will deliver the dose required within the specified maximum and minimum limits.

Equipment qualification focuses on four discrete topics: suitability of the design, correctness of installation, operation and maintenance, and dose mapping, which determines the $D_{max}/D_{min}$ ratios for different densities of product. The most important outcome of equipment qualification is the determination of the *minimum process dose* while respecting the maximum process dose. This documentation will be a part of the validation program documentation.

It must be ascertained that the radiation facility fulfills three requirements:

1. The ability to process materials within specified dose and time requirements consistently

2. Sufficient processing capacity for the product

3. Operation in a manner consistent with current good manufacturing practices (cGMPs).

The essential items for equipment qualification are well delineated in the ANSI/AAMI/ISO 11137 guidelines. This document can be obtained from AAMI.[4] It must be emphasized that regardless of who owns/operates the equipment, it is the responsibility of the *product manufacturer* to ensure that all appropriate documentation exists and is being followed.

Dummy product is usually used to assess the suitability of the equipment of a particular service firm or in-house facility. There are many different configurations of equipment. Most have been designed to handle a variety of material densities. It is unlikely that difficulties in finding the proper configuration for a product will be encountered.

For gamma irradiation facilities, most $D_{max}/D_{min}$ problems are resolved by altering the product loading configuration within the carrier. For electron beam facilities, the product loading within the exterior carton (i.e., shipping configuration) may need to be changed. Understanding the limits of each technology and determining the product's limitations/packaging during the product qualification stage will greatly reduce any chances of facility/technology mismatches.

---

4. Association for the Advancement of Medical Instrumentation, 3330 Washington Blvd., Suite 400, Arlington, VA 22201-4598, USA; Tel: 703-525-4890 or 800-332-2264.

## Process Qualification

Process qualification for the irradiation of pharmaceutical products should include, but is not limited to, a consideration of the following subjects: the sterilization approach, dose distribution studies, product loading patterns, biological challenge reduction studies, cycle interruptions, and product temperature control.

### Sterilization Approaches

The sterilization approaches are those methods found in ANSI/AAMI/ISO 11137. The approaches are "bioburden based" (method 1), with a subapproach of "species specific bioburden based." The "small lot/infrequent production based" method is the subject of a separate ISO document and is in a trial mode at the present time. One can no longer utilize the old 25 (or 32) kGy dose as the "accepted" sterilization dose without proper validation. To achieve validation, it is necessary to use the ANSI/AAMI/ISO 11137 methods, so why not start with them in the first place. Method 1 is the least expensive method to determine the most appropriate radiation dose. With pharmaceuticals, it is imperative to keep the radiation dose as low as possible to minimize changes in the product.

The most important step is to determine the product bioburden. It is the single, most critical factor in having a validated radiation process. To achieve this goal, sufficient time must be given for microbial growth. Incubations should be carried out for at least 14 days and preferably 21 days. This will allow time for the "slow growers" to appear. In this regard, AAMI has published a reference document (1991) with guidelines for determining the true microbial population. If a quarterly audit fails, it will most likely be a result of an inaccurate determination of the true product bioburden.

Before determining the sterilization dose, the maximum process dose must have been determined from the product qualification program. From this value and the loading pattern, the minimum process dose is measured with appropriately placed dosimeters.

Using the minimum process dose, and the bioburden per unit sample data, one can use the Method 1 table in the ANSI/

AAMI/ISO 11137 guidelines to determine the log kill provided by the minimum process dose. If a 6 or more log reduction in the bioburden is achieved by this $D_{min}$ dose, then a sterile product has been produced. If less than a 6 log reduction has been achieved, then the product possesses "enhanced sterility assurance." The verification step should also be done to ensure that the distribution of radiation-resistant organisms in a product's bioburden is not significantly different from that used for the ANSI/AAMI/ISO dose determinations.

The bioburden-based approach is the basis of the ANSI/AAMI/ISO 11137 Guidelines. In this approach, the process is validated to prove that the product's bioburden is similar in nature to that used for the ANSI/AAMI/ISO calculations. These methods are eloquently detailed in those guidelines. The reader is strongly encouraged to obtain a copy, because an understanding of this approach is necessary before attempting the species-specific bioburden approach. The species-specific bioburden-based approach is particularly relevant where the microbial population is known and in control. It can result in the determination that a significantly lower dose is acceptable to sterilize the product.[5]

The species-specific bioburden approach is particularly suited to the pharmaceutical industry. It is applicable for those products that are very sensitive to radiation and, therefore, have a low maximum process dose. It relates the delivered radiation dose to the most resistant organism in the bioburden population found in the manufacturing area. This population should be significantly skewed in the direction of radiation-sensitive organisms. It is helpful if the product can be manufactured in a Class 100 or better facility. This would result in a much lower dose of radiation being needed to achieve sterilization.

---

5. The reader should understand the critical difference between SAL for medical devices and pharmaceuticals. The SAL for medical devices is the probability of a single *organism* surviving on a device. For pharmaceuticals, it is the probability of finding a contaminated *unit*. For this reason, an SAL of $10^{-3}$ from a media fill cannot be combined with a radiation dose deemed to provide an SAL of $10^{-3}$ to give an overall SAL of $10^{-6}$. The SAL as defined in the ANSI/AAMI/ISO 11137 guidelines is an absolute number. The SAL from a media fill is a measure of the rate of contamination. The two cannot be combined.

The "small lot or infrequent production" method is presently under evaluation by AAMI and ISO. It is an attempt to utilize a radiation dose of 25 kGy ($D_{min}$) to treat a very expensive product produced in small lots (less than 1,000 units), or which is infrequently produced (once per year). The problem with the method at present is that it is very difficult to gather enough statistical data to demonstrate the true bioburden of the product.

It has been suggested in various workshops that one way to overcome this problem is to substitute dummy product for the real product. The dummy product must have all of the characteristics of the real product to allow it to undergo the same manufacturing process(es), and thus be exposed to the same microbial environment and opportunities for contamination. If this can be done, then sufficient statistical data on product bioburden can be obtained to justify the use of a 25 kGy dose. The problem with this approach is that there must be exceptional control on the bioburden of the real raw materials used to make the product. If, for example, the product bioburden were to vary by 3-fold from lot to lot, there is no guarantee that the 25 kGy dose would be sufficient to achieve the SAL claimed for the product.

It often happens that products processed with this approach fail on quarterly audits for this reason. If one goes to the trouble of using dummy product, it makes more sense to work with method 1 or 2. Others have also wrestled with this problem, and the reader is directed to Genova et al. (1987b), who may offer other alternatives to the AAMI guidelines. What needs to be remembered is that regardless of the method used, it must be statistically and scientifically sound and capable of being validated.

## Biological Challenge Reduction Study

A biological challenge reduction is not required when using methods 1 or 2. However, for those products that are bacteriostatic or bacteriocidal, this extra step may be required. Occasionally, it might be used as a means to ensure that the product does not demonstrate a radioprotective effect on the microbial population in the product.

The specifics of the biological challenge selected for the study should consider product lot-to-lot variation in the bioburden (species and number). A worst-case bioburden challenge using a microorganism with the highest $D_{10}$ value occurring in the product or manufacturing area is acceptable.

The biological challenge should be performed at a sublethal dose using one of the ANSI/AAMI/ISO 11137 methods as a guideline. Biological challenge tests may be conducted simultaneously with the dose distribution study. The placement of the biological challenges should be defined in writing. The biological challenges should be located as close as possible to the $D_{min}$ position. They should also be placed as close as possible to any dosimeters if conducted concurrently with dose distribution studies. A minimum of three test runs should be performed. Records of the organism type, $D_{10}$ value, number of organisms, lot number, placement, and growth results should be available.

*Note: Biological challenges are never used during routine processing. The validated process relies solely on the results of the dosimeters.*

## Dose Distribution Determination

A dose distribution study is conducted to determine the distribution of the radiation field throughout the product when processed in the actual production system. This study, usually done only once, identifies the maximum ($D_{max}$) and minimum ($D_{min}$) dose positions for a given product in the product transport mechanism of the irradiation facility. Knowing the exact location of the $D_{max}$ and $D_{min}$ positions permits routine process control monitoring. This monitoring is done with dosimeters (see above). Dosimeters are typically analyzed using an optical density measuring device.

The dose distribution study must be performed according to written procedures, and the results must be documented. A competent irradiation facility quality assurance person routinely plans and conducts these studies. The data will be collated into a dose-mapping profile that identifies the location of the $D_{max}$ and $D_{min}$ positions, the dose ratio, and the expected variation. A dose distribution study must be performed for each product loading pattern and each product size.

## Product Loading Patterns

The manner in which the product is loaded into the irradiator's product transport mechanism is critical to achieving the required $D_{max}$ and $D_{min}$ and, therefore the required microbial kill. A detailed "map," called a loading pattern, of how to place the product in the transport mechanism is a part of the process validation documentation.

## Treatment Cycle Interruptions

For mechanical, safety, or operational reasons, the irradiation treatment cycle could be interrupted. A procedure must be in place to direct the irradiator operator to the appropriate contact person. This person must have standard operating procedures (SOPs) to follow and direct the irradiation facility operator as to how to proceed. These SOPs must be in place to define how the product will be handled (i.e., allowed to continue, to be restarted, or to be rejected). For products that can sustain microbial growth, define the maximum tolerable duration of a cycle interruption and its point of occurrence in the treatment cycle (for example, what to do if the interruption occurs before 50 percent of the dose has been delivered; a cycle interruption occurs; there is a delay in starting the processing of the product).

## Product Temperature Control

Certain products can be sterilized by gamma radiation if they are cooled or frozen before irradiation. One example is insulin (Soboleva et al. 1981; Nordheim et al. 1985). Other products are "naturally" temperature sensitive. For both situations, it will be necessary to have documentation stating the acceptable temperature limits on arrival and the time available for irradiation. It may be necessary to provide cooling of the product before, during, and after the irradiation treatment cycle. The manner in which this is to be done (e.g., dry ice or ordinary ice) and how it is to be loaded with the product must be specified. This type of information will form part of the process validation documentation.

## EFFECTS OF IONIZING RADIATION ON PHARMACEUTICAL PRODUCTS

Many pharmaceuticals have been investigated over the past 30–40 years. They have been irradiated with gamma, electron, and X rays, at different dose rates, and to various total doses. Some antibiotics, such as penicillin, were irradiated to extremely high doses (e.g., 1000 kGy). However, the one common theme is the finding of no unique radiolytic products. This is a very important finding. It states that despite the popular notion of free radicals, they are not so "free" after all. They must conform to the laws of chemistry and physics. Radiation sterilization is not a process to make gold from lead, especially at the energies used for sterilization.

Another important feature of these older experiments is the lack of sophisticated analytical techniques. Many products were examined with only thin layer chromatography (TLC) as the most sensitive analytical method. Thus, it is important to revisit these products using the modern techniques of high performance liquid chromatography (HPLC) and HPLC–mass spectrometry. Other physical measurements such as ultraviolet absorption and polarimetry, were used in these earlier experiments, but should be verified. A very good example of this type of occurrence is the drug lidocaine. It was reported to "lose 5% activity at 25 kGy, but is otherwise unaffected" and "Breakdown products are not toxic" (Pandula et al. 1967). However, from recent personal experience, this is now known not to be true. Even at dose rates of ~1 kGy/hour and a total dose of only 5 kGy, extensive breakdown occurs, generating an unwanted product. The amount produced does not exceed the pharmacopeial limit, but it is much higher than the control.

Still another important point to remember is that all of the key parameters of irradiation that have been discussed in this chapter are frequently not recorded. Thus, it is very difficult to compare the results obtained from one laboratory with those of another. It also explains some of the confusion regarding the successful application of gamma radiation processing to many products. Sometimes the dose is reported in units other than in Mrad or kGy, making it very difficult to relate the information to current work. The moral is, use the literature as a signpost or starting point. It may be right or it may be wrong. There is no substitute for doing the work on your product.

## Radiolysis of Water

The radiolysis of water has been studied over the years by many investigators. Some 28 different species of ions have been identified and characterized to varying degrees. The most important concept to remember is that water should be eliminated from pharmaceutical products as much as possible for the most stable response to radiation. However, even if this is impossible, there are other ways to minimize the effects of the free radicals generated in water. Another important fact to remember is that despite the large number of radicals that have been identified, virtually all react with each other to "self-neutralize" their effect. In addition, the speed with which they react (or their half-life) is extremely short ($10^{-12}$ to $10^{-3}$ sec), rendering the possibility of diffusion to react with the pharmaceutical almost negligible.

The five basic products that are left to react with the pharmaceutical are $H_2$, $O_2$, $H_2O_2$, and the free radicals •H and •OH. The abundance of any one of these products is dependent on the irradiation conditions and the total dose delivered. For example, it has been reported that irradiation of water to 25 kGy produces $10^{-4}$ M $H_2O_2$ (Jacobs 1991). An excellent treatment, if somewhat dated, appears in Antoni (1973). The reader is referred to this text for details that are beyond the scope of this chapter.

One of the more obvious ways to negate the effects of water on a pharmaceutical product is to treat the product while in the frozen state (Zagorski 1968). One of the critical parameters to develop for this is the freezing process. It is important to squeeze the molecule out of the ice crystal. In this way the radicals generated within the crystal lattice structure of the water will react with each other and not with the pharmaceutical (Soboleva et al. 1981). This technique has been successfully applied to the irradiation of sera for cell culture work. It has not been successful as a means to treat human blood for readministration.

## Color Centers

The deposition of energy can most easily be observed through a phenomenon known as "F centers" or color centers (Prasil and Marlin 1991b; Safarov et al. 1980). The irradiation of starch or talc or any other material with a crystalline structure will display this phenomenon. As the total dose of irradiation increases, so does the degree of yellowness.[6] This can also be a time-dependent process requiring the translocation of the energy into the crystal lattice. No change in the physical or chemical properties of the compound will be detected by ultraviolet absorption, gas chromatography, mass spectrometry, or HPLC. It is simply an altered energy state. When the material is placed in a suitable solvent, the color generally disappears, and the recrystallized material regains its natural white color.

When irradiated, glass darkens. This can be removed or reduced by exposure to ultraviolet light or by annealing. Again, it centers around perturbations in the crystal structure surrounding chemical impurities in the glass. By annealing the glass, the crystal structure is relaxed—the energy is released and the glass regains, for the most part, its normal color (Prasil and Marlin 1991a).

## Proteins

There exists in the minds of many the notion that proteins cannot be treated by irradiation without causing significant

6. Yellow is the most common color reported. The intensity and chroma may vary depending on the particular compound. Sodium chloride will turn purple when irradiated (Zagorski 1986).

degradation. This statement is indeed true if one examines some of the early literature where very high radiation doses were used. It also applies to more recent experiences where samples were "sent off to be irradiated" and returned for analysis. Because of the lack of understanding on the part of the experimenter about the nature of commercial gamma radiation facilities and on the part of the contract facility operator of the nature of the product, the proteins actually received excessive doses.

Certain changes may be induced in the conformation of proteins by ionizing radiation, and the changes occurring in their biological function may be studied simultaneously. Accordingly, there is a possibility of favorably influencing their structural conformation by ionizing radiation. In other words, it is possible to modify the structure of proteins and, by doing so, alter their existing biological activity or initiate new useful biological functions.

This concept of the deposition of energy and micro heating of certain areas is most important for sensitive biological molecules. While all of the existing literature indicates that there is no dose rate effect from various gamma-ray sources, with the notable exception of the difference between electron beams and gamma rays, a recent patent teaches that this is not so (Sterways 1993). The dose rate can be a very important part of the process and process validation for sensitive biological molecules. In normal gamma radiation processing, as the product moves through the cell and passes the source, it travels through a number of different dose rate fields.

One only needs to reflect on the controversy surrounding food irradiation and the difficulty in detecting irradiated food to realize that biological molecules, including proteins, can safely be irradiated without altering their primary functions. While small commercial facilities are unlikely to produce enough heat in the products to denature the proteins, with the recent development of very large irradiation facilities, there has been a significant increase in product temperature. In addition, the dose rate has more than doubled. This has the potential to have a significant impact on a protein product in a limited aqueous environment. The localized heat generated by the rapid deposition of energy may be sufficient to denature the protein.

It should be remembered that much of the early work on the irradiation of foodstuffs and biologically active proteins was done in low curie content gamma cells or commercial facilities. This concept was not grasped by those active in the field as the focus shifted to the sterilization of disposable medical products and the requirement to sterilize very large volumes of products in as short a time as possible. Hence, the move to large commercial facilities holding many megacuries of $^{60}$Co. The failure to understand the importance of this principle has enabled one company to obtain a patent on a process (Sterways 1993) that critically controls the dose rate delivered to a product. Other patents also exist on the exposure of products while frozen.

For the sterilization of pharmaceutical products, however, the principle of large product volumes with rapid throughput does not usually apply. Certain products need to begin the sterilization step within 24 hours of finishing the packaging operation. However, the volume of product produced, in terms of cubic feet, is almost insignificant next to that of disposable medical products. Hence, very different radiation conditions can be used. Therefore, it is important to check pharmaceutical products, especially proteinaceous ones, for their sensitivity to dose rate.

## Carbohydrates

Carbohydrates are important as part of the complex liquids used for intravenous infusion as well as their incorporation into drug delivery molecules. Again, different changes will occur in compounds when irradiated as solids or as solutions and the degree of oxygenation of the solutions. Color centers will develop when these materials are irradiated in the solid state, the degree of color depending on the degree of crystallinity.

Several fine articles have been written on the degradation products found when irradiating different sugars (e.g., Gopal 1978). A very interesting article comparing the difference between irradiation and steam sterilization for dextrose demonstrates the greater purity of the product after gamma irradiation. There are no furfurals produced with gamma radiation (Berry et al. 1965).

## Excipients

Excipients cover a wide range of materials, from gums, to talcs, to starches, to clays. In fact, an excipient is any material that can be added to a product to enhance its pharmacological action. What is important is that these materials be free of pathogens and other gram-negative bacteria. There have been numerous studies on these materials. All demonstrate that at the doses required for most applications, there is little or no chemical or physical change to the material (Bor 1982). The major ones to be aware of are a drop in pH for starch and a loss of viscosity for poly(vinyl pyrrolidone) (PVP) and some gums. The degree of change is totally related to dose: the larger the dose, the greater the change.

It is important when using thickening agents or others, such as PVP or carboxymethyl cellulose (CMC) where viscosity is a critical parameter, to start with higher molecular weight material, knowing that there will be some degradation due to radiation. Further, it is important to have a very narrow weight range of the material. Under nonirradiation conditions, a small amount of high molecular weight material will be sufficient for the viscosity needs. However, there is often a large proportion of the material with low molecular weight contributing to the final viscosity. It is this fraction that is lost with irradiation. If the low molecular weight component is of the order of 30–40 percent of the total weight, then a large change in viscosity will be observed on irradiation.

MDS Nordion undertook specific studies designed to illustrate the effects, or lack thereof, of gamma radiation on specific bulk materials. The original materials selected were starch, talc, gelatin, and bentonite (Reid 1993a). These materials are widely used in both the pharmaceutical and cosmetic industries. The effects of gamma radiation on StayRx™, Avicell™, and lactose were examined as they are widely used in tableting (El-Bagory et al. 1993). In summary, no changes in the USP or tableting properties of these materials were found after exposure to 25 kGy of radiation initially or after two years of storage.

Other materials that have been investigated privately, include the parabens, methylcellulose, hydroxymethylcellulose, and povidone. The information that is allowed to be shared publicly is that these materials can be safely irradiated as powders.

## Colors

The use of FD&C colors is one area that is very strictly controlled. All organic colors are susceptible to some breakdown after exposure to radiation. The success of irradiation for products containing organic colors will largely depend on the dose required to treat the material containing the color. The lower the dose required, the less likely negative effects will be found. Nevertheless, appropriate testing is required to determine whether the product and/or the color is affected. Thus, the use of organic colors in capsules, for example, may limit the use of radiation processing for these products. There is none of the above concern for inorganic colors. These materials are essentially stable to irradiation, even at high doses.

## Preservatives

Very few preservatives, especially in a liquid finished product, will survive exposure to radiation (Krabbenhoft et al. 1964; McCarthy 1978). This is an important point to note. Failure to select the correct radiation-resistant preservative will result in the loss of protection of the product from contamination by the consumer. Therefore, if a finished pharmaceutical product containing a preservative is to be sterilized by irradiation the effect of the irradiation treatment on the preservative must be investigated.

## Specific Drugs

In the "modern" age, post-1989, the following materials have been investigated privately, including anesthetics, antibiotics, the parabens, methylcellulose and hydroxymethylcellulose, povidone, mastitis products, and some narcotics. Some have been treated as solutions; others as the dry or lyophilized powder (Patil 1984, Phillips 1973). Some products have shown changes at higher radiation doses (25 kGy), and some products have shown an unacceptable degree of change. Each product and formulation must be individually tested. The reader's attention is drawn to the excellent work of Jacobs (1985) on the irradiation response of the cephalosporins.

Tables 10.1–10.3 list certain products and a recommendation for a maximum tolerated dose which have been documented in the literature.

*Note: This is not a recommended sterilization dose. That can be obtained only through the use of the AAMI/ANSI/ISO 11137 procedure for gamma sterilization. The tables should serve as a guideline only.*

## Biologicals

Radiation processing is ideally suited to treat very sensitive materials (i.e., biologicals). Because many of these products come from microbial or viral processes, it is necessary to ensure that no viable organisms are present in the product. Radiation is currently the only technology available to achieve this result. Modest doses of radiation will inactivate most viruses, and certainly kill any bacteria present. Other biologicals, such as monoclonal antibodies, may also be irradiated.

## Enzymes

Contrary to popular opinion, enzymes can be successfully irradiated. Information on the irradiation of certain enzymes in solution can be done, although this is less common. The most stable form is as a lyophilized powder (Nordheim et al. 1985).

## Finished Product (Other Than Biologicals)

Finished products (other than biologicals) represents the largest group of products available to radiation processing. The ability to treat a product in its final container, after aseptic assembly, provides the highest degree of sterility assurance possible. If the

**Table 10.1. Safe to Irradiate to 25 kGy**

| Drug | Comments |
|------|----------|
| Sodium Chloride | Safe to irradiate, but some $H_2O_2$ will be generated. The amount is small, $10^{-4}$ M. The assay value for this must be subtracted from that for oxidizable substances, or the latter test will fail. |
| Hydrochloric Acid (10%) | Safe to 25 kGy |
| Isopropyl Alcohol | Safe to 25 kGy |
| Sodium Citrate | Safe to 25 kGy |
| Water for Injection | Safe to 25 kGy (Note: must subtract $H_2O_2$ value from oxidizable substances test or it will not pass.) |
| $Na_2HPO_4$ (anhydrous) | Safe to 25 kGy |
| Edetate Disodium | Safe to 25 kGy |
| Phenol | Safe to 25 kGy |
| Chlorocresol | Safe to 25 kGy |
| $H_2SO_4$ | Safe to 25 kGy |

**Table 10.2. Safe at a Reduced Dose**

| Drug | Comments |
|------|----------|
| Creatinine | Safe below 8 kGy |
| Metronidazole | Compounds with a benzene ring are generally stable to radiation. This compound is expected to be stable below 15 kGy. Be aware that other ingredients, especially the solution for IV use, may require special precautions before irradiation. |
| Amikacin Sulfate | Probably safe below 15 kGy |
| Sodium Bisulfate | Probably safe to 20 kGy |

## Table 10.3. Likely Unstable Unless Irradiated in a Frozen State

| Drug | Comments |
| --- | --- |
| Naloxone HCl | This solution contains methyl and propyl parabens as preservatives. If it were not for these ingredients, the Naloxone would likely be stable to 20 kGy in the liquid state. It will be stable to this dose when irradiated in the frozen state. |
| Methyl Paraben | As the dried powder, but not in solution unless frozen. |
| Propyl Paraben | As the dried powder, but not in solution unless frozen. |
| Nandrolone & Nandrolone Decanoate Phenyl-propionate | Both of these solutions are in sesame oil with benzyl alcohol as the preservative. |
| Benzyl Alcohol | If used as a preservative, it will decompose unless frozen. |
| Ciprofloxacin Lactate | This preparation normally contains benzalkonium chloride as a preservative. Therefore, it must be irradiated in a frozen state and an assay for benzalkonium chloride used. |
| Dexamethasone Sodium | The benzyl alcohol in this formulation requires that it be irradiated in the frozen state. Dexamethasone itself is stable below 15 kGy. |
| Na Metabisulfite | Not stable; it will act as a free radical scavenger. |
| Betamethasone | This product also contains benzalkonium disodium phosphate chloride as a preservative. This ingredient is unstable to radiation unless done in the frozen state. Betamethasone is stable. |
| Benzethonium Chloride | This quaternary ammonium compound will act as a radioprotectant. Thus, it is likely to be destroyed during normal irradiation. It must be irradiated in the frozen state. It may also be necessary to increase the amount in the formulation to account for radiation losses. |

total dose is sufficiently high, then a measure of viral inactivation is also achieved. This cannot be done with gaseous treatments.

## EFFECTS OF IONIZING RADIATION ON PHARMACEUTICAL PACKAGING

For those pharmaceuticals that must be sterile, there is a need to treat either the container, the pharmaceutical, or both. This sterilization step puts additional stress on the plastic and can be the source of many problems, from leaking containers or poorly fitting caps to an increase in the level of extractables (Reid 1993b).

Currently, there are several means to sterilize the inner and outer surfaces of a container or closure when the pharmaceutical is absent, including radiation, sterilant gases, gas plasmas, and hydrogen peroxide washes. However, once the pharmaceutical is enclosed in the container, the choice of methods for sterilization becomes more limited: steam (autoclaving) or ionizing radiation (gamma or electron beam). Many older plastics were unsuited to these methods as they deformed with steam or became yellow or brittle with radiation.

Packaging for pharmaceuticals that will be terminally sterilized by ionizing radiation brings a new set of considerations. Glass, the preferred packaging material for most parenteral pharmaceutical applications, darkens on exposure to radiation and is a fragile container. However, glass will not work for a transdermal patch or some of the other newer forms of delivery devices for pharmaceutical products.

Both heat and ionizing radiation can cause the level of extractables to increase. This is the major concern regarding the use of these containers for injectable pharmaceuticals. These concerns about the reaction of a compound with its plastic container are not new (Landfield 1980; Duggin 1989; Gopal et al. 1973; Charlesby 1960). Plastic resin manufacturers have become very inventive, developing new resin formulations that can withstand both forms of sterilization (Sipos and Adamis 1990; Sparacio and Amini 1987; Guise 1989). In addition, the plastics industry has developed a technology for coating the surfaces of plastics with a thin film of glass (as silicone) to reduce the potential for product/container interaction (Johansson 1993). This can minimize or prevent the migration of extractables from the plastic to the inner surface of the container and thus limit their release and/or reaction with the contents.

With the single exception of the glass coating on plastics, which is not yet widely used, the most important area on which to focus is the extractables. The pharmaceutical industry needs resins with low extractables. While these extractables have so far proved to be nontoxic, in clinical trials (Danielson 1992; Gopal 1978), they are higher than those found without irradiation. This is a general statement of fact based on the resins that have been tested. However, new resins are appearing almost daily and they need to be tested for this property.

The pharmaceutical industry also needs to learn about the lubricants and stabilizers used in the plastics industry. Lubricants and other additives, or their radiation breakdown products, are a source of extractables and have the potential to interact with pharmaceuticals at some point in time. The pharmaceutical industry is presently less knowledgeable about what lubricants or additives are used than they perhaps should be. It is incumbent on both industries to keep each other abreast of their particular needs.

Thus, it is easy to understand why the selection of the correct container for the final pharmaceutical product is so important for product qualification, stability studies, and process validation. The use of controlled environments for the production of the containers and the pharmaceutical will reduce the bioburden. This, in turn, will reduce the radiation dose required to achieve the desired SAL and the potential for radical production and interactions.

In addition to the container, there is also the closure. There are three good sources of information for pharmaceutical septa. The first is a very good study published by the PDA (Kiang et al. 1992). The second is a technical brochure published by American Stelmi (Merceille and Le Gall 1993) and the third is a similar publication by The West Company (Farley and Drummond 1978). Both deal with the effects of gamma radiation on various elastomeric formulas. The work done by the PDA was at higher doses than is anticipated being used for parenteral products.

Certain forms of pharmaceutical packaging encompass what is best termed a "gray area." It involves the intimate coupling of both the pharmaceutical and the device. Examples would be the patch-type drug delivery system used for hormone replacement therapy or for providing a substitute for nicotine. Here the device is both the drug package and, to some degree, the delivery system. The package must have a special type of adhesive that protects the drug during routine handling, yet will stick to the skin without causing dermal reactions. This is a very tough adhesive problem. In addition, while these drugs/ devices are meant to be applied to the intact dermal surface, this may not always be the case. Therefore they must also have some degree of sterility. Because the drug is contained in a viscous base or the adhesive itself, for timing of delivery, gamma radiation is the method of choice for sterilizing these products. However, gamma radiation has the potential to adversely affect the adhesive if care is not taken in selecting the appropriate formulation.

Packaging qualification is an integral part of product qualification. In addition to any and all of the attributes discussed above, it also involves testing for the following[7]:

- Physical changes

- Chemical changes

- Extractables

- Lubricants

- Stabilizers

The needs of the pharmaceutical industry are varied and complex. The importance of the raw materials used in plastics cannot be overemphasized. The molding process is the second

---

7. Include control samples when product is shipped to the irradiator for treatment. It enables transportation/handling changes to be separated from those due to irradiation.

step in a series of steps leading to the manufacture of the product. Only when all of the ingredients are known can the effects of exposure to gamma radiation be estimated with any accuracy. Knowing the potential reactants allows the skillful formulator to add or delete the correct ingredients to offset any potential radiation effects.

It is helpful if information regarding the extractables and their composition can be provided. (A confidentiality agreement may be required in some cases.) Packaging qualification studies include stability or shelf-life determinations. However, it is possible to market a drug with a reasonable shelf life (at least six months) while additional stability data is generated.

## REGULATORY CONSIDERATIONS

### Product Related

#### Material Effects

Presently, there are few regulatory concerns regarding the irradiation of most pharmaceutical raw materials. Raw materials, including excipients, tend to be particularly stable to radiation processing. The product user or supplier must show that the material being irradiated is usable after treatment; that is, any changes induced are manageable from a formulation point of view. "Know your raw materials" cannot be emphasized enough.

Many products are subjected to a variety of processes before they come to the final manufacturing facility. Most dry, finished products (e.g., antibiotics, freeze-dried products) should respond well to irradiation treatment (Soboleva et al. 1981; Jacobs 1982).

Those finished products that are in solution or suspension are less stable. This is due to the fact that water is the major source of free radicals. The potential for radiation to produce minute amounts of breakdown products that may adversely affect the stability and/or toxicology of the product are of concern to regulators. An appropriate guideline to consult on how to address the presence of any impurities is the recent ICH Guideline (ICH 1994). It provides a very convenient "decision tree" for determining a course of action.

However, to date, no unusual breakdown products have been discovered after irradiation (Tsuji et al. 1983). The products found are typically the initial starting materials, known breakdown products (found on stability studies), and, occasionally, the metabolites of the starting material. Radiation can produce different breakdown products from those produced by steam sterilization. This has been demonstrated for dextrose (Berry et al. 1965). These products are less toxic than those produced by steam sterilization.

Radiation introduces only energy, not new science. Therefore, the reactions that take place are those along known or predictable pathways. Radiation, after all, is simply another form of energy that enhances certain chemical reactions. It is the impurities in materials, that normally have been thought to be insignificant or inert, that may be the source of unwanted or unexpected product degradation. This is why it is important to know fully the origin, source, and prior treatment of raw materials.

## Sterility Assurance Levels

The SAL required for terminal sterilization is a 6 log reduction below one surviving organism. For many parenteral products and others coming from an aseptic, barrier, or isolation manufacturing environment, the bioburden may well be one or less. It is perhaps prudent to consider the initial bioburden as one organism per vial. Then the log reduction required is $10^{-6}$. In some cases where the pharmaceutical is very sensitive to radiation, then a dose that will not produce 6 logs of inactivation may be used. However, the product cannot be labeled as "sterile." However, it can be stated to have an "enhanced SAL" based on the number of log reduction achieved.

Where there is a particular organism present with a low radiation resistance, it may be possible to achieve an SAL of $10^{-6}$ with only 2 kGy. While this is very rare, it is not unknown to achieve this level with 8–10 kGy.

It is also important to remember that the SAL obtained from a media fill cannot be added to that obtained by ionizing radiation. The first is a "frequency" number; the latter an absolute number. When the statement is made that the frequency of contamination is 1:1000 or 1:10,000, nothing is said about how many organisms are involved in the contamination of that unit. When the ANSI/AAMI/ISO 11137 document is followed, it relates the number of organisms per unit to the dose required to inactivate (kill) those organisms.

## Pyrogen Assurance Levels

The pyrogen assurance level (PAL) is a concept developed by French researchers working at BioWhittaker at the time (Guyomard et al. 1987). It involves the well-known phenomenon of pyrogen destruction by gamma rays, but not by electron beam. It would appear to be due to lipid peroxidation. While gamma radiation cannot be recommended as a means of depyrogenation, there is still some added "comfort level" achieved knowing that residual pyrogens may be inactivated with this process.

## Viral Inactivation Assurance Levels

The viral inactivation assurance level (VIAL) is the newest concept. It has evolved from the need to provide a level of viral inactivation for the products of all plasma- and biotechnology-derived products. It is similar to an SAL in measurement and concept.

The viral inactivation method should be capable of inactivating both lipid and non-lipid enveloped viruses. For products coming from animal or human cell lines or microorganisms using various tissue culture media, the method should also be capable of inactivating mycoplasmas. It is important to have sufficient viral inactivation data to determine the slope of the curve with good statistical confidence. This will permit extrapolation of the curve to higher doses with a greater degree of confidence. The product formulation and other parameters must be strictly controlled to ensure that the level of viral inactivation will be the same from lot to lot.

## Personnel Related

Gamma radiation leaves no residues and imparts no radioactivity to the product. This makes it one of the safest technologies to use. There are no concerns for personnel contacting irradiated product or working in a radiation processing facility. The latter is very strictly regulated. Radiation facilities are designed so that radiation levels are not detectable above the normal background radiation in the area in which they are sited. An irradiation facility is also a more pleasant environment in which to work, as compared to facilities using steam sterilization.

## Environment Related

The disposal or recycling of medical devices and pharmaceutical containers involves a different set of regulations and regulators. Concerns about the environmental impact of packaging only serve to complicate further the selection of the right plastic formulation. Does one design for recycle or reuse? If the latter, how will molders control the blends of resins? When will pharmaceutical residue levels reach sufficient concentrations to cause surprising reactions? These and many similar questions will play an ever-increasing role in the future.

## CONCLUSION

The authors have attempted here to provide a basic background for the rational application of gamma radiation processing to pharmaceuticals, their raw materials, and packaging. The application of these concepts should assist the reader in achieving the microbial/viral assurance goal desired. It is strongly suggested that the reader contact the nearest gamma irradiation facility for assistance and guidance. They have access to expert assistance for material(s) effects and can assist in the judicious use of the irradiation facility to reduce the cost while achieving the maximum information from an irradiation experiment. Finally, the authors would strongly recommend that this technology be considered as the safest, most economical means of assuring the sterility or viral inactivation of pharmaceutical products or raw materials.

## ACKNOWLEDGMENTS

The authors would like to thank their respective spouses for the long evening hours devoted to the production of this chapter. In addition, Brian Reid would like to acknowledge the support of MDS Nordion; Barry Fairand would like to acknowledge the support of SteriGenics International.

## REFERENCES

Alper, T. 1961. *Mechanisms in Radiobiology*. New York: Academic Press.

AAMI. 1991. *Microbiological methods for gamma irradiation sterilization of medical devices*. Arlington, VA: Association for Advancement of Medical Instrumentation.

ANSI/AMMI/ISO 11137-1994. *Sterilization of health care products—requirements for validation and routine control-radiation sterilization*. Arlington, VA: Association for Advancement of Medical Instrumentation.

Antoni, F. 1973. *The effect of ionizing radiation on some moelcules of biolgical importance*. In Manual on Radiation Sterilization of Medical and Biological Materials, Technical Report Series 149, STI/DOC/10/149. Vienna: International Atomic Energy Agency, pp. 13–36.

ASTM E1205. 1993. *Practice for use of a ceric-cerous sulfate dosimetry system*. West Conshohocken, PA: American Society for Testing and Materials.

ASTM E1261. 1994. *Guide for selection and calibration of dosimetry systems for radiation processing*. West Conshohocken, PA: American Society for Testing and Materials.

ASTM E1275. 1993. *Practice for the use of a radiochromic film dosimetry system*. West Conshohocken, PA: American Society for Testing and Materials.

ASTM E1276. 1993. *Practice for the use of a polymethylmethacrylate dosimetry system*. West Conshohocken, PA: American Society for Testing and Materials.

ASTM E1400. 1994. *Practice for characterization and performance of a high dose radiation dosimetry calibration laboratory*. West Conshohocken, PA: American Society for Testing and Materials.

ASTM E1401. 1994. *Practice for use of a dichromate dosimetry sytsem*. West Conshohocken, PA: American Society for Testing and Materials.

ASTM E1607. 1994. *Practice for the alanine-epr dosimetry system*. West Conshohocken, PA: American Society for Testing and Materials.

ASTM E1650. 1994. *Practice for use of a cellulose acetate dosimetry system*. West Conshohocken, PA: American Society for Testing and Materials.

ASTM E1707. 1995. *Guide for estimating uncertainty in dosimetry for radiation processing*. West Conshohocken, PA: American Society for Testing and Materials.

Baldy, R. W., et al. 1970. Recovery of viability and radiation resistnace by heat-injured conidia of *Penicillium expansum* Lk. Ex Thom. *J. Bacteriol.* 102:514.

Barricelli, N. A. 1956. *Acta Biotheoret.* 11:107.

Bender, E., J. Fritzsche, M. Bar, and W. Nordheim. 1989. Inactivation of Mycoplasmas and Other Bacteria in Calf Serum by Irradiation with Gamma-Rays from Radiocobalt. *Archiv fur Experimentelle Veterinarmedizin* 43(5):783–788.

Benzer, S. 1952. *J. Bacteriol.* 63:59.

Berry, R. J., et al. 1965. *Int. J. Radiation Biol.* 9:559.

Bor, C. 1981. *Gamma radiation sterilization of pharmaceutical active ingredients, adjuvants, and packaging materials*. 2nd Working Meeting on Radiation Interaction, Selected papers—Part 2, edited by O. Brede and R. Mehnert. Leipzig: Academy of Sciences of GDR, Central Institute for Isotope and Radiation Research. pp 444–458.

Brinston, R. M. 1995. The economics of sterilization. *Medical Device Technology* (June):16–22.

Brinston, R. M. 1991. Gaining the competitive edge with gamma sterilization. *Medical Device Technology* (June):28–33.

Charlesby, A. 1960. *Atomic radiation and polymers*. New York: Pergamon Press, pp. 229–236.

Danielson, J. W. 1992. Toxicity potential of compounds found in parenteral solutions with rubber stoppers. *J. Parenteral Sci. Technol.* 46 (2):43–47.

Dessauer, F. 1922. *Z. Physik* 12:38.

Dorman-Smith, V. 1991. Considerations when using ethylene oxide for the sterilization of medical devices. *Medical Device Technology* (June): 42–47.

Duggin, G. 1989. Drug packaging. *Manufacturing Chemist* (February): 37.

Dzieglielewski, J., B. Jezowska-Trzebiatowski, et al. 1973. Gamma radiolysis of 6-aminopenicillanic acid and its derivatives. *Nucleonika* 18:513–523.

El-Bagory, I., B. D. Reid, and A. G. Mitchell. 1993. The effect of gamma irradiation on the tableting properties of some pharmaceutical excipients. *Int. J. Pharm.* 105 (3):225–258.

Epstein, R. H. 1958. *Virology* 6:382.

Farley, J. J., and J. N. Drummond. 1978. *The effect of gamma irradiation on various elastomeric formulations.* Promotional Material by The West Company, Phoenixville, PA.

Fluke, D. J., and E. C. Pollard. 1955. *Ann. N.Y. Acad. Sci.* 59, 484.

Genova, T. F., R. A. Hollis, C. A. Crowell, and K. A. Schady. 1987a. Procedure for validating the sterility of individual gamma radiation sterilized production batches. *J. Parenteral Sci. Technol.* 41 (1):33–36.

Genova, T. F., R. A. Hollis, C. A. Crowell, and K. A. Schady. 1987b. A procedure for supplementing the AAMI b1 method for validating radiation sterilized products. *J. Parenteral Sci. Technol.* 41 (4):126–127.

Gergov, P., et al. 1988. Sterilizing effect of ionizing radiation on mycoplasma contamination of calf serum. *Beterinarna Sbirka.* 86 (8):25–27.

Ginoza, W. 1968. Inactivation of viruses by ionizing radiation and by heat. In *Methods in Virology,* vol. IV, edited by K. Maramorosch and H. Koprowski. New York: Academic Press, pp. 139–209.

Gopal, N. G. S. 1978. Radiation sterilization of pharmaceuticals and polymers. *Radiat. Phys. Chem.* 12:35–50.

Gopal, N. G. S., S. Rajagopalan, and G. Sharma. 1973. Chemical effects of radiation on plastics and pharmaceuticals. *Radiation Sterilization of*

*Medical Products*. Bombay: Report on the Colloquium held at Bhaba Atomic Research Centre, 17–18 August.

Guise, B. 1989. Plastic containers. *Manufacturing Chemist* (July): 32–34.

Gunter, S. E., and H. I. Kohn. 1956. *J. Bacteriol.* 71:571.

Guyomard, S., V. Goury, J. Laizier, and J. C. Darbor. 1987. Defining of the pyrogenic assurance level (PAL) of irradiated medical devices. *Int. J. Pharmaceutics.* 40:173–174.

Guyomard, S., V. Goury, and J. C. Darbord. 1988. Effects of ionizing radiations on bacterial endotoxins: Comparison between gamma radiations and accelerated electrons. *Radiat. Phys. Chem.* 314–6:679–684.

Hoxey, E. 1991. Validation of sterilization procedures. *Medical Device Technology* (June): 25–27.

ICH. 1994. Impurities in new drug substances: Draft consensus guideline. Released for Consultation 30 June at Step 2 of the ICH process.

ICRU Report 14. 1969. Radiation dosimetry: X-rays and gamma rays with maximum photon energies between 0.6 and 50 MeV. Bethesda, MD: International Commission on Radiation Quantities and Units.

ICRU Report 35. 1984. Radiation dosimetry: Electron beams with energies between 1 and 50 MeV. Bethesda, MD: International Commission on Radiation Quantities and Units.

Jacobs, G. P. 1981. The effect of radiation on drugs. *Proceedings of the PMA Seminar Program on Radiation Sterilization.* (November 30): 139–150.

Jacobs, G. P. 1991. Radiation in the sterilization of pharmaceuticals. In *Sterile Pharmaceutical Manufacturing*, vol. 1. Buffalo Grove, IL: Interpharm Press, Inc., pp. 57–78.

Jacobs, G. P. 1985. A review: Radiation sterilization of pharmaceuticals. *Radiat. Phys. Chem.* 26 (2):133–142.

Jacobs, G. P. 1995. A report on $D_{10}$ Values for gamma and electron irradiated microorganisms. Private communication.

Johansson, K. 1993. Overview of recent glass coatings developments: Emphasis on plasma techniques. *Proceedings of the European Conference on Pharmaceutical and Medical Plastics Packaging*, edited by H. R. Skov. pp. 4.1–4.11.

Kiang, P. J., chairperson., et al. 1992. Effects of gamma irradiation on elastomeric closures: Technical Report No. 16. *J. Parenteral Sci. Tech.* Supplement 46(S2).

Krabbenhoft, K. L., A. W. Anderson, and P. R. Elliker. 1964. *Appl. Microbiol.* 12:424.

Krieg, D. R. 1959. *Virology* 8:80.

Landfield, H. 1980. Radiation effects on device and packaging materials. *MD & DI* (May).

Latarjet, R. 1964. In cellular control mechanisms and cancer. edited by P. Emmeot and O. Mühlbock. Amsterdam: Elsevier, p. 326.

Laughlin, J. S., and S. Genna. 1996. *Radiation dosimetry*, 2nd ed. New York: Academic Press, pp. 389–442.

Lea, D. E. 1961. *Actions of radiation on living cells.* London: Cambridge Univ. Press.

Lewis, H. 1961. Effects of hydration on the radiation sensitivity of bacteriophage. M.S. Thesis, University of California at Los Angeles.

Ley, F. J. 1973. The effect of ionizing radiation on bacteria. In *Manual on Radiation Sterilization of Medical and Biological Materials,* Technical Report Series 149, STI/DOC/10/149, Vienna: International Atomic Energy Agency, pp. 37–64.

McCarthy, T. J. 1978. The effect of gamma irradiation on selected aqueous preservative solutions. *Pharmaceutisch Weekblad* 113:698–700.

McLaughlin, W. L., A. W. Boyd, K. H. Chadwick, J. C. McDonald, and A. Miller. 1989. *Dosimetry for radiation processing.* London: Taylor and Francis.

McLaughlin, W. L., and N. W. Holm. 1973. Physical characteristics of ionizing radiation. In *Manual on Radiation Sterilization of Medical and Biological Materials.* Technical Report Series 149, STI/DOC/10/149, Vienna: International Atomic Energy Agency, pp. 5–12.

Merceille, J. P., and P. Le Gall. 1993. *Radiosterilization of rubber stopper for injectable preparations.* Stelmi Technical Article.

Mortimer, R. 1954. Studies on the effects of X-rays on yeast cells of different ploidy. *Radiat. Res.* 1:225.

Nam, J. W. 1983. Assuring good irradiation practice. *IAEA Bulletin* 25 (1):37.

Nordheim, W., S. Brauniger, G. Petzold, J. Reinhardt, M. Bar, and K. C. Bergmann. 1985. Application of ionizing radiation for production of drugs, vaccines, and biochemicals. *Isotopenpraxis* 21 (11):375–379.

Pandula, E. L., E. Farkas, and I. Rácz. 1967. *Effects of radiosterilization on sealed aqueous solutions.* Radiosterilization of Medical Products, Proceedings of a Symposium in Budapest, 5–9 June. Vienna: International Atomic Energy Agency, SM–92/6, pp. 83–89.

Patil, S. F., D. Ravishankar, P. Bhatia, and I. B. Chowdhary. 1984. Chemical effects induced by gamma-irradiated salts in aqueous medium. *Int. J. Appl. Radiat. Isot.* 35 (6):459–462.

Phillips, G. O. 1973. Medicines and pharmaceutical base materials. In Manual on Radiation Sterilization of Medical and Biological Materials, Technical Report Series 149, STI/DOC/10/149, Vienna: International Atomic Energy Agency, pp. 207–228.

Pollard, E. C. 1973. The effect of ionizing radiation on viruses. In *Manual on Radiation Sterilization of Medical and Biological Materials,* Technical Report Series 149, STI/DOC/10/149, Vienna: International Atomic Energy Agency, pp. 65–72.

Polley, J. R. 1962. The use of gamma radiation for the preparation of virus vaccines. *Canadian Journal of Microbiology* 8:455–459.

Pope, D. G., K. Tsuji, J. H. Robertson, and M. J. DeGeeter. 1978. Raw material microbial count reduction via cobalt-60 irradiation. *Pharmaceutical Technology* (Oct): 31–41.

Prasil, Z., and T. Marlin. 1991. Radiation coloration of glass—state-of-the-art. *Beta-Gamma* 4:11–12.

Prasil, Z., and T. Marlin. 1991. Two colours out of one. *Beta Gamma* 2+3:18–19.

Reid, B. D. 1996. *The sterways process: A new approach to inactivating viruses using gamma radiation.* Presented at the International Conference on the Virological Safety of Plasma Derivatives, 20–22 November, in Bethesda, MD.

Reid, B. D. 1993a. *Sterilization of four cosmetic materials: A Canadian study.* Published by MDS Nordion. (Also available from the author.)

Reid, B. D. 1993b. *Gamma radiation processing for pharmaceuticals: Implications of container selection.* Presented to the Society Plastics Engineers, Copenhagen.

Reid, B. D. 1995. Gamma processing technology: An alternative Technology for Terminal Sterilization of Parenterals. *J. Parenteral Sci. Technol.* 49 (2):83–89.

Safarov, S. A., O. L. Grigor'eva, V. T. Kharlamov, V. V. Sedov, and E. A. Tyrina. 1980. Stabilization of nikethamide solution for injection with the aim of sterilization with ionizing radiation. *Pharma. Chem. J.* (USSR) 13 (7):747–750.

Sauerbier, W. 1962. *Virology* 16:398.

Sipos, M., and Z. Adamis. 1990. The effects of radiation sterilization on different types of plastics. *Korhaz-es Orvostechnika* 28(6):164–168.

Soboleva, N. N., A. I. Ivanova, V. L. Talrose, V. I. Trofimov, and V. P. Fedotov. 1981. Radiation resistivity of frozen insulin solutions and suspensions. *Int. J. of Appl. Radiat. and Isotopes* 32:753–756.

Sommer, N. 1973. The effect of ionizing radiation on fungi. In *Manual on Radiation Sterilization of Medical and Biological Materials,* Technical Report Series 149, STI/DOC/10/149, Vienna: International Atomic Energy Agency, pp. 73–80.

Sparacio, D. A., and M. A. Amini. 1987. Effect of gamma radiation on the permeability of ophthalmic preservatives through fluorine surface-treated low-density polyethylene bottles: Influence of radiation on material properties. *Proceedings of the ASTM 13th International Symposium,* Philadelphia, Part II, pp. 688-700.

Sterways Pioneer Inc. 1993. U.S. Patent 5,362,442.

Symonds, N., and E. W. McCloy. 1958. *Virology* 6:649.

Tsuji, K., P. D. Rahn, and K. A. Steindler. 1983. [60]Co-irradiation as an alternate method for sterilization of penicillin g, neomycin, novobiocin, and dihydrostreptomycin. *J. Pharmaceutical Sciences* 72 (1):23–26.

Winkler, U. M. 1964. *Virology* 24:518.

Wyatt, D. E., J. D. Keathley, C. M. Williams, and R. Broce. 1993a. Is there life after irradiation? Part 1: Inactivation of biological contaminants. *BioPharm* (June): 34.

Wyatt, D. E., J. D. Keathley, C. M. Williams, R. Festen, and C. Maben. 1993b. Is there life after irradiation? Part 2: Gamma-irradiated FBS in cell culture. *BioPharm* (July–August): 46.

Zagorski, Z. P. 1968. Low temperature irradiations. In *Radiation chemistry and its applications,* Technical Report Series, No. 84, Vienna: International Atomic Energy Agency, pp. 70–71.

# 11

## *PureBright®* Pulsed Light Sterilization

*Joseph Dunn, Ph.D.*

PurePulse Technologies, Inc.
San Diego, CA

*PureBright®* pulsed light uses powerful, short duration flashes of broad spectrum white light to kill all exposed microorganisms.

A new sterilization method, which carries the tradename *PureBright®*, has been developed by PurePulse Technologies, Inc. in San Diego, CA. The method uses high power pulses of light for sterilization in situations where light can access all of the important surfaces and volume of the product or material. Air, water, and package surfaces are examples of areas of applicability. Other significant uses, such as the treatment of sealed products through the package, become possible when packaging materials that transmit light are used.

In this chapter, this new processing and sterilization method is introduced. Because this is a relatively new method and much of the work presented has not been previously documented, some degree of technical detail will be included, hopefully without sacrificing readability. After the fundamentals of the process are described, a few applications are discussed in detail to demonstrate the potential of the method. A brief discussion of process chemistry, mechanisms, and economics then follows.

## PUREBRIGHT® PULSED LIGHT

The *PureBright®* pulsed light method of sterilization uses very brief, intense pulses, or flashes, of broad spectrum white light to sterilize packaging, medical devices, pharmaceuticals, parenterals, water, and air. High levels of all microorganisms exposed to the light are killed, including bacteria, fungi, spores, viruses, protozoa, and oocysts. Each *PureBright®* flash is similar to sunlight (Figure 11.1, Thekaekara, NASA Technical Report); both are broad spectrum white light with peak output at about 450 nm. They are different in that *PureBright®* is rich in ultraviolet at

**Figure 11.1.** Comparison of *PureBright®* pulsed light and sunlight.

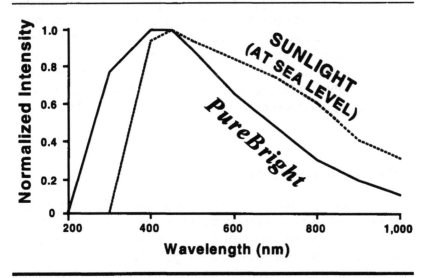

wavelengths less than 300 nm, which are filtered by the earth's atmosphere and therefore not present in sunlight at sea level, and *PureBright®* is much more intense. Each pulse lasts only a few hundred millionths of a second yet is very intense, being about 20,000 times brighter than sunlight. The light contains nonionizing wavelengths from the far ultraviolet, through the visible, to the infrared; approximately 25 percent of the light is in the ultraviolet range.

The sterilizing effects of *PureBright®* are due to the rich, broad spectrum ultraviolet content, the short duration, and the very high peak power of the light pulse. Unique bactericidal effects are produced that are not observed when the same total energy is applied at low intensity in a sustained or continuous mode over a relatively longer period of time. Although the peak power of each pulse of light is very high, the total energy is relatively low and the average power requirement ("wall power") modest. As a result, the process is relatively economical compared to conventional sterilization processes.

The method can sterilize products on-line at high throughput, since one to a few flashes yield high levels of microbial kill and tens of flashes can be produced per second. An important feature of the pulsed light method is its ability to be monitored and controlled. During processing, it is possible to monitor the important treatment parameters electronically in real time on a pulse-to-pulse basis and to verify the level of treatment, adjust key parameters, or shut down the system depending on parametric measurements and preset specifications. Important uses of pulsed light sterilization will be in the manufacture of medical devices and pharmaceuticals and for the production of very high quality water.

## A Simple Demonstration

The high killing effects obtained using pulsed light are easily demonstrated using a simple method in which different concentrations of a test microorganism are treated as small droplets dried onto the surface of medium (Figure 11.2). To produce the plates shown, an overnight *Staphylococcus aureus* shake-culture

**Figure 11.2.** *PureBright®* kill of *Staphylococcus aureus* on the surface of medium.

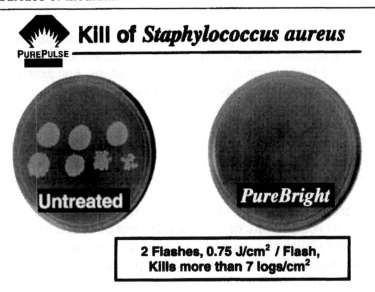

KIll of *Staphylococcus aureus*

PUREPULSE

Untreated

*PureBright*

2 Flashes, 0.75 J/cm² / Flash,
Kills more than 7 logs/cm²

suspension and its 10-fold serial dilutions were dried as a series of 25 mL droplets onto the surface of standard methods agar. The untreated dish on the left shows about 20 colonies/cm² growing in the most dilute spot, and ~2 × 10⁷/cm² in the most concentrated. Two *PureBright®* flashes (0.75 J/cm²/flash) produced no surviving organisms in the identically inoculated dish on the right. This killing effect was produced using a relatively small unit of energy. For example, a watt is equal to 1 J/sec; therefore, a 100-watt light bulb would operate for 0.015th of a second with the amount of energy used to treat the dish on the right.

This simple demonstration of the killing effects of *Pure-Bright®* pulsed light on organisms on the surface of medium illustrates the kinds of effects seen when organisms are sprayed onto surfaces or packaging materials. A single pulsed light flash at an energy per unit area (or fluence) of about 1–2 J/cm² will kill 6 log colony forming units (CFU) of bacterial spores per square centimeter of inoculated surface. Treatment with a few flashes to fluences of 4–6 J/cm$^2$ will kill 7-8 log CFU/cm$^2$ of bacterial or mold spores, and more than 9 log CFU/cm$^2$ of vegetative organisms.

This killing effect is significantly greater than that obtained using conventional continuous ultraviolet generated by low-, medium-, or high-pressure mercury vapor lamps. For example, *Aspergillus niger* is relatively resistant to sterilization using continuous wave ultraviolet light. When an *Aspergillus niger* spore preparation is sprayed onto packaging, continuous wave ultraviolet light produces about a 2.5 to 4.5 log reduction in CFU during the first 6–10 seconds of treatment (Cerny 1977). Longer exposure times do not significantly increase the kill obtained (Figure 11.3); consequently, continuous wave ultraviolet light is recognized as falling short of the effectiveness required for sterilization. By comparison, one to a few flashes of pulsed light applied in a fraction of a second can produce greater than 7 log CFU/cm$^2$ *Aspergillus niger* reductions on packaging. Pulsed light produces much a higher kill in a shorter time and provides the effectiveness required for high levels of sterility assurance.

**Figure 11.3.** Comparison of the effectiveness of *PureBright®* with that of a high intensity conventional ultraviolet source.

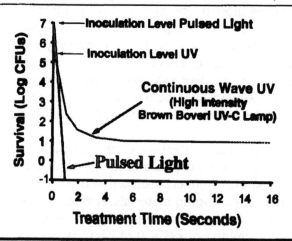

## Application of Pulsed Light

*PureBright®* pulsed light sterilization is applicable to situations and products where light can access all of the important surfaces and volume. Air, water, and many packaging situations meet this criterion and permit the use of pulsed light. For example, Tetra Laval (Lausanne, Switzerland) is actively pursuing the commercial development of *PureBright®* for use in certain packaging and food processing equipment. Tetra Laval is the world's leading designer and manufacturer of aseptic food packaging machines, such as those that produce the Tetra Brik® milk and juice boxes; consumers throughout the world are familiar with brikpak aseptic products. Tetra Laval has tested *PureBright®* pulsed light on packaging materials and foods in San Diego and other test facilities, and will market products incorporating *PureBright®* technology worldwide.

The criterion for light access is met in many other packaging applications, such as the treatment of cups and lids, packaging or products for aseptic or extended shelf-life filling applications, and products through the package. Although pulsed light will not penetrate opaque materials, it is transmitted efficiently through many plastics and may be used to treat a final filled and sealed product through its plastic packaging. The olefins (polyethylene, polypropylene), acrylics, many polyamides and nylons, ethylene acrylic acid, ethylene vinyl acetate, ethylene vinyl alcohol, and similar plastics that transmit in the ultraviolet generally transmit *PureBright®* pulsed light well, and products can be effectively treated through the package. As will be discussed in more detail later, absorption at the wavelengths primarily responsible for the killing effects of *PureBright®* treatment occurs predominantly in aromatic hydrocarbon molecules. This means that plastics with aromatic hydrocarbon backbones, side groups, or additives [i.e., plastics that have a high content of aromaticity in the final material such as polystyrene, polyesters, polycarbonate, polyethylene terephthalate, and common poly(vinyl chloride) formulations] generally absorb too strongly to treat products through the package. Also, labels or printing on a package can obstruct treatment and are best applied after *PureBright®* processing.

Many liquids and solutions can be effectively treated to depths that allow the bulk treatment of a final filled container. The primary limitation on the depth of treatment relates to the degree of aromatic compounds or color present. Color hinders pulsed light transmission, since most colored compounds are, by definition, composed of molecules that interact strongly with light and are also often aromatic in molecular structure. Many colored or absorptive materials, though not treatable at final filled container depths, can be effectively treated as thin layers and then aseptically filled (for example, into *PureBright®* sterilized packages). Since pulsed light is readily transmitted through air, water, saline, dextrose, opthalmic and many other pharmaceutical solutions, high microbial kill and sterilization can be attained.

*PureBright®* is effective for the through-the-package treatment of many finished products and packages as they currently exist. Increasingly, however, the design of the entire manufacturing process, from conception and initial development to production and sterilization, is becoming an integrated effort. Increasingly rare, and rightly so, are instances where the method of sterilization to be used is considered only after the product and package design have been developed. *PureBright®*'s advantages will lead to its wider use as manufacturers begin to integrate pulsed light sterilization into product and package design and development.

## PULSED LIGHT TERMINAL STERILIZATION OF FILLED BLOW/FILL/SEAL CONTAINERS

Many pharmaceutical and medical products are not terminally sterilized; they are prepared by a series of aseptic steps. This group of products is growing in importance and market presence, as many new therapeutic agents cannot be terminally sterilized by conventional means. While recognizing that many products are damaged by current terminal sterilization methods, the U.S. Food and Drug Administration (FDA) has nevertheless proposed that wherever possible drugs and devices should be terminally sterilized. Accordingly, there is a growing need to develop alternative terminal sterilization technologies.

Blow/fill/seal (BFS) manufacturing systems, such as those manufactured by Automatic Liquid Packaging Inc. (Woodstock, IL) produce aseptically filled packages for dispensing pharmaceutical, opthalmic, parenteral, and other liquid products. BFS packaging systems operate with high levels of asepsis and are currently validated by process simulation tests (media fills) to demonstrated high levels of sterility assurance and safety. However, BFS packaging situations represent an ideal model system for demonstrating the ability of pulsed light to act as a terminal sterilization process for aseptically filled and sealed products, thereby providing added levels of protection, safety, and quality assurance. A study was performed in order to investigate the potential for pulsed light to act as a terminal sterilization method for aseptically filled and sealed BFS products (Dunn et al. 1997).

A laboratory *PureBright*® system and static batch treatment arrangement were used in this study to demonstrate the terminal sterilization potential of pulsed light. Ten replicates of four sizes of Water for Injection (WFI) filled and sealed BFS containers were inoculated with 6 log CFU of four biochallenge organisms. Three of the tested organisms were selected because of their demonstrated resistance to well-characterized sterilization processes. *Bacillus pumilus* spores (recommended by the U.S. Pharmacopeia [USP] as the indicator spore for gamma ray or electron beam ionizing radiation sterilization) were chosen because of their known resistance to radiation. *Bacillus subtilus* strain *niger* variety *globigii* spores (recommended by the USP as the indicator spore for ethylene oxide, hydrogen peroxide, and dry heat sterilization) were selected because of their known resistance to alkylating agents, strong oxidizing chemicals, and heat. *Bacillus stearothermophilus* spores (recommended by the USP as the indicator spore for autoclave steam sterilization) were used because of their known resistance to steam heat. In addition, a fourth organism, *Aspergillus niger* spores, known for high resistance to ultraviolet light, were tested.

The results obtained in this laboratory study were highly successful and demonstrated the ability of *PureBright*® to kill

6 log CFU per container in multiple samples of various sizes. Since the publication of the results of this study, commercial *PureBright®* systems have been designed and fabricated for interface with BFS packaging machines for on-line automated treatment of BFS containers. A description of an automated BFS treatment system and the test results obtained during its use are presented next.

## Automated Treatment of Blow/Fill/Seal Containers

The *PureBright®* on-line automated treatment system has three components:

1.   A sterilization tunnel with an integrated conveyor system for receiving and treating containers directly from a BFS production machine

2.   A pulser module that operates and controls sterilization tunnel treatment

3.   A validation module and data logger for verifying and recording the treatment parameters for every flash

The *PureBright®* sterilization tunnel shown in Figure 11.4a–b uses two lamp units with integral reflectors, one on each side of a conveyor, directed at the product and flashing simultaneously to treat cards containing multiple BFS containers of up to 50 mL in volume. The card of containers is carried on the conveyor with the path of travel directly between the lamps. The conveyor servo and *PureBright®* flash rate are synchronized to provide a constant level of treatment. The number of flashes delivered per container can be programmed and controlled, it is generally within the range of 1 to 3 flashes.

The *PureBright®* pulser and validation module are shown in Figure 11.5. The pulser module shown (PurePulse Model PBS-2) can flash four 14-inch lamps at up to 5 times per second (or eight 14-inch lamps at 2.5 times per second); however, it is operated here with two 9-inch lamps. The validation module/data logger is designed to monitor and measure *PureBright®* treatment. With

**Figure 11.4a.** *PureBright*® terminal sterilization tunnel for BFS containers: An overall view of the system.

this system, treatment quality is monitored by interrogating the energy and spectral properties of each *PureBright*® flash to verify that proper treatment was produced.

Before operation, the system was calibrated using a surface absorbing calorimeter (Gentec, ED-200L) with a 1 cm$^2$ aperture to determine the broad spectrum energy of the *PureBright*® flash. With the treatment chamber opened, the total incident fluence measured at the surface of a test 20 mL BFS container is about 1.7 J/cm$^2$ (i.e., broadband light energy at the surface of the container facing each lamp is at least 1.7 J/cm$^2$). Since the package and product are relatively transparent to the light, and the

**Figure 11.4b.** *PureBright®* terminal sterilization tunnel for BFS containers. The main door has been opened to show one of the two opposing lamp/reflector assemblies.

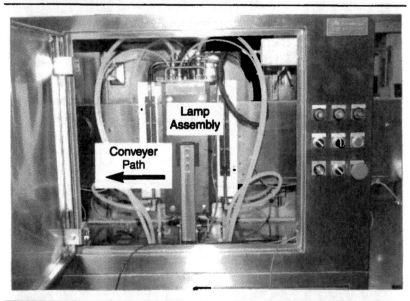

**Figure 11.5.** A *PureBright®* pulsed light system for full-scale, high throughput commercial production. The control module can be placed up to 100 feet away from the treatment site.

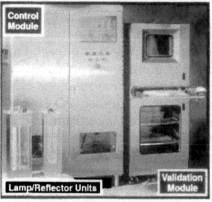

**PBS-2**

**16 kW**

**4 lamps
at 5 Hz**
or
**8 lamps
at 2.5 Hz**

**PureBright System**

simultaneous flash treatment is arranged within a reflective semi-cavity formed by the two opposing reflector units, the total delivered full spectrum energy at the outer surface of each treated container with both lamps flashing simultaneously is increased by multiple reflection of the light. It was calculated by computational analyses to be relatively uniform around the container surface at about 4.5 J/cm$^2$ per flash.

Before discussing the results of the sterility tests, some discussion of procedures is in order. The methods were the same as those used in the earlier published laboratory static treatment study except that for automated treatment, cards containing 20-unit 20 mL containers were inoculated and treated on the conveyor as though they were samples exiting a BFS machine during production. The biochallenge organisms were inoculated into individual 20 mL containers in a BFS card with a syringe, the card placed on the conveyor, and *PureBright®* treated in an automated mode. On exit from the system, the contents of each container were removed in a high efficiency particulate air (HEPA) filtered laminar flow cabinet using a sterile syringe, plated on tryptic soy agar, and incubated at 32°C for 14 days. The microbiological results using this on-line automated system are presented in Table 11.1.

Contaminants other than the inoculated test organism were noted in six samples; these samples were removed from further consideration in the study. The occurrence of these contaminants is noted in Table 11.1 by an asterisk. In the *Bacillus subtilus* sterility assay plates, five instances of contaminating colonies were noted. They were subsequently identified as *Staphylococcus epidermidis* (four instances as a single colony on an assay plate) and one *Bacillus pumilus* (one colony on one plate identified as the *Bacillus pumilus* strain used in another portion of the biochallenge tests but also used routinely at high numbers as a test organism in our facility). One contaminating *Aspergillus niger* colony was recovered on one of the *Bacillus pumilus* sterility assay plates. These contaminants appear to have been introduced after treatment and during the sterility assay procedures. This level of contamination represents a background contamination rate of 6/250 samples or 2.4 percent and is not unexpected in laboratory tests (using laboratory aseptic procedures but with the worker intruding into the aseptic environment) carried out under nonbarrier isolation conditions. It is for this reason that the author chose to assay the contents of the treated containers rather than to attempt to assay both the contents and the container. Such tests are more appropriately performed using controlled aseptic conditions either as media fills using the actual aseptic filling machine environment or as assays performed within a barrier isolation environment.

In the results of the sterility tests shown in Table 11.1, two sterility failures were noted (bold type in the results column). One out of 9 samples inoculated with 8.5 log CFU of *Bacillus subtilis* spores per the 20 mL BFS container volume and treated with 2 *PureBright®* flashes demonstrated survival. The level of survival could not be determined since a swarm of growth was noted on the surface of the survival assay medium. Similarly, 1 out of 10 samples inoculated with 6.6 log CFU of *Bacillus stearothermophilus* spores and treated with a single *PureBright®* flash showed 1 surviving colony. These results verify the results observed in previous laboratory BFS treatment tests. *PureBright®* pulsed light treatment seems to have significant potential for use as a terminal sterilization method for transmissive BFS containers and products.

## *PureBright®* Inactivation Kinetics

Though demonstrating the highly effective nature of *PureBright®* treatment, the inactivation seen in Table 11.1 using the four biochallenge organisms does not provide insight into process kinetics. To generate this insight, the same 20 mL polyethylene WFI filled BFS containers were tested using the automated on-line system under reduced treatment conditions. Reductions in treatment were produced by decreasing the charge voltage for the capacitor energizing the lamp to obtain less than specified normal operating conditions by decreasing the energy of the pulse into the lamp. This reduces the emitted energy per flash; the emission losses are expressed most strongly at the shorter wavelengths of the spectrum (i.e., in the ultraviolet range). Thus, a reduction of the per flash electrical

Table 11.1. PureBright® Test Results Treating 20-Unit Sample Cards of 20 mL BFS Polyethylene Containers Filled with WFI. Automated On-line Treatment Was Simulated Using a PureBright® Treatment System Interfaced with a Conveyor.

| Organism | Inoculation Level (log CFU recovered from controls) | Treatment (number of flashes[a]) | Fraction Positive Results (number positive per number tested) |
|---|---|---|---|
| Aspergillus niger Spores | 6.04 | 1 | 0/10 |
|  | 6.04 | 3 | 0/40 |
| Bacillus subtilis Spores | 6.33 | 1 | 0/10 |
|  | 6.33 | 2 | 0/10 |
|  | 6.61 | 3 | 0/37*** |
|  | 8.47 | 1 | 0/9* |
|  | 8.47 | 2 | 1/9* |
|  | 8.47 | 3 | 0/10 |
| Bacillus pumilus Spores | 7.38 | 1 | 0/10 |
|  | 7.38 | 2 | 0/10 |
|  | 7.38 | 3 | 0/10 |
|  | 8.21 | 1 | 0/10 |

Table 11.1 continued on next page.

*Table 11.1 continued from previous page.*

| Organism | Inoculation Level (log CFU recovered from controls) | Treatment (number of flashes[a]) | Fraction Positive Results (number positive per number tested) |
|---|---|---|---|
| *Bacillus pumilus* Spores | 8.21 | 2 | 0/9* |
|  | 8.21 | 3 | 0/10 |
| *Bacillus stearothermophilus* Spores | 5.48 | 1 | 0/10 |
|  | 5.48 | 3 | 0/10 |
|  | 6.6 | 1 | 1/10 |
|  | 6.6 | 2 | 0/10 |
|  | 6.6 | 3 | 0/10 |

[a]Normal operating conditions employed (i.e., relative energy input of 1 and relative UV content of 100%).

*Each asterisk denotes the presence of a contaminant (not the inoculated challenge organism) in the final sterility assay medium; such contaminated samples were removed from consideration in the results of the study.

Bold type denotes presence of sterility failure.

energy into the lamp causes a reduction in light output. The light output does not decrease uniformly across the pulsed light spectrum, but occurs mainly in the ultraviolet. As will be discussed later, the antimicrobial effects of pulsed light treatment relate strongly to the ultraviolet content of the flash at wavelengths less than about 320 nm; reductions in electrical pulse energy into the lamp produce significant reductions in killing potential. The results of tests using *Aspergillus niger* and *Bacillus pumilus* spores at reduced and below normal operating conditions are summarized in Table 11.2.

The results in Table 11.2 show the effects of treatment using a single treatment flash at reduced energy per flash into the lamp. Each of the reduced energy per flash treatments is quantitated relative to the normal specified lamp operating energy (with the normative condition assigned a value of 1) and relative to the ultraviolet content of the flash. The ultraviolet content of each flash condition was monitored as a broad window from 240 to 280 nm centered at 260 nm; the normative operating measured ultraviolet content value was designated as 100 percent.

The *Aspergillus niger* spore preparation showed no survival in multiple samples when samples inoculated with 6 log CFU were treated using a single flash under normal operating conditions (relative energy input of 1 and relative ultraviolet output of 100 percent, see Table 11.1), nor when operating with a 0.86 relative energy input resulting in a measured relative ultraviolet content of 88 percent (Table 11.2). However, when the *PureBright®* treatment system was operated using a single flash at a relative energy input of 0.72 and 79 percent measured relative ultraviolet content, surviving *Aspergillus niger* was isolated in 2 out of 10 tested samples (Table 11.2, of the positive samples, 1 CFU was recovered after treatment from one container and 5 CFU from the other). Greater reductions in lamp operating energy and measured ultraviolet output resulted in higher levels of single flash treatment *Aspergillus niger* survival. The observed *Aspergillus niger* spore preparation inactivation kinetics are shown graphically in Figure 11.6. Note that the inactivation kinetics observed appear to be biphasic with a sharp increase in dose response between 65 percent and 79 percent relative ultraviolet content.

When *Bacillus pumilus* spores were treated at 6 log CFU per container, no survivors were noted in multiple samples treated using a single flash at relative input energies of 0.86, 0.72, 0.57, and 0.48, which resulted in measured relative ultraviolet content values of 88 percent, 79 percent, 65 percent, and 56 percent, respectively (Table 11.2). Moreover, *Bacillus pumilus* spore inoculations at more than 8 log CFU per container yielded no surviving organisms when treated with a single flash at normal operating conditions (Table 11.1). However, a low level of *Bacillus pumilus* spore sterility failures were noted using the two reduced operating condition treatments tested at this high inoculation level (Table 11.2). A relative energy input of 0.57 producing a relative measured ultraviolet content of 65 percent, resulted in 1 surviving organism found in 9 treated samples. Reductions resulting in a relative measured ultraviolet content of 56 percent, yielded 1 surviving organism in 10 treated samples. The 8 log CFU per container *Bacillus pumilus* spore inactivation kinetics observed are shown graphically in Figure 11.7.

The data in Table 11.2 and Figures 11.6 and 11.7 describing the inactivation performance of single flash treatments on *Aspergillus niger* and *Bacillus pumilus* spore preparations begin to provide some insight into treatment kinetics. Combined with the results shown in Table 11.1, the data suggest that for the 4 biochallenge organisms (*Bacillus pumilus* [ATCC 27142] spores, *Bacillus subtilus* var. *niger* spores [AMSCO/Steris commercial preparation, Mentor, OH], *Bacillus stearothermophilus* spores [AMSCO/Steris commercial preparation], and *Aspergillus niger* [ATCC 16404] spore preparations) a *PureBright®* treatment of 2 or 3 flashes applied to the 20 mL BFS polyethylene WFI container is an effective treatment providing added protection, safety, and quality assurance to the aseptic BFS manufacturing process.

## PULSED LIGHT FOR HIGH QUALITY WATER— CRYPTOSPORIDIUM DISINFECTION

Cryptosporidium is a protozoan parasite that has been recently implicated in multiple, waterborne outbreaks of diarrhea and

Table 11.2. *PureBright®* Test Results Using Reduced Operating Parameters. Normative Operating Conditions Are a Relative Energy Input of 1.0 and a Measured Relative Ultraviolet Content of 100 Percent.

| Organism | Inoculation Level (log CFU recovered from controls) | Relative Energy Input (measured relative UV content | Treatment (number of flashes | Results | |
|---|---|---|---|---|---|
| | | | | Fraction Positive (number positive per number tested) | Count Reduction (inoculation minus mean log survival) |
| *Aspergillus niger* Spores | 5.94 | 0.86 (88%) | 1 | 0/10 | NA* |
| | 5.94 | 0.72 (79%) | 1 | 2/10 | NA |
| | 5.94 | 0.57 (65%) | 1 | 10/10 | 2.3 |
| | 5.83 | 0.48 (56%) | 1 | 10/10 | 1.7 |
| *Bacillus pumilus* Spores | 6.12 | 0.86 (88%) | 1 | 0/10 | NA |
| | 6.12 | 0.72 (79%) | 1 | 0/10 | NA |

*Table 11.2 continued on next page.*

Table 11.2 continued from previous page.

| Organism | Inoculation Level (log CFU recovered from controls) | Relative Energy Input (measured relative UV content | Treatment (number of flashes | Results | |
|---|---|---|---|---|---|
| | | | | Fraction Positive (number positive per number tested) | Count Reduction (inoculation minus mean log survival) |
| *Bacillus pumilus* Spores | 6.12 | 0.57 (65%) | 1 | 0/9 | NA |
| | 5.98 | 0.48 (56%) | 1 | 0/9 | NA |
| | 8.22 | 0.57 (65%) | 1 | 1/9 | NA |
| | 8.22 | 0.48 (56%) | 1 | 1/10 | NA |

*NA = Not applicable; mean log survival values are shown only for those instances in which surviving organisms were recovered from a majority of the samples.

**Figure 11.6.** *Aspergillus niger* spore preparation inactivation kinetics using a single *PureBright®* flash.

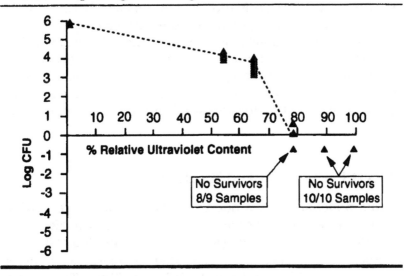

**Figure 11.7.** *Bacillus pumilus* spore inactivation kinetics using a single *PureBright®* flash.

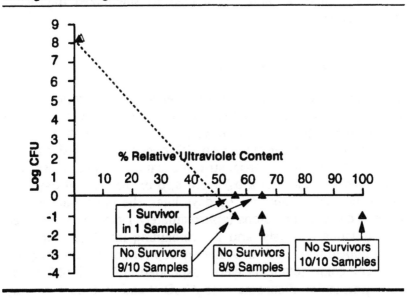

associated mortalities. *Cryptosporidium parvum* oocysts are notoriously resistant to disinfection, especially in the treatment of drinking water by chlorination or other means. Because of the emerging threat this organism poses to international water supplies, tests were conducted to determine the potential for pulsed light to disinfect waterborne suspensions of *Cryptosporidium parvum* oocysts. Oocyst preparations and in vivo mouse infectivity assays were performed by Michael Arrowood and Long-Ti Xie, at the Centers for Disease Control and Prevention, Atlanta, GA.

## Oocyst Preparation and Treatment

*Cryptosporidium parvum* oocysts (IOWA isolate) were purified from calf feces via discontinuous sucrose and microscale cesium chloride gradient methods (Arrowood and Sterling 1987). The oocysts were washed and suspended in phosphate buffered saline (PBS, 0.01 M, pH 7.2) at $10^6$/mL and $10^7$/mL. The suspensions were *PureBright®* treated as 1 mm thick fluid volumes confined between two quartz plates using 1, 6, or 12 flashes at 1 J/cm$^2$ per flash. The suspensions were recovered from the treatment chamber and transferred to siliconized microcentrifuge tubes.

## In Vivo Infectivity Assays

The control and treated oocyst suspensions were concentrated by a single-step centrifugation (16,000 × g, 3 min) and resuspended to $4 \times 10^5$ or $4 \times 10^6$ oocysts/mL. This concentration was chosen to allow delivery of $10^5$ or $10^6$ oocysts (per 25 μL aliquot) to neonatal mice via orogastric intubation. Neonatal BALB/c mice (Jackson Laboratories) were approximately 4–6 days old when inoculated. Six and $^1/_2$ days later, the mice were euthanized. The terminal colons (approximately 1 cm) were removed for individual analysis, and the remaining intestinal segments were pooled from each treatment group for composite analysis.

The colon segments were suspended in 400 μL of 2.5 percent potassuim dichromate ($K_2Cr_2O_7$) and homogenized by application of a disposable wooden applicator and extensive vor-texing. The pooled intestinal segments were suspended in 20 mL $K_2Cr_2O_7$ and homogenized in a VirTishear® tissue homogenizer for 40 sec at a power setting of 70. Oocyst quantitation (a mea-sure of infection success) in aliquots of the colon and pooled in-testinal homogenates was performed by flow cytometry analysis as previously described (Arrowood et al. 1995). Briefly, the sam-ples were vortexed and allowed to stand approximately 15 sec to allow large debris to settle. Supernatant aliquots (200 $\mu$L) were overlaid onto microscale sucrose gradients prepared by layering 750 $\mu$L of a 1.103 specific gravity sucrose solution beneath 750 $\mu$L of a 1.064 specific gravity sucrose solution in 2 mL mi-crofuge tubes. Each tube was centrifuged at 1,000 × g for 20 min. The interface between the 2 sucrose solutions was then collected, washed by centrifugation (16,000 × g, 3 min) with saline, and suspended in 300 $\mu$L PBS supplemented with 0.1 percent bovine serum albumin. The partially purified stool concentrate was in-cubated for 30 min at 37°C with 5 $\mu$L of an oocyst-specific monoclonal antibody conjugated with fluorescein isothio-cyanate (OW50-FITC, 1/50 dilution in PBS). Samples were ad-justed to 600 $\mu$L with PBS, stored at 4°C, and protected from light until analyzed by flow cytometry.

The samples were evaluated by using a 102 sec sampling in-terval (approximately 100 $\mu$L) with logical gating of forward and side light scatter and OW50-FITC fluorescence signals on the flow cytometer. Data files were stored on floppy disk and sub-sequently analyzed with software provided with the FACScan (LYSYS II, Becton Dickinson). Each experimental run included positive and negative controls that were used to calibrate the re-gion settings necessary to discriminate the labeled oocyst popu-lation from background debris.

## Results and Discussion

Oocyst recoveries from mouse colon segments and from pooled intestinal homogenates are presented in Table 11.3. No evidence of parasite colonization was observed in any of the mice receiv-ing oocysts treated with pulsed light. All three treatment levels

Table 11.3. Results of In Vivo *Cryptosporidium parvum* Oocyst Infectivity Assays. Oocysts at $10^6$ and $10^7$/mL were Pulsed Light Treated at 1 J/cm$^2$/Flash. Values Indicate Recovered Oocysts per 100 mL of Mouse Colon Tissue.

| Pulsed Light Treatment | Treated Oocysts/mL | Inoculated Oocysts/Mouse | Mice per Group | Mean Oocysts in Colon | Oocysts in Pooled Intestines |
|---|---|---|---|---|---|
| 1 Flash | $10^7$ | $10^6$ | 5 | 5* | 9* |
| 6 Flash | $10^7$ | $10^6$ | 7 | 3* | 10* |
| 12 Flash | $10^7$ | $10^6$ | 4 | 4* | 10* |
| Control | $10^7$ | $10^6$ | 5 | 1188 | 510 |
| 1 Flash | $10^6$ | $10^5$ | 4 | 3* | 14* |
| 6 Flash | $10^6$ | $10^5$ | 4 | 3* | 24* |
| 12 Flash | $10^6$ | $10^5$ | 4 | 2* | 11* |
| Control | $10^6$ | $10^5$ | 4 | 2119 | 441 |

*Values below 25 are below background (i.e., negative for oocysts).

(1, 6, or 12 flashes at 1 J/cm$^2$/ flash) resulted in significant oocyst inactivation.

Estimates of the minimum infectious dosage necessary to infect 50 percent of a group of mice (MID$_{50}$) are approximately 79 (Finch et al. 1993) and 100–500 (Ernest et al. 1986). Unpublished data from Dr. Arrowood's laboratory indicate the MID$_{50}$ is between 50 and 100 oocysts (personal communication). The lack of detectable infection in the neonatal mice indicates significant oocyst disinfection activity ($\geq$ 3–6 log reduction) even at the minimum 1 J/cm$^2$ treatment level.

## Summary

The initial results demonstrate that pulsed light treatment effectively eliminates the infectivity of high concentration suspensions of *Cryptosporidium parvum* oocysts when assayed using in vivo mouse infectivity assays. Subsequent tests have confirmed the high degree of effectiveness of pulsed light against *Cryptosporidium parvum* oocysts in further in vivo and in vitro assays. In addition, tests performed to titrate the effects of pulsed light at lower treatment levels show the method is highly effective against *Cryptosporidium parvum* oocysts, even at relatively low doses.

## A Pulsed Light Point-of-Entry Water Treatment System

These previous results provided the initial impetus for the development and testing of a pulsed light water treatment system. The design concept was a point-of-entry system capable of producing high quality water free of potential protozoan, bacterial, or viral pathogens. Pulsed light dose response tests were conducted using multiple strains of a wide range of potentially pathogenic waterborne organisms, including *Cryptosporidium parvum*, *Escherichia coli*, *Salmonella* sp., *Klebsiella* sp., poliovirus, and rotavirus, in order to determine appropriate treatment parameters. For use in follow-on in-flow treatment trials, tests were also performed using *Bacillus pumilus* (ATCC 27142) spores to index the kill of the tested pathogens to this nonpathogenic surrogate.

These studies led to the design and fabrication of the 4 gal/min (5 gal/min maximum peak flow) point-of-entry *PureBright®* water treatment unit shown in Figure 11.8. This unit is currently undergoing extensive testing on an international basis. The first use of this unit will be for restaurant applications; the system is relatively inexpensive and can be adapted for in-home use. The pulsed light system will serve as an in-line, on-demand unit for treating municipal or groundwater prior to entry into homes or

**Figure 11.8.** A *PureBright®* pulsed light 4 gal/min point-of-entry water treatment system.

restaurants to enhance and assure water quality and safety. The technology can also be easily scaled to higher flow rates for industrial or municipal applications. The use of pulsed light in systems for the production of ultrapure water for pharmaceutical or electronics manufacturing is also under development.

## SOME COMMENTS ON MICROBIAL INACTIVATION MECHANISMS

### The Peak Power and Broad Spectrum of Pulsed Light Are Unique

*PureBright®* pulsed light works so effectively for killing high levels of a broad range of microorganisms because of its very high peak power and the broad spectrum nature of the light flash. The pulsed light treatment parameters important to achieving sterilization are the spectral content of each flash, the energy of each flash (or fluence in J/cm$^2$/flash), and the number of flashes delivered. For sterilization applications, it is important to keep the flash number low and the power (energy per flash) high. Therefore, the two most important treatment parameters for sterilization are the broad spectrum ultraviolet content of the flash and the total light energy per flash. One or both of these characteristics of the *PureBright®* pulsed light treatment distinguish it from other sources, such as conventional ultraviolet systems (i.e., low-, medium-, or high-pressure mercury vapor lamps and eximer lasers).

### *PureBright®* Ultraviolet Effects

*PureBright®* pulsed light is a rich ultraviolet light source. The ultraviolet portion of the pulsed light spectrum contains those wavelengths that are most responsible for the high killing efficiency of pulsed light (Dunn et al. 1990). *PureBright®* light, unlike some conventional ultraviolet sources, has a continuous spectrum well into the ultraviolet range. It is a rich source for all wavelengths greater than 200 nm that interact strongly with biological matter and are detrimental to cell growth and division.

More than 50 years of research and literature relate to the biological effects of ultraviolet light. In fact, modern molecular biology owes much of its roots to some of the findings and understandings originating from work in the 1940s, 1950s, and 1960s on the effects of ultraviolet light on cells. Many excellent reviews are available; the reviews by Jagger (1967, 1985) remain outstanding sources and reviews of ultraviolet light photobiology for students and researchers alike. It is important to draw on this work and understandings, rather than repeat them. However, most of that work was performed using relatively low power, monochromatic sources and long exposure times and does not address the effects on microorganisms of broad spectrum ultraviolet light. To the author's knowledge, PurePulse's work with *PureBright®* pulsed light systems represents the first use and examination of the biological effects of very high power pulses of broad spectrum ultraviolet-rich light.

## Molecular Interactions of Ultraviolet Light

The wavelengths of light included in the *PureBright®* spectrum are classically referred to as "nonionizing wavelengths." The intent of this classical definition is to distinguish high energy wavelengths, such as the photons present in X-ray or gamma sources, from more benign, lower energy photons, such as those present in sunlight or ultraviolet light. The basis for this classical definition is that sunlight or ultraviolet light wavelengths do not have sufficient photon energies to penetrate the window of, or ionize the detector gases in, a Geiger counter tube. In fact, the wavelengths of light in *PureBright®* do not have sufficient single photon energy to ionize atoms, but are instead only absorbed by specific resonant molecular bonding structures. The wavelengths of light in a *PureBright®* flash cannot, for example, interact with water molecules to produce hydroxyl radicals, which are produced by and represent the primary mode of action of more energetic ionizing photons.

A molecule can absorb light quanta only of specific wavelengths. The absorbed photon's wavelength and energy must be equal to or greater than the energy required to perturb an electron, raising it to an orbit more remote from the atomic nucleus than the ground state. The absorbed photon energy is then part of the molecule, which is said to be "activated" or in an activated state. It is this activated state that can lead to intermolecular or intramolecular rearrangements and chemical change.

For the wavelengths present in *PureBright®* pulsed light and for the molecular structures in biological systems, the primary and predominant absorption occurs in aromatic (or alternating single and double carbon bond systems in conjugated polyaromatic hydrocarbons or benzene-based molecules) carbon-carbon double bond systems. In structures of this type, the electrons in the bonds formed by overlap of carbon p orbitals are delocalized and assigned to the entire molecular structure rather than to specific interatomic bonds; this lowers the bond energy (and raises the wavelength of absorption) to a degree roughly proportional to the degree of aromaticity. When ultraviolet light is absorbed in biological systems, it is an electron in such a $\pi$ electron system that is activated or excited to a higher energetic state. The half-life of this transient excited state (or "singlet") is very short. It either loses the excess energy (via fluorescence) and falls back to the ground state, transitions to other similarly excited yet unstable states (such as "triplet," metastable, intermolecular "cage," or intramolecular charge transfer states), or rearranges to produce chemical change within half-life time scales of nanoseconds to fractions of a millisecond.

Thus, the aromatic nucleotide bases in DNA absorb broadly in the ultraviolet range, with an absorption maximum at about 260 nm. It is this absorption that has been primarily associated with cell reproductive death. For example, the ultraviolet action spectrum for microbial death generally peaks at about 260 nm. This association of cell death with changes in DNA, such as the production of thymine dimers, primarily relates to the sensitive nature of DNA to ultraviolet light and the profound effects small changes in DNA structure exert on a cell's reproductive viability. However, ultraviolet absorption spectra in general (and the ultraviolet absorption of DNA specifically) are broad, and the killing effects of ultraviolet light are expressed only slightly less efficiently at wavelengths above or below 260 nm. For example, the killing effects of ultraviolet light at 280 nm are, on a log scale, about 0.3–0.5 of those observed using 260 nm ultraviolet light. DNA absorption and the formation of pyrimidine dimers in double-stranded DNA has been observed at wavelengths up to 365 nm (Tyrrell and Webb 1973). In addition, there are many other aromatic structures within a cell, besides DNA, that absorb ultraviolet strongly to produce a resulting chemical change. Some of these other events may function synergistically with DNA specific events. Investigators of near-ultraviolet effects in bacteria have observed greater levels of inactivation using polychromatic rather than monochromatic treatments or sources and this synergy has been at least partially attributed to damage to repair systems and inhibition of repair (Tyrrell 1973; Tyrrell and Peak 1978).

## Kinder, Gentler Chemistry— No Measurable Oxidative Cascade

As discussed above, the consequences of *PureBright®* pulsed light treatment are predominantly chemical changes in aromatic or conjugated carbon-carbon double bond molecular systems. From a photochemical viewpoint, these chemical changes are classified into two broad categories: "direct" and "oxidative cascade" events and effects. The primary event of both pathways is the same (i.e., the absorption of light in and the activation of carbon-carbon double bond molecular systems). For the "direct" ultraviolet events and effects, the activated carbon-carbon double bond system rearranges in some way, either within the molecule or with a proximate molecule. The "oxidative cascade" effects are a particular subset of intramolecular bonding events in which an activated carbon-carbon double bond system reacts with a proximate oxygen molecule. The first product of this reaction is a peroxide formed by the insertion of the oxygen molecule into the activated site. The peroxide so formed is usually unstable and decomposes to yield a variety of potential final products, including aldehydes, ketones, and carboxylic acids.

This range of potential pathways and products is lumped together under the "oxidative cascade" name, and separated from the direct effects category, because of its importance from a product and historical perspective. For example, there has been a long history of the use of ultraviolet light for the treatment of meats. Product color changes and/or changes in lipid oxidation have been noted from such treatments. These changes are related to oxidative cascade events and effects during and after ultraviolet treatment. The author has not been able to measure similar changes when meats are treated with *PureBright®* pulsed light. This is in contrast to the clear evidence of color or lipid oxidation changes observed and normally associated with the treatment of meats with conventional ultraviolet systems.

This difference between *PureBright®* and conventional ultraviolet light treatments has led to the following hypothesis: the short pulse times that *PureBright®* pulsed light employs (for most applications on the order of a 100 microseconds in duration), combined with the short half-lives of excited $\pi$-bonds (in the range of nanoseconds to fractions of milliseconds), prevent efficient coupling of free or dissolved oxygen into the reaction sequence. Only oxygen immediately available to the activation site during a $\mu$sec-msec time window could potentially interact. This amount of oxygen would be relatively small compared to that available during conventional, continuous wave ultraviolet treatments, where (it is proposed that) the relatively long exposure times (many seconds) continuously produce activation sites to which oxygen can freely migrate and react. Ancillary to this hypothesis is the suggestion that oxidative events during *PureBright®* treatment as normally employed are "diffusion limited" and are of no consequence at low flash numbers (normal treatments generally use 1–3 flashes). This nypothesis is supported by several other observations that do not need to be detailed here. The author wishes to reemphasize the observation that oxidation reactions and "oxidative cascade" chemistries seem to be rare with *PureBright®* pulsed light treatment.

### *PureBright®* Peak Power Effects: Fluence Per Flash Is Important

The importance of the peak power, or energy delivered per unit area during each flash (fluence in $J/cm^2/flash$, as distinguished from the total fluence delivered over some number of flashes) is seen in the results of bacterial and/or fungal spore survival tests in which the ability of different pulsed light treatments to produce sterilization levels of kill is evaluated. For example, the sterilization capabilities of several different treatments on *Bacillus pumilus* spores inoculated onto packaging material is shown in Table 11.4.

The test results in Table 11.4 were generated using *Bacillus pumilus* (ATCC 27142) spores sprayed onto 6 $cm^2$ (2 cm by 3 cm) samples of packaging material with an air-shrouded ultrasonic spray nozzle interfaced with a syringe pump (SonoTek, Poughkeetsie, NY). For the samples here, 7 log CFU of spores in a 5 µL volume were sprayed to cover a circular 1.75 $cm^2$ area

### Table 11.4. Effect of Different Pulsed Light Treatment Conditions on Fraction Positive Assay Results Using *Bacillus pumilus* (ATCC 27142) Spores Sprayed at 7 log CFU per Packaging Material Sample

| Fluence per Flash ($J/cm^2/F$) | Total Treatment Fluence ($J/cm^2$) | | | | | |
|---|---|---|---|---|---|---|
| | 2.4 | 2.5 | 2.56 | 3.2 | 4 | 6 |
| 0.3 | 13/25* | | | | | |
| 0.5 | | 11/25 | | | 6/19 | 1/20 |
| 0.64 | | | 7/25 | | | |
| 0.8 | | | | 1/10 | | |
| 1 | | | | | **0/10** | |
| 3 | | | | | | 2/30 |

*Number of samples positive for growth/total number of samples tested.

Bold type indicates tests results interpreted as indicating sterilization within the limits of assay sensitivity (1 false positive per 10 samples).

with droplets from approximately 5–100 μm in diameter and a mean droplet diameter size of ~20 μm. After *PureBright®* treatment (as specified in Table 11.4), a presterilized glass cylinder 2.5 cm in diameter and ~4.9 cm² in area was bonded (using a presterilized paraffin wax/mineral oil mixture, mix ratio adjusted for set consistency) to the surface of the packaging material so the glass cylinder surrounded the inoculation spray pattern, the sealed packaging/cylinder apparatus placed in a sterile dish, and the cylinder volume filled with tryptic soy broth. The results were read as growth positive or negative after 14 days at 32°C.

In one set of samples, a treatment of 8 flashes at an incident fluence of 0.5 J/cm² (total treatment fluence of 4 J/cm²) is seen to yield surviving organisms in sterilization assays in 6 out of 19 tests (i.e., approximately 1/3 of the samples were not sterilized by this treatment). However, identically inoculated samples are seen to yield surviving organisms in only 1 out of 10 tests when treated with 4 flashes at an incident fluence of 0.8 J/cm² (total treatment fluence of 3.2 J/cm²); there were no surviving organisms (0 survival) in each of 10 tests when treated with 4 flashes at an incident fluence of 1 J/cm² (total treatment fluence of 4 J/cm²). Thus, it is seen that higher levels of per flash treatment energy produce higher levels of sterility assurance.

It should be noted that the sterility test results documented here do have an inherent background failure rate due to contamination of the samples after treatment. This background false positive rate is associated with posttreatment sample manipulations required to complete the assays, such as handling, media filling, and so on, and the environment in which these manipulations are performed. The results of control tests using sterile, uninoculated samples treated with high levels of pulsed light and subsequently assayed suggest the false positive rate in the Pure-Pulse laboratory is equal to or less than about 1 sample in 10. Therefore, the limit of accuracy for the procedures used is approximately 90 percent, and experimental sample set results with 1 or less growth positive samples per 10 total test assays are interpreted as indicating a treatment resulting in the maximum detectable level of sterility assurance for the test methodology used.

The importance of fluence per flash is further demonstrated in the results seen in Figure 11.9, where the killing effects of pulsed light treatments are shown on *Aspergillus niger* (condiospores, hyphal debris, and sporangia recovered by a crude wash of a mature culture) inoculated by spraying the preparation onto the surface of white plastic packaging material. The inoculum was a crude aqueous 0.05 percent Tergitol-7 wash of a mature *Aspergillus niger* yeast morphology agar culture (>7 days at 28°C) that was sprayed onto the surface of small, approximately 3 × 4 cm samples cut from a white food grade polyethylene tub. The spray pattern covered ~1.75 cm² of surface and approximately 1 log CFU of the spray concentration was lost between spraying and recovery (the majority of this loss is attributable to recovery inefficiencies). Three control samples and three treated

**Figure 11.9.** Log CFU recovery from control (0 Flashes) and *PureBright®* treated white plastic tub packaging material spray inoculated with *Aspergillus niger.*

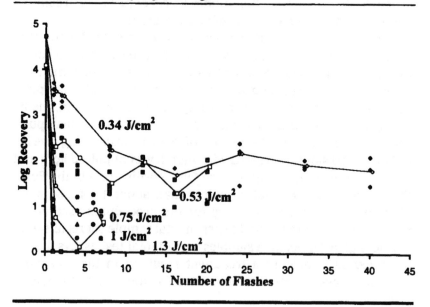

samples were assayed for each data point; the charted lines connect the mean values of each data set. The results for samples treated at a variety of fluences per flash and with a variety of total flash numbers are shown. Samples demonstrating no recovered viable CFUs were plotted as 0 log recovery.

The log survival results are documented for treatments of (0.34, 0.53, 0.75, 1, and 1.3 $J/cm^2$ per flash) used as a single flash, or after various total numbers of flashes (and, therefore, various total treatment fluences). When 0.34 $J/cm^2$/flash is used, the resulting kills obtained are relatively constant after about 7 flashes and through treatments ranging up to 40 flashes (40 flashes equals a total accumulated treatment fluence in excess of 13 $J/cm^2$). However, a single flash treatment at 1.3 $J/cm^2$ is seen to result in no recovered viable organisms for each of three identically inoculated samples tested.

Thus, the peak power of the treatment (i.e., fluence per flash) clearly affects the overall killing efficiency achieved using either bacteria or fungi spores. This is observed on the surfaces of packaging in Table 11.4 (*Bacillus pumilus* spores) and Figure 11.9 (*Aspergillus niger* spores). It is also apparent treating organisms in solution; the results seen in Figure 11.6 combined with other observations to be published elsewhere confirm the importance of treatment fluence per flash in liquids.

The importance of fluence per flash effects are more prominent in *Aspergillus niger* spore than in bacillus spore populations. For both bacterial and fungal spores, the importance of fluence per flash is most apparent at high levels of kill. This is, of course, logical since high fluence (i.e., high fluence per flash) treatment is synonymous with high kill levels. This contributes added complexity to the study of this phenomenon because large sample sizes must be tested since survival is a very rare event. For example, the tailing phenomenon observed at high exposure levels with conventional ultraviolet sources (see Figure 11.3), does not generally become apparent until kill levels greater than about 4.5 log CFU inactivation are observed. Thus, to study this event requires large numbers of samples since it represents an event occurring in only a small portion of the treated population, for example, in less than 1 cell per $10^4$ to $10^5$.

It is also best to study this phenomenon using barrier isolation equipment or aseptic filling machines rather than conventional laboratory methods and laminar flow hoods or clear room environments.

A potential hypothesis relating to the importance of fluence per flash effects have been proposed; certainly some of the effects seen seem to be related to thermal events. Just as a surface is heated by exposure to sunlight, an absorbing surface will experience a thermal transient when flashed with an intense short duration broad spectrum light source such as the *PureBright®* flash. It is proposed that a thermal transient on an absorbing surface is coupled to organisms on the surface so they experience both ultraviolet and thermal effects. Similarly, the surface of absorbing organisms or clumps of organisms may experience thermal transient events capable of increasing killing efficiency. In addition, specific bonds or absorption sites in organisms (chromophores) may experience thermal and vibrational effects which heighten ultraviolet effects and killing potential.

The measurement of the properties of a *PureBright®* induced thermal cycle on an absorbing surface are, however, not trivial. Since a typical *PureBright®* flash is about 100 microseconds in duration, cyclical thermal events on an absorbing surface are confined to the surface and near-surface regions and are of very short duration, probably lasting on the order of less than one to, at the most, several milliseconds. Surfaces do not feel heated or hot after normal *PureBright®* treatment. *PureBright®* associated thermal transients cannot be measured by conventional means, such as by touch, thermometers or resistive thermal devices, but require relatively sophisticated equipment and experimentation for accurate measurement. The thermal transients measurements discussed here relate to surface absorption events on packaging material. However, as mentioned above, similar considerations can be envisioned on smaller scales, such as on the surface of organisms or within specific chromophores. Small-scale absorption effects are not amenable to measurement and require theoretical assessment, which will be addressed elsewhere.

## Ozone Production Is Minimal and Plays No Role in Kill

Oxygen absorbs in the ultraviolet light range with an absorption maximum centered at about 180 nm. The oxygen absorption band is, however, very broad (it extends, for example, on the upper end to about 240 nm). The consequence of oxygen absorption across this range of wavelengths is the production of ozone ($O_3$). The *PureBright®* pulsed light flash, therefore, does have some potential to produce ozone. This potential is greatest in the lamp/reflector unit itself. *PureBright®* lamp units have a cleanable 316 stainless steel housing and thick quartz window; the lamp is contained in a reflector within the lamp unit. *PureBright®* lamp units are generally operated with a trickle flow of nitrogen through the housing to minimize ozone production in the unit.

*PureBright®* lamp units are relatively small and easy to integrate into existing equipment and lines. If a *PureBright®* unit is installed in a small compartment (for example, a 2 ft$^3$ cube) with no ventilation, ozone will build up to part per million levels during continuous operation. Such levels can be detectable by sensitive instruments, including the human nose. In a typical installation, however, such as with a *PureBright®* pulsed light system integrated into an aseptic packaging machine or line with attendant sterile airflow, ozone production within the machine is small and easily vented. Outside the machine or line, ozone is almost undetectable and well below U.S. Occupational Safety and Health Administration (OSHA) exposure limits. For use in ventilated systems, the attendant airflows are such that ozone levels are insignificant and well below long-term exposure limits for personnel. It is important to note that ozone plays no role in the normal *PureBright®* pulsed light killing process; the same level of effectiveness is observed when treating organisms in an anaerobic environment (such as under $N_2$ gas).

## PUREBRIGHT® ECONOMICS

*PureBright®* pulsed light is a rapid, high rate, and high throughput processing method that produces no residuals. In most instances it can be easily integrated for on-line sterilization. Estimates of the cost of *PureBright®* use are on the order of 0.1–0.3 cent per square foot (or 1–3 cents per square meter) of treated product surface area (the units are surface area—rather than volume—because depth of penetration is often material or product specific and treatment must, of course, penetrate all of the necessary volume). These cost-of-use estimates include the initial investment cost amortized over 10 years, and the costs of lamps, maintenance, and electrical energy. A more in-depth analysis of the comparative costs of pulsed light versus conventional sterilization methods is in preparation. Though this evaluation is not yet complete, it is apparent in early assessments that an on-line method gains significant economic advantage over batch processing methods simply as a result of reduced handling and other expenses associated with batch processing. Estimates suggest that 25 percent or more of the manufacturing cost of some products, for example, parenteral solutions, might be saved by replacing a batch sterilization process, such as autoclaving, with a continuous on-line sterilization method, particularly if the new process favors parametric release.

## SUMMARY

A new method has been developed that uses brief intense flashes of broad spectrum white light for sterilization. All organisms and microbial types (bacteria, fungi, vegetative, spores, oocysts, and viruses) are killed when exposed using one to several flashes at sufficient intensity. To be effective for sterilization, *PureBright®* pulsed light must be able to access all the important surfaces and volume where microbial contamination may exist. The pulsed light parameters important to achieving sterilization are the spectral content of the flash, the fluence of the flash, and the number of flashes delivered. Opportunities exist for using this method in the sterilization of air, water, packaging materials, medical devices, and pharmaceuticals.

The criteria for the use of pulsed light for sterilization is met in many BFS packaging situations. Laboratory and on-line test results using a range of biochallenge spores suggest that filled, sealed products can also be effectively sterilized through the package. Automatic Liquid Packaging and PurePulse Technologies have teamed to develop the use of *PureBright®* as a terminal sterilization method for BFS packaged products and thereby provide an extra level of quality, safety, and sterility assurance. Other manufacturing sterilization applications will emerge as use of the method and regulatory status develop further.

In addition, pulsed light has been shown to be highly effective for killing *Cryptosporidium parvum,* and a pulsed light point-of-entry water treatment system for home or restaurant use has been fabricated and successfully tested. Larger volume water treatment systems and systems for the production of high quality sterile water are under development.

Pulsed light processing and sterilization is an emerging new method that offers a high degree of effectiveness for on-line, high throughput processing at an economical cost. As acceptance and use of the method grows, it will become a significant new pharmaceutical and device manufacturing technology.

# REFERENCES

Arrowood, M. J. 1997. Personal communication. Dr. Arrowood is in the Division of Parasitic Diseases, Centers for Disease Control, Atlanta, GA.

Arrowood, M. J., and C. R. Sterling. 1987. Isolation of *Cryptosporidium* oocysts and sporozoites using discontinuous sucrose and isopycnic Percoll gradients. *J. Parasitol.* 73:314–319.

Arrowood, M. J., M. R. Hurd, and J. R. Mead. 1995. A new method for evaluating experimental cryptosporidial parasite loads using immunofluorescent flow cytometry. *J. Parasitol.* 81:404–409.

Cerny, G. 1977. Sterilization of packaging materials for aseptic packaging 2. Investigations of the germicidal effects of UV-C rays. *Verpackungs-Rundschau,* 28 (10), Techn-wiss. Beilage, S. 77–82.

Dunn, J., D. Burgess, and F. Leo. 1997. Investigation of pulsed light for terminal sterilization of WFI filled blow/fill/seal polyethylene containers. *Parenteral Drug Assoc. J. of Pharm. Sci. & Tech.* 51 (3):111–115.

Dunn, J. E., R. W. Clark, J. F. Asmus, J. S. Pearlman, K. Boyer, F. Painchaud, and G. A. Hofmann. 1990. Methods for aseptic packaging of medical devices. U.S. Patent 4,910,942.

Ernest, J. A., B. L. Blagburn, and D.S. Lindsay. 1986. Infection dynamics of *Cryptosporidium parvum* (Apicomplexa: Cryptosporidiae) in neonatal mice (*Mus musculus*). *J. Parasitol.* 72:796–798.

Finch, G. R., C. W. Daniels, E. K. Black, F. W. Schaefer, and M. Belosevic. 1993. Dose response of *Cryptosporidium parvum* in outbred neonatal cd-1 mice. *Appl. Env. Microbiol.* 59:3661–3665.

Jagger, J. 1967. *Introduction to research in ultraviolet photobiology.* Englewood Cliffs, NJ: Prentice-Hall.

Jagger, J. 1985. *Solar-UV actions on living cells.* New York: Praeger Publishers.

Thekaekara, M. P., ed. *The solar constant and the solar spectrum measured from a research aircraft,* NASA Technical Report R-351, Goddard Space Flight Center, Greenbelt, MD.

Tyrrell, R. M. 1973. Induction of pyrimidine dimers in bacterial DNA by 365 nm radiation. *Photochem. Photobiol.* 17:69–73.

Tyrrell, R. M., and M. J. Peak. 1978. Interaction between UV radiation of different energies in the inactivation of bacteria. *J. Bacteriol.* 136:437–440.

Tyrrell, R. M., and R. B. Webb. 1973. Reduced dimer excision in bacteria following near-ultraviolet (365 nm) radiation. *Mutation Res.* 19:361–364.

# 12

## Virus Inactivation and Removal

*Wayne P. Olson*

Oldevco,
Beecher, IL

Viruses are commonplace in human-source proteins (e.g., hepatitis B virus or HBV); large plasma pools from which clotting factors, immunoglobulins (gamma globulins), and serum albumin are prepared often contain several hepatitis viruses and the human immunodeficiency virus (HIV). Half of the hemophiliacs in France seroconverted to the AIDS (acquired immunodeficiency syndrome) virus (HIV–1) because they received HIV-contaminated human clotting factor VIII (hF-VIII) (reviewed in Olson 1995). Most hemophiliacs die not from bleeding episodes but as an indirect result of HIV infection received from injected clotting factors (Ehmann et al. 1995).

The preferred strategies detailed in this chapter impart viral safety to proteins for injections. Hemophiliac seroconversions to HIV–positive in France (Olson 1995), Canada, and Japan (Anonymous 1996) seem attributable to the poor judgment of government officials controlling the acquisition, distribution, and use of clotting factors, and not to the state of the art in viral inactivation.

In the first half of the 1980s, shortly after AIDS had surfaced as a menace, the Hyland Division of Baxter Healthcare (now Baxter Biotech) developed methods for the dry heat inactivation of HIV and other viruses (Griffith et al. 1991). Other methods for virus inactivation have since been developed, and many human proteins have been cloned in prokaryotic and eukaryotic cells. However, regulatory officials in Europe, Japan, and the United States continue to be concerned about viruses in the blood-source proteins still in use, viruses in the formed elements (cells and platelets) from human blood, and viruses that may be harbored and expressed by recombinant animal cells in culture that produce human proteins. Therefore, methods for viral elimination and inactivation are appropriate to a pharmaceutically oriented text on sterilization.

This chapter has been written from the viewpoint of the manufacturer of injectables. The best review, which focuses primarily on chemical means of viral inactivation, is that by Prince et al. (1991).

## VIRUSES OF CONCERN IN NATURE AND IN CELL CULTURE

HIV is of greatest concern because, with a handful of exceptions that remain to be understood, HIV infection leads to AIDS, which invariably appears to be fatal. This is the reason that self-donations have become commonplace. Self-donations are blood donations made by the patient for personal use during planned surgery.

The human hepatitis viruses A, B, C, D, and E—best known by the acronyms HAV, HBV, HCV, HDV, and HEV—can cause or participate in fatal liver disease. Other human-source viruses of concern are cytomegalovirus (CMV) and Epstein-Barr virus (EBV). Table 12.1 details the murine- and human-source viruses that are prohibited by the U.S. Food and Drug Administration (FDA) in therapeutic monoclonal antibodies (MAbs) and products produced by means of murine (m) MAbs. The sizes

## Table 12.1. Viruses (Murine- and Human-Source) and Antigen That Are Forbidden in mMAbs or hMAbs[a]

|  | Acronym | Diam. (nm) |
|---|---|---|
| Human-source viruses | | |
|   Cytomegalovirus | CMV | 100–200 |
|   Epstein-Barr virus | EBV | 100–200 |
|   Hepatitis B surface antigen | $HB_sAg$ | 42 |
|   Retroviruses | e.g. HIV | 80–120 |
| Murine-source viruses | | |
|   Hantaan | | 90–100 |
|   Ectromelia | | 230–400 |
|   Lactate dehydrogenase-elevating virus | LDH | 150–300 |
|   Lymphocytic choriomeningitis | LCM | 150–300 |
|   Minute virus of mice | MVM | 18–20 |
|   Mouse adenovirus | | 65–80 |
|   Mouse encephalomyelitis | GDVII | 22–30 |
|   Mouse hepatitis | | 75–160 |
|   Mouse salivary gland virus | murine CMV | 100–200 |
|   Pneumonia virus of mice | | 150–300 |
|   Polyoma | | 45–55 |
|   Reo type 3 | | 75–80 |
|   Sendai | | 150–300 |

[a]Adapted from Hoffmann (1987) and Prince et al. (1991)

of these viruses, which relate to their filterability, are also shown.

    Animal viruses are of concern because a rapidly increasing number of biologics are produced in mammalian cell culture, and because the use of mMAbs in the purification of human proteins is widespread (Olson 1995). Because diagnostic methods are much improved in recent years, and because humans increasingly are destroying animal habitats, a number of animal viruses appear to be emerging as sources of human infection (Le Guenno 1995).

Improved analytical methods reveal well-known diseases as virally induced. For example, hemorrhagic fever with renal syndrome, a human disease, was known in China 1,000 years ago, and in 1951–1953 it infected more than 2,000 UN troops during the Korean War. In 1976, the viral agent was identified in the lungs of a Korean field mouse. Hantaviruses have since been implicated in human diseases in Japan, Russia, Finland, the Balkans, New Mexico, Colorado, and Nevada. Mice are minimally affected.

A variety of animal viruses are infectious to humans. The Marburg filovirus had been introduced into Germany with vervet monkeys from Uganda. HIV likely is a variant of the simian immunodeficiency virus and probably was transmitted to an African male sometime before 1959. In Zaire in 1976, 280 of the 318 patients infected with a new virus, Ebola, died; at least 14 Ebola cases have died in Gabon in 1996. The Ebola filovirus likely has an animal source. Most viruses infect one or two hosts, but influenza viruses can prey on humans, horses, pigs, seals, and a variety of birds.

Unknown animal viruses are a concern in human proteins produced in mammalian cell culture. This is part of the driving force for conversion to insect cell culture. The presumption is that any viral DNA (deoxyribonucleic acid) or cDNA (copy DNA) sequestered in the genome of an insect will not be infectious in humans if the virus is expressed. The promoters of this art would be wise to remember that certain Togaviridae multiply in arthropod and in mammalian hosts, hence diligence in detecting virus expression, even by recombinant insect cells, is important. The naked single-stranded RNA (ribonucleic acid) of arboviruses is infectious to cells.

Known human viruses are an issue in those materials that, at this writing, must be human sourced. Those human-sourced materials that cannot presently be simulated with approved products available in the marketplace include packed erythrocytes, platelets, white cells, and cells of the bone marrow, such as stem cells. As will be explained, human serum albumin (HSA) and (with one noted exception) the clotting factors now are virus safe by any reasonable measure. The viral safety of gamma globulins from human plasma *not specifically treated for viral inactivation* presently is in question. Viruses (e.g., adenoviruses), sometimes used as vectors in human gene therapy are conceivably infectious, although their restructure should preculude that.


## MATERIALS REQUIRING VIRAL INACTIVATION AND ELIMINATION

Injectables that may contain infectious virions are human whole blood, human blood fractions, formed elements (erythrocytes, platelets, leukocytes), and recombinant human proteins produced in eukaryotic cells (e.g., Chinese hamster ovary [CHO] cells, which may contain a parvovirus).

Equipment or matrices used in the fractionation or transfer of such materials require decontamination. These include chromatographic matrices, columns, pipettes, flasks, and the like (Kuwahara and Chuan 1995). So far as is possible, such materials usually are disposable or of stainless steel or Pyrex® glass so that they can be steam autoclaved and made risk free.

Viruses as aerosols have not been included herein. Data and methods are available on the survival of aerosolized viruses but these do not appear to be a manufacturing concern. For example, the incidence of hepatitis among plasma workers appears to be no higher than the average in the population at large, although aerosols form during plasma manipulation (including centrifugation). Aerosols likely were formed during the manufacture and arming of biological weapons that now are outlawed internationally.


## WHERE VIRUSES AREN'T

Small-volume parenterals (SVPs) other than biologics (i.e., organic drugs such as antimicrobials, cancer chemotherapeutics, any secondary metabolite or synthetic of less than 2–3 kDa) and simple large-volume parenterals (LVPs) such as 5 percent dextrose in water (D5W), saline, solutions for parenteral nutrition (containing salts, small peptides, individual amino acids, vitamins, phospholipids, etc.) are virus free unless deliberately contaminated by an act of industrial sabotage, which is extremely rare but not unknown.

*Bacterial* viruses, known as bacteriophages (eaters of bacteria), or more simply as phages, conceivably might be found in solutions of recombinant proteins produced in bacteria; however, phages are pathogenic only to particular bacteria and never have been a source of concern in injectable drugs. Human antibody titers to phages have not been reported (injection of amounts sufficient to stimulate the immune system would give rise to anti-phage antibodies). It is not reasonable to expect to find phages in biologics. The biologic manufacturer who produces human protein(s) in a bacterium in fermentation tanks need not screen for these routinely; a phage infection in the fermentation tank is obvious as the bacterial host cells lyse, literally exploding and generating sufficient foam to interrupt fermentation.

## METHODS FOR VIRAL INACTIVATION AND ELIMINATION

The basic techniques for viral inactivation and elimination may be categorized as follows:

1. Decontamination of materials not intended for injections

    a. Moist heat sterilization

    b. Dry heat sterilization

    c. Exposure to strong oxidizers or alkylating agents

    d. Gamma sterilization

    e. Other means

2. Decontamination of materials intended for injections

    a. Exclusion of virus-containing material by prior testing

b.   Dry heating of freeze-dried material

c.   Moist heating of freeze-dried material

d.   Pasteurization of material in solution

e.   Virus removal by hydrophobic interaction

f.   Virus removal by ultrafiltration

g.   Solvent/detergent disintegration of enveloped viruses

h.   Gamma sterilization

i.   Psoralen derivative + ultraviolet irradiation

j.   Cold ethanol fractionation

Of these, all except 2e and 2h currently are in widespread use. The hydrophobic interaction method is not entirely reliable and recently has been discontinued in use by the one firm that used it in process. Gamma sterilization may become a preferred method because, for the sterilization of devices, it is well-established, cost-effective, and accepted everywhere. However, the effectiveness of gamma against viral targets is not as widely known as gamma use in the kill of bacteria and fungi.

For parenterals, none of these methods (other than 1a–d) are used alone. For example, the solvent/detergent method (see below), although highly effective against enveloped viruses (e.g., HIV) and many others, proved ineffective against HAV (Mannucci et al. 1994). To provide very high levels of viral safety and to avoid potential difficulties should one method prove ineffective, combinations are employed (Feldman et al. 1995; Griffith et al. 1991).

## Decontamination of Materials Not Intended for Injections

All surfaces that are exposed repeatedly to precursor or product liquids require thorough cleansing by validated methods. This author does not suggest that these methods, singly or in combination, are satisfactory in the absence of cleaning with detergents and/or disinfectants prior to inactivation steps. Cleaning is essential prior to sterilization procedures because detritus protects the organisms to be destroyed.

## Moist Heat Sterilization

Steam autoclaving presently is the method of choice for destroying viruses. Any autoclave cycle that has been validated for the kill of any type of bacterial spore will destroy any described virus. A commonplace cycle is 5–15 minutes at 121.1°C (250°F), once thermal equilibrium has been established. Eight minutes at temperature usually is more than adequate for spore kill, provided steam penetration is thorough. Owens (1993) provides excellent background on what is required. The deterrent is the heat-lability of many solutions and devices.

In some viruses, the DNA is double stranded (e.g., adenovirus, hepadnavirus, herpesvirus, papovirus, poxvirus) and the RNA of reoviruses is double stranded. Heating dissociates double-stranded nucleic acids and causes the helical structure to become a random coil. Capsid proteins, which require specific folding to assume the three-dimensional shape required of them, unfold in some time-temperature continuum as does any enzyme within the virion (e.g., reverse transcriptase in retroviruses). The lipid envelopes of some viruses (e.g., HIV) also lose structure when heated sufficiently; loss of envelope structure renders the virus unable to bind to host cells.

Viruses contain only minimal amounts of nucleic acid and protein (with, perhaps, a carbohydrate ligand that may impart a degree of thermal protection) and, in the case of enveloped viruses, lipid. There are only a few key structures, so many viruses are less heat labile than many vegetative forms of bacteria. Nonetheless, steam autoclaving destroys viral integrity irreversibly. Higher forms of life contain proteins dedicated to the folding/refolding of other, complex proteins. Virions are monastic structures without such luxuries.

All containers and matrices that have been used in the preparation of human proteins or cells should be steam autoclaved after use. All liquid and solid waste that may contain infectious material should be moist heat sterilized before discarding, without exception. Incineration may be acceptable provided any particles released to atmosphere have been charred to the point of viral destruction. Material suspected of containing HIV, any hepatitis virus, or radionuclides (especially $^{32}P$, $^{125}I$, $^3H$, $^{14}C$, or $^{51}Cr$), if steam autoclaved, must be contained inside a sturdy autoclave bag. This is a safe and adequate procedure (Stinson et al. 1990).

## Dry Heat Sterilization

Currently, glassware used as final containers are dry heat sterilized so as to depyrogenate. Dry heat is far less efficient as a lethal factor than moist heat. Common temperatures would be 185°C for sterilization or 250°C for sterilization and depyrogenation. A variety of time-temperature conditions are suitable (e.g., for depyrogenation, 350°C for times ranging from about 1.5 to 11.5 minutes may be satisfactory). Such conditions destroy spores and invariably are rapidly lethal to viruses on glassware. For an excellent review on dry heat methods, see Wood (1993).

## Exposure to Strong Oxidizers or Alkylating Agents

One of the most common oxidizers that is used widely as a sterilant is hydrogen peroxide. Other sterilizing oxidants in widespread use include chlorine dioxide and peracetic acid. An oxidized element or compound loses electrons to an oxidizer; the oxidizer is itself reduced (i.e., gains electrons). Hence oxidation and reduction invariably are paired and referred to as *redox reactions*. A classic example is the oxidation of manganous ion, such that

$$Mn^{2+} + 6H_2O \rightarrow MnO_2 + 4H_3O^+ + 2e^-$$

Here, $Mn^{2+}$ as a divalent cation has been oxidized by water to a tetravalent cation (the possible valence states for Mn are +2, +3, +4, and +7). Another example of a redox reaction is the oxidation of ascorbic acid, in which case a double bond forms between covalently coupled carbons as an electron and a proton are lost to an oxidizer.

The structure of hydrogen peroxide is HO:OH and the active species is the hydroxyl radical (as distinct from the hydroxyl ion); the hydroxyl radical extracts a proton and an unpaired electron from the compound that is oxidized. Another example would be the extraction of an electron and a proton from a compound with a hydoxyl group, to form a ketone (in this case, actually an aldehyde):

$$R\text{-}CH_2\text{-}CH\text{-}OH + \cdot OH \rightarrow R\text{-}CH_2\text{-}CH=O + H_2O$$

In the example given, the aldehyde so generated also has the capability of acting as an alkylating agent.

An alkylator has the capacity to form a covalent bond with another molecule. Consider the aldehyde $RCH_2CH=O$ reacting with lysine (an amino acid with an amino residue). If we represent the lysine as $R'\text{-}NH_2$, then

$$R\text{-}CH_2\text{-}CH=O + H_2N\text{-}R' \rightarrow R\text{-}CH_2\text{-}CH(OH)\text{-}NH\text{-}R' + H^+$$

This occurs because the aldehydic carbon is somewhat cationic as a result of electron propensity for the keto oxygen. The nitrogen of the lysine is relatively electron rich. Ergo, the reaction goes to the right. For practical purposes, most alkylations are irreversible.

The most widely used alkylating agent is glutaraldehyde, also known as bis-glutaraldehyde since the molecule has bilateral symmetry and an aldehyde function at either end. Glutaraldehyde (usually 2 percent in slightly alkaline aqueous solution) cross-links proteins or polysaccharides rich in glycosyl amides (invariably beta-GlcNAc residues if there is to be reactivity). Two minutes of exposure to glutaraldehyde is sufficient to inactivate $3.5 \times 10^6$ pg HIV antigen/mL and to sterilize, for example, endoscopes (Hanson et al. 1990).

Formaldehyde (HCHO) is an excellent alkylator and chemical sterilant when used in the gas phase in low-temperature steam (LTS) but is a suspected carcinogen. The odor is quite unpleasant. Ethylene oxide (EtO) is used very widely as an alkylator and general sterilant but has become a target for the U.S. Occupational Safety and Health Administration (OSHA) and the U.S. Environmental Protection Agency (EPA), although EtO is very, very practical for device sterilization, especially for packaged devices on pallets for storage or shipment. Its use is in slow decline.

An alkylating agent no longer in use is beta-propiolactone (BPL). BPL use has been discontinued as it is a suspected carcinogen and has been shown **ineffective** in the inactivation of HIV. One possible reason is that BPL partitions almost exclusively to the lipid envelope of the virus and does not react to a significant degree with the protein capsid, RNA, or the reverse transcriptase (see Figure 12.1).

Oxidizing agents and alkylating agents give rise to essentially irreversible changes in proteins, lipids, and nucleic acids as well as small organic compounds. Hence, the viral capsid (always comprised of proteins or glycoproteins) and the viral DNA or RNA are inactivated or irreversibly altered. Enveloped viruses are surrounded by an organized lipoid layer that is essential to the penetration of enveloped viruses into a host cell. Destruction of the integrity of the lipid envelope, the capsid, or the genome destroys the virus as an infective entity. This may be done by oxidizers or alkylators, but can also be done with much gentler-acting compounds, such as detergents (see below).

## Gamma Sterilization

There are several advantages to gamma sterilization with $^{60}Co$:

- Gamma-sterilized product does not require a sterility test and, all other qualifications being met, product can be released immediately for sale. Twenty-five kilo-Gray (2.5 Megarads) is accepted as a sterilizing dose **for radiation-resistant bacterial spores.**

- The process is amenable to either continuous or batch operation. For continuous operation, boxed product passes into and out of the gamma field several times and in several aspects.

- Gamma radiation shares an important advantage with EtO; the product can be packaged before it is sterilized. After sterilization, the product is palletized and shipped for sale. No degassing is required as is the case with EtO.

There are some **disadvantages** as well.

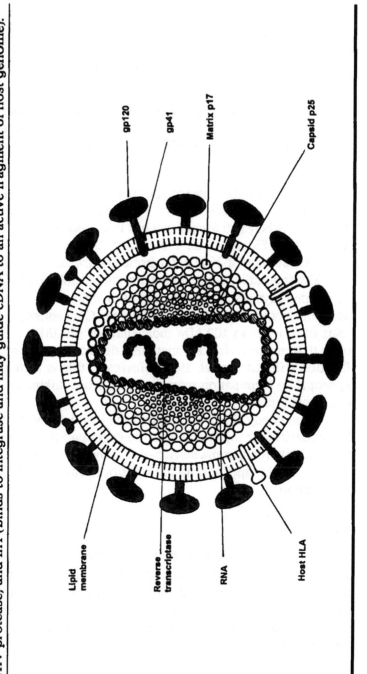

**Figure 12.1.** Structure of HIV-1. Enzymes now known to be present in HIV include reverse transcriptase (which makes cDNA from the RNA of the virus), integrase (integrates cDNA into the host genome), HIV protease, and In1 (binds to integrase and may guide cDNA to an active fragment of host genome).

- Gamma sterilization is an ionizing radiation that leaves a track of ions, free radicals, and excited molecules in the device being sterilized as well as in the microorganisms being killed. This sometimes results in brittleness or discoloration, or both, in plastics.

- Viruses, and viral genomes, are much smaller targets than bacteria or fungi and may require higher radiation doses for a kill than are required to eliminate bacteria.

Human and animal viruses are not present in unused medical devices, but may be present in body fluids and tissues and used devices. Thirty kGy (3.0 Megarads) has been recommended for HIV inactivation (Fideler et al. 1994). Campbell et al. (1994) recommend a *minimum* of 25 kGy for bone allografts.

The free radicals and ions generated in solutions of proteins by gamma rays result in some loss of protein structure and function, and some degree of cross-linking. Consequently, for proteins of more than 100 kDa molecular weight ($M_r$), the gamma dose that ensures viral kill may also destroy significant protein function (Hiemstra et al. 1991). However, this is a caution, and appropriate formulation may eliminate cross-linking as a significant consideration.

Waste body fluids may be gamma irradiated with impunity, but materials for injections must be examined by experiment as to suitability for gamma sterilization. Low levels of gamma irradiation may be appropriate for solutions, but this, too, requires verification.

## Other Means

Viruses can be difficult to eliminate by filtration (see below), in part because conditions required to eliminate the virus also may alter the product. A possibility is the use of two or more methods in concert (synergistic sterilization), each at a dose that, used alone, would be insufficient to sterilize. Two examples of synergies as follows:

1.  Hydrogen peroxide + $CuCl_2$ or $FeCl_2$ are applied syn-
    ergistically to devices. The XJ-clone 3 of Junin virus, in
    0.9 g percent serum, can be inactivated in a solution of
    0.2 g percent $CuCl_2$ + 5 g percent $H_2O_2$ for 60 minutes
    at 20°C and at a slightly alkaline pH. Ferrous salts are
    less efficient per unit weight. This approach clearly is
    inapplicable to products that are labile to oxidants or
    to which metallic cations may not be added.

2.  Hydrogen peroxide oxidation is catalyzed by ultravio-
    let radiation, which promotes generation of the hy-
    droxyl radical.

## Decontamination of Materials Intended for Injections

Sterilization is readily achievable with heat, at a cost. Solutions
of most peptides and all proteins are denatured irreversibly.
Proteins, such as insulin ($M_r$ = 5 kDa), become therapeutically
useless.

Polysaccharides usually are relatively stable to autoclaving,
but monosaccharides (e.g., dextrose) heat sterilized in the final
container (i.e., terminal sterilization) may polymerize and undergo other reactions (e.g., caramelization). Aldehydes form,
and some aldehydes (e.g., acetaldehyde) are toxic. In SVPs, a
very modest amount of caramelized material and aldehyde probably would cause no harm in the patient, but these are of concern
with LVPs such as dialysis solutions (Holmes 1993). Heat alone
may not be the best method to sterilize solutions other than
saline.

### Exclusion of Virus-Containing Material by Prior Testing

At this writing (1997), the tests employed by Baxter Biotech and
many of the other plasma protein fractionators for the screening
of donated bloods are as follows:

*   A, B, and O for cell compatability between donor and
    recipient

*   *Treponema pallidum* (syphilis organism)

- anti-HIV antibodies

- HTLV-1

- Hepatitis B-surface antigen ($HB_sAg$)

- Hepatitis B-core Ag (HBC)

- HCV

Fractionators do not use the formed elements (cells), but purify individual plasma proteins, primarily using Cohn's cold ethanol process (see Olson 1987). Fractionators do not test for hepatitis B core antigen or HTLV-1, but examine plasma for the following:

- Alanine amino transferase (ALT), an enzyme that is a surrogate indicator for non-A, non-B and non-C hepatitis infections

- anti-HCV Abs

- anti-HIV-1 and HIV-2

- HIV Ag by ELISA (enzyme-linked immunoassay)

Such screening signficantly reduces, but does not eliminate, the inclusion of virus-hazard material in donated plasmas, formed elements, and plasma for fractionation. The process is not foolproof because viruses have a latent period in the host, during which time they may be undetected. For example, HIV DNA, prepared from HIV RNA (when HIV enters a cell) by means of reverse transcriptase from the virus, may enter the genome of a human T4 cell, where it is undetected by simple screening methods until the virus multiplies in the cell and emerges. That latency may extend from three months to a number of years.

Since December 1994, Immuno-U.S. has screened *pooled* human plasma (from which HSA, IgG [gamma globulin], hF-VIII, hF-IX [human clotting factor IX]) for HBV, HBC, and HIV using the polymerase chain reaction (PCR). This is a test for the various virus genomes, not host antibodies directed against the viruses. PCR is qualitative but far more sensitive than ELISA. The cost of a PCR test is estimated to be less than $200 (Lewis 1996).

Tests have sensitivity limits. For example, HBV–positive material may be infectious at 1 to 2 log dilutions *below the limit of sensitivity of the test.* The implication is that a single unit of whole blood or packed red cells or platelets may be contaminated with insufficient virus for detection. The use of PCR dramatically increases test sensitivity, but at tremendous cost where applied to small amounts of injectable material, such as single units of packed red cells. The merit of a red cell substitute is obvious.

Screening for viruses in peptides or proteins for injections is particularly important because patients with malignancies are among the major recipients of whole blood. Chemotherapeutic drugs destroy (in part) the ability of the tumor to reproduce but also diminish the capacity of stem cells in the bone marrow to produce red and white cells and platelets for the circulating plasma. Among the white cells that are killed in smaller numbers during chemotherapy are certain of those (e.g., plasma cells) that participate in the immune response.

## Dry Heating of Freeze-Dried Material

As is apparent from Table 12.2, the heating of freeze-dried clotting factors is an important part of antiviral strategies. Prior to the implementation of highly effective viral inactivation methods in pooled cryoprecipitates, 80 percent of German hemophiliacs were HCV positive, 71 percent were HBV positive, and 44 percent were HIV positive (Schramm et al. 1989).

An early heating regimen, not shown in Table 12.2, was 60°C for 10 hours, which was insufficient to inactivate HBV. According to the FDA (Cruzan 1995), as recently as the fourth quarter of 1995, 6 Canadian patients receiving product from any of 3 lots of hF-VIII (lot numbers B71308, B71309, B71508) of a U.S. manufacturer had seroconverted to HIV positive. According to the Internet report, the records of the manufacturer showed that product in the dry state had been heated to 60°C for 10 hours, a cycle shown in 1978 to be inadequate for the inactivation of HBV, which is less labile than HIV. Even 60°C for 32 hours had been shown insufficient to inactivate particular strains of HBV, HBC, and HDV (Purcell et al. 1985). However, what the Internet report does not provide is information about the control of blood supplies in Canada.

## Table 12.2. Virus Inactivation or Removal Methods Currently in Use in the Preparation of Human Clotting Factors for Injection[a]

| Method | Manufacturers |
|---|---|
| Dry heating, 80°C, 72 h | BPL[b]; Cutter division of Bayer |
| Dry heating, 60°C, 144 h | Baxter Biotech (Hyland Labs) |
| Vapor (moist) heating, 60°C, 10 h | Immuno |
| Pasteurization, 60°C, 10 h | Armour (recently Rhone-Poulenc-Rorer, now Centeon); Behringwerke; Sclavo |
| Heating a suspension in n-heptane, 60°C, 20 h | Alpha Therapeutics |
| Other heating regimens for dry hF-VIII | |
| 60°C, 30 h | not stated |
| 60°C, 72 h | not stated |
| Solvent/detergent | Alpha Therapeutics; American Red Cross; AIMA; Baxter Biotech; Biotest; Biotransfusion; Cutter (Miles) division of Bayer; Centro de Hematologia Santa Catarina; NY Blood Center; Octapharma |
| Solvent/detergent + dry heating, 80°C, 72 h | Novo Nordisk (Copenhagen) |
| Solvent/detergent + dry heating, 100°C, 30 min | AIMA |
| Sodium thiocyanate + ultrafiltration | Armour (recently Rhone-Poulenc-Rorer, now Centeon) |

[a] Modified from Mannucci (1993)
[b] Bio-Products Laboratory, Elstree, U.K.

Mark Kennedy (1996) reports on the preliminary findings of a commission headed by Justice Horace Krever.

*Krever heard that many of those players [Canadian Red Cross, Canadian Blood Agency, federal Health*

> *Department] moved slowly to clear infected blood from the system so it could be destroyed and replaced by newly produced, safe, heat-treated products. As well, they had access to newly designed kits to screen out tainted blood shortly after donation, but delayed using them.*

Consequently, depending on the age and status of the products, the manufacturer(s) may not have been remiss. Furthermore, although the Internet report cites the seroconversion of 6 patients, the Krever Commission in Canada has been studying the causes of the HIV-1 seroconversion of approximately 1,200 hemophiliacs. It would appear that the Canadian system for the treatment of hemophiliacs has had some difficulties over an extended period.

The $D$ value at 60°C for HIV is 32 minutes for the freeze-dried material, but 24 seconds when the virus is in a liquid suspension (heating hF-VIII in aqueous suspension inactivates the clot promotion function of the protein). Although the heating of freeze-dried proteins is successful for the inactivation of many viruses (Table 12.2 provides details of viral inactivation strategies applied by various manufacturers of human plasma fractions), it is not consistently a successful strategy with HCV (Olson 1987), nor with HAV (Mannucci et al. 1994). $D$ values, which represent the time for destruction of one log of an organism, decrease exponentially with increasing temperature. Therefore, the survival of HIV drops like a stone at higher temperatures. Limited studies of hemophiliacs receiving hF-VIII dry heated at 80°C for 72 hours (Rizza et al. 1993) indicated that that regimen reduced the risk of HCV transmission from about 90 percent to 0–11 percent. Similar (but apparently unpublished) results had been obtained in the United Kingdom several years earlier.

Heating of freeze-dried materials is done because the heat denaturation (loss of structure, most often irreversible) of proteins in solution is well known. The classic example is a hard-boiled egg. Ovalbumin, surrounding the viscous yolk, and the yolk, solidify irreversibly if heated while in solution inside the eggshell. At a given temperature, there is a limiting time at which irreversible denaturation occurs. Similarly, at any given time, there is a limiting temperature at which irreversible denaturation of a protein occurs.

Aside from time and temperature as variables, there are two other essential considerations about the heat denaturation of proteins (including plasma proteins and cytosol from ruptured cells). Although a protein may unfold and lose the shape in which it is functional, in solution the proper refolding can occur, with highest efficiency if other proteins, termed *chaperonins*, are present. Proper refolding does not occur if the unfolded protein encounters other unfolded proteins to form dimers, tetramers, and polymers. Polymerization is by the interaction of the hydrophobic residues that normally are shielded from the exterior of the properly folded molecule. The addition of a heat-stable background protein, such as albumin, increases the heat stability of clotting factors suspended therein.

The other consideration is salt concentration and the dielectric constant of the solution. As conductivity of the aqueous phase increases, the aqueous phase becomes less and less accommodating to hydrophobic side chains of amino acids such as tyrosine. At high salt concentrations, hydrophobic residues are driven to associate, forming a water-free hydrophobic phase and visible particles in the suspension.

## Moist Heating of Freeze-Dried Material

Moist heating is the Immuno (Vienna, Austria, and Detroit, MI) process, done at 60°C for 10 hours. Moist heat is well known to be far more lethal than dry heat. To that end, Immuno employs heat and high relative humidity to freeze-dried hF-VIII (Kryobulin TIM 3) and a variety of other clotting factors. The initial assumption of the dry heat art was that moisture would cause the rapid irreversible denaturation of a large proportion of the freeze-dried plasma proteins. This assumption was in error.

Mannucci et al. (1992) were successful in inactivating HBV, HCV and HIV, using the Immuno process. A recent study (Shapiro et al. 1995), which included 6–15 month follow-ups, indicated the effectiveness of the Immuno method against HBV, HCV, and HIV.

## Pasteurization of Material in Solution

Fractionators of human plasma proteins are familiar with the pasteurization of 5 percent, 20 percent, and 25 percent HSA at 60°C for 10 hours. HSA (68 kDa) contains a number of disulfide bridges that stabilize the globular shape of the molecule. The addition of caprylate (an alkyl lipid with a carboxylic terminus; caprylate binds in the hydrophobic clefts of HSA) further stabilizes the molecule to heat. No known virus tolerates such a time-temperature continuum in solution. As temperature is increased, time to virus inactivation decreases logarithmically.

There have been studies (reviewed in Olson 1987) in which various sugars have been added to, for example, clotting factors in solutions that were then spiked with test viruses. The primary problem is that additives that interfere with the irreversible denaturation of a particular plasma protein also tend to protect the target virus. The human plasma fractionators who pasteurize solutions (other than albumin, which everyone pasteurizes) are given in Table 12.2.

Products in the late stages of development at this writing are cross-linked human hemoglobin (hHb) from erythrocytes and from recombinant Hb. The cross-linked product is amenable to pasteurization and, properly made, regardless of source, should be entirely virus safe. Platelet membranes for the treatment of bleeding also are well along in development (PRP Inc., Watertown, MA; Rhone-Poulenc-Rorer [now Centeon], Collegeville, PA, and Bradley, IL) and are treated for viral inactivation.

## Virus Removal by Hydrophobic Interaction

A Swedish group incorporated alkyl groups into various filters and found that enveloped viruses were removed with good efficiency by columns of, for example, C18 ligands immobilized on relatively stable biocompatible supports in columns (Einarsson et al. 1981). The concept is that an alkyl group, extending from a membrane or a column matrix, penetrates into the lipid coat of enveloped viruses. The problem is that lipoproteins, and serum proteins that have lipid-binding sites (e.g., HSA), mask alkyl ligands on the resin or may displace enveloped viruses. If the plasma or the selected clotting factor concentrate is quite lipemic (i.e., contains high concentrations of lipoproteins), the method probably is ineffective. To the best of this author's knowledge, the method was employed by one European firm that has since discontinued its use.

## Virus Removal by Other Adsorptive Methods

Viruses have been concentrated from various aqueous sources for purposes of identification and quantitation (Hurst et al. 1989). One of the methods applicable to virus-contaminated proteins is the precipitation of protein(s) together with virions, achieved by the addition of polyethylene glycol and salts. Because virtually all pharmaceutical makeup water is Water for Injection (WFI), and all WFI produced in the U.S. is produced by distillation (so far as this author is aware), virions in the feedwater to a plant are not a concern.

However, where there is a desire to reduce the viral content of relatively simple liquids (e.g., supernatants from cell culture), several highly adsorptive types of microporous membrane filter matrices (e.g., nylon) should be highly successful in reducing by many orders of magnitude the viral burden of aqueous solutions that carry a low particle burden. However, all adsorptive processes are saturable and displacements are commonplace as influent quality changes. Such methods are unsuitable for use with a protein in solution.

## Virus Removal by Ultrafiltration

Pores of nanometer (nm) diameter present a tremendous resistance to liquid flow, regardless of pore density. Membrane filters with large surface areas, containing a high density of nm-sized pores, have been manufactured from a variety of polymers. The polymeric membranes are of two forms—largely isotropic—consisting of foam-like cells (e.g., cuprammonium cellulosic membranes) and highly anisotropic membranes, the bulk of which have very open, nonretaining pores overlaid by a very thin, dense layer containing nm-sized pores.

Cuprammonium membranes are in widespread use in blood dialyzers (artificial kidneys), and isotropic membranes are in widespread use in the concentration and diafiltration (counterpart to static dialysis across a membrane) of various macromolecules (primarily proteins) by means of cross-flow systems (Michaels et al. 1995). Viral reductions in tangential flow systems are commonplace (e.g., Michaels et al. 1995). However, most (if not all) of the anisotropic membranes have occasional imperfections in the surface. They reduce viral burden significantly but, in this author's experience with 1980s products, could not be relied on to eliminate viruses fully.

Roberts (1995) has reviewed the situation as follows: Filters with cutoffs of 15, 35, 40, and 70 nm are used in the removal of viruses from various human clotting factors and now from immunoglobulins (IgGs). IgG ($M_r$ = 150 kDa) can be filtered through the 35 nm membrane and hF-VIII ($M_r$ = 350 kDa) through the 70 nm filter; hF-IX and hF-XI pass readily through the smallest ultrafilter pore sizes. HIV is quite large (80–100 nm) and is readily removed by filtration. HAV is quite small (25–30 nm); hence, from proteins of high $M_r$, HAV cannot be removed by filtration.

Asahi Chemical Industry (Tokyo and Osaka) has produced in its membrane division a cuprammonium cellulosic follow fiber filter cartridge composed of 15 nm (PLANOVA15) or 35 nm pores in a hollow fiber system. A cross-section of the membrane is shown in Figure 12.2 and detail of the pore structure is shown in Figure 12.3.

When that system is operated in the dead-end mode, it is claimed to remove viruses of diameters greater than the pore size. Asahi claims, and comments from one user, are consistent. Asahi data show a correlation between the removal of Japanese encephalitis virus (JEV, 45–50 nm diameter) and the removal of 35 nm diameter gold nanoparticles from suspension (Table 12.3). JEV is a member of the Flaviviridae (among the arboviruses) and is infectious to pigs, horses, and birds.

The PLANOVA filters are in use but almost certainly at low flux since any dead-ended filtration system is flux sensitive, low fluxes being preferred. Data on throughputs and product recovery at a variety of differential pressures doubtlessly exist, but are held proprietary by the users.

Testing of the filters for integrity (prior to use) is, in effect, the pressure-hold test that is familiar to microporous membrane

**Figure 12.2.** Cuprammonium cellulosic nanofilter (PLANOVA) cross-section; the upstream side of the filter is on the lower right and the dark particles are stained HIV virions.

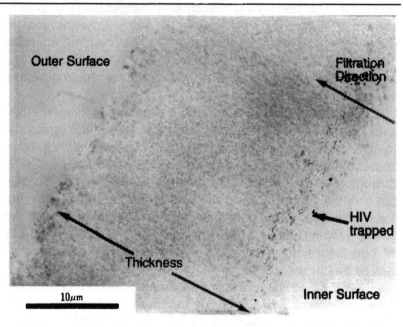

filter users and is a leak test. The Asahi filters are function tested *following* product filtration. The advantage of testing with nano-gold is that analysis is spectrophotometric, hence rapid. An alternative challenge test might be with a suitable bacteriophage. Candidate phages are given in Table 12.4. The preference of the U.S. FDA in another application is PhiX174 because it shows minimal hydrophobic and electrostatic interactions with various matrices (Lytle and Routson 1995).

Cross-flow systems (which usually provide far higher fluxes than dead-ended systems) that remove over 4 logs of Sindbis virus (40–60 nm) are the Amicon YM100 and the Millipore Viresolve (Feldman et al. 1995).

In the recent past, one of the issues in relying largely on ultrafiltration for the removal of virions is the absence of a nondestructive test of ultrafilter pore size and effectiveness. The bubble

**Figure 12.3.** Detail of the pore structure of the PLANOVA filter.

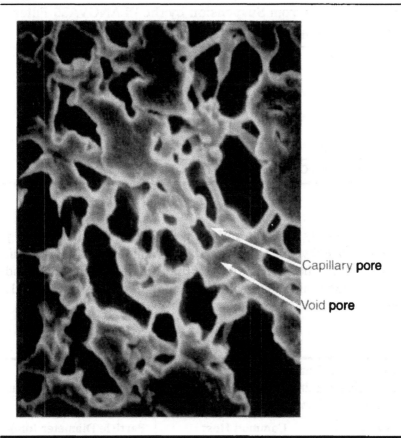

Capillary **pore**

Void **pore**

point for an nm pore membrane would be beyond the strength of materials in the system. However, Feldman et al. (1995) have reported their screening with poly(vinyl pyrrolidone) (PVP). PVP, like most synthetic polymers, contains a wide range of $M_r$. Figure 12.4 shows the $M_r$ distribution of several PVPs. Although the molecule is largely linear, the $M_r$ fraction that penetrates an nm pore ultrafiltration membrane is proportional to the pore size. Centeon (nee Armour, and subsequently Rhone-Poulenc-Rorer) uses only the membranes that retain PVP-K60. Small amounts of PVP are tolerated well and excreted by humans. Other polymers

**Table 12.3. Correlation Between the Log Retention Value for JEV (45–50 nm diameter) and the Retention of 35 nm Gold Nanoparticles from Suspension by the PLANOVA35 Filter[a]**

| JEV Log Retention Value | Log Retention, 35 nm Gold |
|:---:|:---:|
| 5.5 | 2.23 |
| 5.0 | 2.03 |
| 4.5 | 1.84 |
| 4.0 | 1.65 |
| 3.5 | 1.45 |

[a]Data from Asahi Chemical Industry, Tokyo and Osaka

that are readily available and are well-tolerated by humans (e.g., PEGs [polyethylene glycols] and dextrans) also should be suitable. This is a nondestructive test that can be done *prior* to product filtration, and 100 percent elimination of the PVP from the system need not be shown.

**Table 12.4. Candidate Bacteriophages for Destructive, Post Use Testing of nm-Pore Size Ultrafilters Used in Virus Removal; All Are Without Tails[a]**

| Phage | Common Host | Particle Diameter (nm)[b] |
|:---|:---|:---:|
| Phi6 | *Pseudomonas phaseolica* | 65 |
| PM2 | *Pseudomonas* Bal-31 | 60 |
| PhiX174 | *Escherichia coli* | 27 |
| S13 | *Escherichia coli* | 27 |
| M12 | *Escherichia coli* | 27 |
| G4 | *Escherichia coli* | 27 |
| MS2 | *Escherichia coli* | 24 |
| f2 | *Escherichia coli* | 24 |

[a]From Joklik (1988)

[b]Approximate diameters, from electron micrographs

**Figure 12.4.** Distribution of molecular weights of PVP polymers.

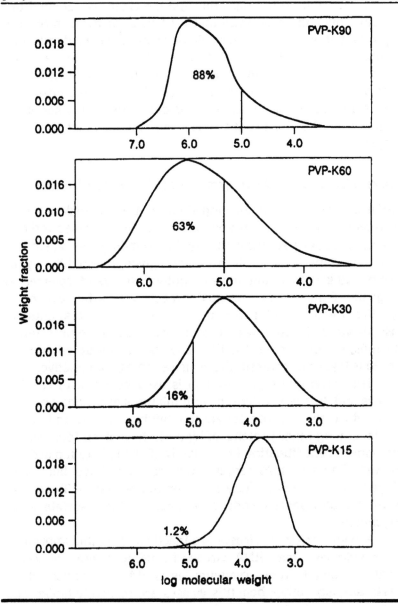

In a dead-ended system (e.g., with the Asahi fibers), as contrasted with a cross-flow system (Michaels et al. 1995), molecules or particles that do not pass through the filter are retained upstream. This author assumes that PVP, PEG, or dextran of high $M_r$, retained on the upstream side of an nm pore deadend filter, would cause severe throughput problems because the retained polymers are protein precipitants. Therefore, the dead-ended systems are leak tested by a pressure-hold test of the type used with some microporous filters, and challenged with nanogold suspension *after* the product has been filtered (post use challenge).

## Solvent/Detergent Disintegration of Enveloped Viruses

HIV is an example of an enveloped virus. A highly ordered, proteinaceous capsid contains within it two RNA molecules and two molecules of reverse transcriptase. Surrounding the capsid is a lipoid envelope in which is embedded glycoprotein 120 (a protein of 120 kDa that contains a carbohydrate ligand, referred to as gp120, Figure 12.1). The gp120 reacts with a fair degree of specificity with receptor proteins on cells (e.g., T4) that endocytose the virus. Reaction with a receptor protein is the first step in the infection of a cell. If the lipid envelope is disrupted and the gp120 is lost or denatured, the virus becomes noninfective.

The solvent/detergent system is used widely with clotting factors and, to a lesser degree, with mixtures of plasma proteins from which the cryoprecipitate (containing hF-VIII and fibronectin, among other proteins of high $M_r$) have been removed by high-speed centrifugation immediately after the frozen plasma is thawed. The solvent/detergent mixtures that have been employed with success include tri(n-butyl)phosphate + Tween 80; tri(n-butyl)phosphate + sodium cholate; and amyl acetate + deoxycholate.

A wide variety of viruses are inactivated by this approach. Picornaviruses (small RNA viruses including polio) are somewhat resistant to these reagents, as is HAV (Mannucci et al. 1994), although other hepatitis viruses are relatively labile. The human plasma fraction manufacturers that use this method are noted in Table 12.2.

When the solvent/detergent method is employed with plasma, either immediately before or during pooling, $10^{12.1}$ tissue culture infectious doses ($TCID_{50}$) of HIV can be inactivated, and the downstream recovery of IgG is 87 percent ± 3 percent (Gao et al. 1993). Although this method does not inactivate HAV, most other viruses are destroyed.

### Psoralen Derivative + Long Wave Ultraviolet Light

Alkylation and oxidation as virus kill methods usually pose problems with materials for injections. Proteins that are altered by alkylation or oxidation may become antigenic to the patient, in which case they cannot be administered repeatedly. Formed elements administered by injections on an emergency basis likely do not pose the same risk, but it would be preferable that they, too, are nonantigenic.

Psoralens (= furocoumarins) are linear, three-ring heterocyclic compounds recovered from tropical fruits and many fungi (Hearst 1981). The psoralens (about 100 are known) are very weakly reactive with proteins but, irradiated with ultraviolet light in the presence of DNA or RNA, result in cross-linking of the DNA and uracil substitution of the RNA. Therefore, psoralens and UV irradiation have been used to inactivate viral nucleic acids in vaccines.

Viruses added to platelet concentrates have been inactivated by means of ultraviolet light-activated 8-methoxypsoralen (8-MOP) (Lin et al. 1989). The issue with 8-MOP is not toxicity (it is a coumarin derivative) but aqueous solubility. Margolis-Nunno et al. (1992) reran the Lin et al. (1989) experiments with ultraviolet A (UVA) at 82.8 $J/cm^2$ for 2 hours into a solution of 4'-aminoethyl-4,5',8-trimethylpsoralen (AMT) at 85 $\mu M$ in an oxygen-depleted solution, in which concentrated human platelets were suspended. The UVA–AMT system inactivated > $10^{6.5}$ vesicular stomatitis virus (VSV, a common viral surrogate)/mL. AMT ($M_r$ 186.16) is found in a variety of legumes and, as mentioned, is a coumarin. It likely is of low immunogenicity and low toxicity. Considering the widespread use of platelet concentrates in bleeding trauma, and the potential for viral transmission via platelet concentrates, this work is of great importance and must be extended to viruses other than VSV.

### Cold Ethanol Fractionation

The larger plasma fractionators no longer rely on ethanol to inactivate the panoply of viruses that may be present in human material. For example, Alpha Therapeutics (Los Angeles)/Green Cross (Osaka) incorporates the solvent/detergent process early in the purification of IgGs (Uemura et al. 1994).

## APPLICATIONS OF VIRAL INACTIVATION AND ELIMINATION

The first rule is: No matter how effective a single process appears to be, use at least two and preferably three or more different methods based on different principles (Feldman et al. 1995; Griffith et al. 1991). This is good sense since all such methods are probabilistic and assume ideal conditions that, in manufacturing, may not be obtained on occasion.

Feldman et al. (1995) provide as an example the methods for the preparation of hF-IX (Mononine). The steps are as follows:

- Adsorption of the hF-IX to an immobilized MAb with washing of the column to remove proteinaceous and viral contaminants (a similar process, reported by Olson [1995], removes over 5 logs of various viruses).

- Desorption of the hF-IX from the immunosorbent column with sodium thiocyanate (chaotrope to interdict the binding of hF-IX to the immobilized MAb) and incubation in the chaotrope at 4°C for 1 hour, which inactivates 8.0 logs of HIV.

- Ultrafiltration with YM100 (Amicon) which removes 11.2 logs of HIV (80–100 nm) and 3.3 to 4.7 logs of encephalomyocarditis virus (EMC [27–38 nm] in the Feldman paper).

## A Viral Reference for Inactivation or Elimination

HAV is very small (30 nm) and hence is difficult to remove by filtration. HAV has high thermal resistance (Mannucci et al. 1994) and is quite resistant to chemical disinfection (Mbithi et al. 1990). The virus is not highly susceptible to gamma irradiation (Scheid et al. 1981).

Therefore, HAV is an appropriate model to which various inactivation or elimination strategies may be applied. The strain of choice is HM-175, which can be titered by plaque assay (Springthorpe 1995). A lytic plaque assay also indicates infectivity and provides more information than assays employing fluorescent, radiolabeled, or enzyme-labeled antibodies, or PCR, which is far more sensitive but less quantitative.

Inactivation data for HAV might then be correlated with the corresponding data for a bacteriophage, such as one of those listed in Table 12.3. Thereafter, all testing could be done with the safer, more easily (and rapidly) quantitated phage at a far lower cost. Other phages without tails, between 27 and 60 nm in diameter, and larger than 65 nm, would be of interest.

## Prions

Inflammatory reports in the British press in 1996 gave rise to public fears (justified, it would appear) that a variation of Creutzfeldt–Jacob Disease (CJD) may be transmitted by the ingestion of beef from cattle fed, in part, on scrapie-infected sheep meat and bone meal. CJD, kuru, fatal familial insomnia (FFI), and Gerstmann-Straussler-Scheinker disease (GSS) are prion-related human diseases. Animal counterparts are scrapie in sheep, and bovine spongiform encephalopathy (BSE) in cattle. The following is background so that the reader can understand why prions are of modest concern in biologicals for sterile injection.

Prions appear to be aggregated proteins of extremely low aqueous solubility. The aggregates appear in brain tissue as abnormal fibrils. Abnormal fibrils also are found in the brains of humans with FFI or GSS.

The CJD infectious agent appears to be a modification of a normally occurring glycoprotein designated PrPc. The normal glycoprotein is present in CJD patients as PrPsc with one of several mutations. The conversion of PrPc to PrPsc has been accomplished in vitro (reviewed by Hope 1995). PrPsc normally is not found in muscle but may be in some tissues of animals fed the insoluble protein.

Iatrogenic cases of CJD, associated with the injection of human growth hormone (HGH) recovered from postmortem human pituitaries, are mentioned by Wilfert (1988) as is CJD in a dentist and two of his patients, and a recipient of a corneal transplant. The latency period for each of these was less than a year. In Britain, 12 young males (x = 27.5 years, none older than 42) have contracted a form of spongiform encephalopathy resembling CJD. Through 1995, the mean for the onset of CJD has been 55 years (Anderson et al. 1996).

There is widespread speculation in the press (Glauber 1996), and recent confirmation (Anderson et al. 1996) that the disease resulted from the ingestion of beef (contaminated between 1986 and 1989 by fodder containing scrapie-infected sheep meat and bone meal). The scrapie prion is transmissible to cattle by ingestion, and from cattle to humans by ingestion. The European Community presently bans the importation of British beef products. The U.S. did so seven years ago. Beef exports from Britain to Europe had been about 3 billions (U.S.$) per annum and now are nil.

Prions as infectious agents resist boiling (but may become noninfectious at 120°C), resist ionizing radiation, and are protease resistant. Table 12.5, modified from Wilfert (1988), indicates the effect of various treatments on prion infectivity.

Kuru occurred in the Fore people of New Guinea and was associated with cannibalism, which included eating the brains of captured enemies. No one born after the cessation of

**Table 12.5. Inactivation Strategies and Resistance of the Agent of Creutzfeldt-Jacob Disease (modified from Wilfert 1988)**

| | |
|---|---|
| **Infectivity destroyed or reduced significantly** | |
| Physical factor | Moist heat at 120°C |
| Chemicals—solvents | Phenol |
| | Diethyl ether |
| Chemical—oxidizer | NaOCl (5 percent, w/v) |
| **Infectivity unchanged or reduced moderately** | |
| Physical factor | Resistant to 80°C |
| | Partial inactivation at 100°C |
| Surface-active agents | Deoxycholate |
| | Sarkosyl |
| | Tritons |
| | NP-40 |
| Chemical—solvent | Acetone |
| Chemical—alkylators | Formaldehyde |
| | Glutaraldehyde |
| | beta-Propiolactone |
| Enzmes—proteases | Trypsin |
| | Pepsin |
| | Pronase (protease mixture) |
| | Proteinase K |
| Enzymes—nucleases | DNase |
| | RNase |
| | Micrococcal nuclease |
| Enzymes—other | Phospholipases |

cannibalism has acquired kuru, and no non-Fore in contact with kuru patients acquired the disease (Wilfert 1988). Kuru is not an issue in human-source material used in the preparation of proteins for injections.

CJD is primarily a disease of Libyan Jews, among whom PrPsc is inheritable. However, no case of the transmission of CJD or a related spongiform encephalopathy by a human blood fraction has been shown.

Where a protein is solubilized in the absence of strong surfactants and is filtered carefully to remove insoluble residues (including, one supposes, prion proteins), the probability of inclusion of PrPc or PrPsc should be close to zero. Plasma proteins, whether by the cold ethanol (Cohn) process or by chromatographic methods, are processed using multiple filtration steps.

## ACKNOWLEDGMENT

Thanks to Andrea L. Olson for the preparation of Figure 12.1. Figures 12.2 and 12.3 are from Asahi Chemical Industry, Tokyo and Osaka. Figure 12.4 is from Feldman et al. (1995).

## REFERENCES

Anderson, R. M., C. A. Donnelly, N. M. Ferguson, M. E. J. Woolhouse, C. J. Watt, H. J. Udy, S. MaWhinney, S. P. Dunstan, T. R. E. Southwood, J. W. Wilesmith, J. B. M. Ryan, L. J. Hoinville, J. E. Hillerton, A. R. Austin, and G. A. H. Wells. 1996. Transmission dynamics and epidemiology of BSE in British cattle. *Nature* 382:779–788.

Anonymous. 1996. News in brief: Blood scandal scientist arrested. *Nature* 383:9 (issue for September 5, 1996).

Barbash, F. 1996. Britain wrangles with human fear over mad cows. *The Washington Post* for March 22, pp. A27, A30.

Brandner, S., S. Isenmann, A. Raeber, M. Fischer, A. Sailer, Y. Kobayashi, S. Marino, C. Weissmann, and A. Aguzzi. 1996. Normal host prioon protein necessary for scrapie-induced neurotoxicity. *Nature* 379:339–343.

Campbell, D. G., P. Li, A. J. Stephenson, and R. D. Dakeshott. 1994. Sterilization of HIV by gamma irradiation. A bond allograft model. *Int. Orthop.* 18:172–176.

Cruzan, S. 1995. Recall of antihemophilic factor. U.S. Food and Drug Administration information release P87-37, copied by Fred Landherr and W. P. Olson on the Internet, December 5, 1995.

Ehmann, W. C., M. E. Eyster, L. M. Aledort, and J. J. Goedert. 1995. Causes of death in haemophilia. *Nature* 378:124.

Einarsson, M., L. Kaplan, E. Nordenfelt, and E. Miller. 1981. Removal of hepatitis B virus from a concentrate of coagulation factors II, VII, IX and X by hydrophobic chromatography. *J. Virol. Methods* 3:213–228.

Feldman, F., S. Chandra, and C. Huang. 1995. Improved safety from plasma derivatives: purification and viral elimination characteristics of Mononine. *Acta Haematol.* 94 (suppl. 1):25–24.

Fideler, B. M., C. T. Vangsness, Jr., T. Moore, Z. Li, and S. Rasheed. 1994. Effects of gamma irradiation on the human immunodeficiency virus. *J. Bone Joint Surg.* 76A:1032–1035.

Gao, F., A. M. Prince, D. Pascual, and B. Horowitz. 1993. Enhancement in the safety of immune globulins prepared from high-risk plasma. *Vox Sang.* 64:204–209.

Glauber, B. 1996. Fearful Britain loses appetite for beef. *The Sun* (Baltimore, MD) for March 22, pp. 1A, 16A.

Griffith, M. J., S.-L. Liu, G. Neslund, and H. S. Kingdon. 1991. Reducing the risk of viral transmission in biotechnology processes. In *Sterile pharmaceutical manufacturing: Applications for the 1990s*, edited by M. J. Groves, W. P. Olson and M. H. Anisfeld. Buffalo Grove, IL: Interpharm Press.

Hanson, P. J., D. Gor, D. J. Jeffries, and J. V. Collins. 1990. Elimination of high titre HIV from fibreoptic endoscopes. *Gut* 31:657–659.

Hearst, J. E. 1981. Psoralen photochemistry. *Ann. Rev. Biophys. Bioeng.* 10:69–86.

Hiemstra, H., M. Tersmette, A. H. V. Vos, J. Over, M. P. van Berkel, and H. de Bree. 1991. Inactivation of human immunodeficiency virus by gamma radiation and its effect on plasma and coagulation factors. *Transfusion* 31:32–39.

Holmes, C. J. 1993. Biocompatability of peritoneal dialysis solutions. *Peritoneal Dial. Intl* 13:88–94.

Hope, J. 1995. Mice and beef and brain diseases. *Nature* 378:761–762.

Hurst, C. J., W. H. Benton, and R. E. Statler. 1989. Detecting viruses in water. *J. Am. Water Works Assoc.* 81(0):71–80.

Joklik, W. K. 1988. *Virology,* 3rd ed. Norwalk, CT: Appleton & Lange.

Kennedy, M. 1996. Government upstages blood probe; Talks to reform system to take place before final Krever report. *The Ottawa Citizen* Tuesday, March 12:A1 et seq.

Kleinschmidt, A. K., G. Klotz, and H. Seliger. 1981. Viroid structure. *Ann. Rev. Biophys. Bioeng.* 10:115–132.

Kuwahara, S. S., and J. H. Chuan. 1995. Process validation of separation systems. In *Separations technology: Pharmaceutical and biotechnology applications,* edited by W. P. Olson. Buffalo Grove, IL: Interpharm Press.

Le Guenno, B. 1995. Emerging viruses. *Scientific American* 273:56–64.

Lewis, R. 1996. PCR finds bioprocess applications in detection of viral contamination. *Genetic Engr. News* 16 (1):10,17.

Lin, L., G. P. Wiesehahn, P. A. Morel, and L. Corash. 1989. Use of 8-methoxypsoralen and long-wavelength ultraviolet radiation for decontamination of platelet concentrates. *Blood* 74:517–525.

Lytle, C. D., and L. B. Routson. 1995. Minimized virus binding for tests of barrier materials. *Appl. Environ. Microbiol.* 61:643–649.

Mannucci, P. M., K. Schimpf, T. Abe, L. M. Aledort, K. Anderle, D. B. Brettler, M. W. Hilgartner, P. B. Kernoff, M. Kunschak, and C. W. McMillan. 1992. Low risk of viral infection after administration of vapor-heated factor VIII concentrate. *Transfusion* 32:134–138.

Mannucci, P. M. 1993. Clinical evaluation of viral safety of coagulation factor VIII and IX concentrates. *Vox Sang.* 64:197–203.

Mannucci, P. M., S. Gdovin, and A. Gringeri. 1994. Transmission of hepatitis A to patients with hemophilia by factor VIII concentrate treated with organic solvent and detergent to inactivate viruses. *Ann. Internal Med.* 120:1–7.

Margolis-Nunno, H., B. Williams, S. Rywkin, N. Geacintov, and B. Horowitz. 1992. Virus sterilization in platelet concentrates with psoralen and ultraviolet A light in the presence of quenchers. *Transfusion* 32: 541–547.

Masison, D. C., and R. B. Wickner. 1995. Prion-inducing domain of yeast Ure2p and protease resistance of Ure2p in prion-containing cells. *Science* 270:93–95.

Mbithi, J. N., V. S. Springthorpe, and S. A. Sattar. 1990. Chemical disinfection of hepatitis A virus on environmental surfaces. *Appl. Environ. Microbiol.* 56:3601–3604.

Mestel, R. 1996. Putting prions to the test. *Science* 273:184–189.

Michaels, S. L., C. Antoniou, V. Goel, P. Keating, R. Kuriyel, A. S. Michaels, S. R. Pearl, G. de los Reyes, E. Rudolph, and M. Siwak. 1995. Tangential flow filtration. In *Separations technology: Pharmaceutical and biotechnology applications*, edited by W. P. Olson. Buffalo Grove, IL: Interpharm Press.

Olson, W. P. 1987. Sterilization of small-volume parenterals and therapeutic proteins by filtration. In *Aseptic pharmaceutical manufacturing: Technology for the 1990s*, edited by W.P. Olson and M. J. Groves. Buffalo Grove, IL: Interpharm Press.

Olson, W. P. 1995. Immunoaffinity purification of proteins for injection, and the issue of animal viruses. In *Separation technology: Pharmaceutical and biotechnology applications*, edited by W. P. Olson. Buffalo Grove, IL: Interpharm Press.

Owens, J. E. 1993. Sterilization of LVPs and SVPs. In *Sterilization technology: A practical guide for manufacturers and users of health care products*, edited by R. F. Morrissey and G. B. Phillips. New York: Van Nostrand Reinhold.

Prince, H. N., D. L. Prince, and R. N. Prince. 1991. Principles of viral control and transmission. In *Disinfection, sterilization, and preservation*, 4th ed., edited by S. S. Block. Philadelphia: Lea & Febiger.

Purcell, R. H., J. L. Gerin, H. Popper, W. T. London, J. Cicmanec, J. W. Eichberg, J. Newman, and M. E. Hrinda. 1985. Hepatitis B virus, hepatitis non-A, non-B and hepatitis delta virus in lyophilized antihemophilic facto: relative sensitivity to heat. *Hepatology* 5:1091–1099.

Rizza, C. R., M. L. Fletcher, and P. B. Kernoff. 1993. Confirmation of viral safety of dry heated factor VIII concentrate (8Y) prepared by Bio Products Laboratory (BPL): A report on behalf of U.K. haemophilia centre directors. *Br. J. Haematol.* 84:269–272.

Roberts, P. L. 1995. Value of virus filtration as a method for improving the safety of plasma products. *Vox Sang.* 69:82–83.

Scheid, R., F. Deinhardt, G. Frosner, and V. Gauss-Muller. 1981. Further characterization of hepatitis A virus. *Med. Microbiol. Immunol.* 169: 134–137.

Schramm, W., M. Roggendorf, F. Rommel, R. Kamerer, H. Pohlmann, R. Rasshofer, L. Gurtler, and F. Deinhardt. 1989. Prevalence of antibodies to hepatitis C virus (HCV) in haemophiliacs. *Blut* 59:390–392.

Schwinn, H., M. Stadler, D. Josic, F. Bal, W. Gehringer, I. Nur, and R. Schutz. 1994. A solvent/detergent treated, pasteurized and highly purified factor VIII concentrate. *Arzneimittelforschung* 44:188–191.

Shapiro, A., T. Abe, L. M. Aledort, K. Anderle, M. W. Hilgarner, M. Kunschak, F. E. Preston, G. E. Rivard, and K. Schimpf. 1995. Low risk of viral infection after administration of vapor-heated factor VII concentrate or factor IX complex in first-time recipients of blood components. *Transfusion* 35:204–208.

Springthorpe, V. S. 1995. Personal communication with Susan Springthorpe, Dept. of Microbiology and Immunology, Faculty of Medicine, University of Ottawa, Ottawa, Ontario K1H 8M5, CANADA. HAV strain HM-175 also is used by Dr. Mark D. Sobsey, Dept. of Environmental Science and Engineering, University of North Carolina, Chapel Hiill, NC 27599.

Stinson, M. C., M. S. Galanek, A. M. Ducatman, F. X. Masse, and D. R. Kurizkes. 1990. Model for inactivation and disposal of infectious human immunodeficiency virus and radioactive waste in a BL3 facility. *Appl. Env. Microbiol.* 56:264–268.

Uemura, Y., Y. H. J. Yang, C. M. Heldebrant, K. Takechi, and K. Yokoyama. 1994. Inactivation and elimination of viruses during preparation of human intravenous immunoglobulin. *Vox Sang.* 67:246–254.

Vogel, G. 1996. Yeast prions: DNA-free genetics? *Science* 273:580.

Wilfert, C. M. 1988. Subacute spongiform encephalopathis/unconventional viruses. In *Virology,* 3rd ed., edited by W. K. Joklik. Norwalk, CT: Appleton & Lange.

Wood, R. T. 1993. Sterilization with dry heat. In *Sterilization technology: A practical guide for manufacturers and users of health care products,* edited by R. F. Morrissey and G. B. Phillips. New York: Van Nostrand Reinhold.

# 13

# Sterility Testing, Microbial Monitoring, and Injured Organisms

*Wayne P. Olson*

Oldevco
Beecher, IL

At this writing (October 1997), the sensitivity of the USP (U.S. Pharmacopeia) sterility test has been found wanting for 48 years (Knudsen 1949). Beloian (1983), Olson (1987), and Groves (see chapter 2) have examined the statistics and came to the same conclusion.

The first consideration is that 10 or 100 organisms per mL, injected in a 1 mL small-volume parenteral (SVP), may be innocuous to a patient, or may put the patient at risk. However, 100 mL of a large-volume parenteral (LVP) containing 100 or more organisms/mL, instilled into any patient, almost certainly entails an element of hazard. Hence, all items in a batch of a parenteral drug are required to be sterile. The existing test is not sufficiently sensitive for determining if only a small number of final containers are contaminated. The probability is too low of selecting a few contaminated items amid many sterile ones.

Two additional issues are the slow growth of certain sterility test positives (Besajew 1992; Bathgate et al. 1993) and the difficulty of coaxing growth from dormant, dysgenic, and otherwise damaged organisms (Olson 1996), which may be the same issue in some instances. Another consideration is that whether the sterility test of choice (by membrane filter) is for 7 or 14 days, product must be quarantined and cannot be sold until the test provides no evidence of microbial contamination.

This chapter is about strategies that may alleviate these problems and improve the speed and sensitivity of microbial testing. The long-term objective of the reader should be full characterization and control of the manufacturing process and, eventually, parametric release (see chapter 14). Those facilities that can validate, and support with routine manufacturing records, a product sterility assurance level (SAL) of 6 logs or better are best served by the parametric release of product and discontinuation of the sterility test.

## Definitions

*Parenterals* are drugs administered to patients by all routes other than oral. They include drugs instilled into the eye or body cavities (e.g., sterile glucose solutions for infusion into peritoneal dialysis patients) via subcutaneous, intracutaneous, intravenous, or intra-arterial injections. For practical purposes, a *drug* is anything (including sterile saline), other than food, that is administered to a patient.

*Sterility* is an all-or-nothing event. It is usually indicated by the absence of growth in defined media of particles (including bacteria and fungi, if present), from a putatively sterile solution, trapped on a sterile membrane filter rated at a pore diameter of 0.45 μm. This also serves as a general definition of the *membrane filter sterility test (MFST)*.

The *SAL* is the probability of nonsterility of a unit from a batch of product in the final container. It is assumed throughout the parenterals industry and by regulatory authorities (e.g., U.S. Food and Drug Administration [FDA]) that where a process appears to have a SAL of $10^{-3}$, approximately 1 vial per 1,000 will be contaminated with microbes.

*Injured organisms* include vegetative and spore forms of bacteria, fungi, and mycoplasmas damaged by heat, shear, pressure, chemicals, or a combination of these, but are still integral and retain the capacity to carry out some metabolic functions, to rejuvenate, or both. These include *dysgenic* organisms with a damaged genome, *dormant* organisms that have dampened reproductive and metabolic activity, and organisms that respire but do not reproduce.

A *barrier/isolator* system is a solid or flexible enclosure, partially or fully enclosed, that can be sterilized using a method that ensures an SAL of $10^{-6}$ of a resistant organism. The sterility of air and surfaces in the barrier/isolator is maintained during use by a constant internal flow of high efficiency particulate air (HEPA), usually under a positive pressure relative to the surrounding room. For details, see Coles (1998).

## STATE OF THE ART IN STERILITY TESTING

### The MFST (Old Style)

The simplest and most effective way to detect organisms is the membrane filter version of the USP sterility test (Olson 1987; USP/NF 1995). This involves putting the drug in solution and filtering the drug through a membrane filter that has a pore size rating of 0.45 μm. Organisms and other particles are retained on or near the surface of the filter and can be washed with sterile water or other sterile solutions to rinse out microbial growth inhibitors (e.g., antibiotics). This is a relatively sensitive test for the volume of liquid that is actually tested. If the particle burden is small (almost always the case), the contents of many final containers can be filtered through a single membrane filter.

In one version of the test, the membrane filter is cut in half with sterile scissors and half (in the grip of sterile forceps) is placed into sterile soybean-casein digest medium (SCD medium, also called trypticase soy broth or TSB, a widely used liquid growth medium for aerobic bacteria and fungi). The other half of the membrane filter is placed in sterile fluid thioglycolate medium (FTM) containing resazurin dye. FTM is an anaerobic medium; oxygen diffusing into the medium oxidizes the dye, which then turns red. Usually, the red color can be seen extending 1 or 2 mm into the medium from the surface in a closed container; strict anaerobes and facultative anaerobes survive and grow in FTM.

The contents of 20 containers may be filtered through a single membrane filter—usually a 47 or 50 mm diameter disc—and divided between the two sterile media as indicated above. The aerobes (in TSB) are incubated at about 25°C and the FTM at 35°C, as has been done for many years. Outside the United States, Sabouraud's medium, for fungi, is used in addition to FTM and TSB. American scientists assume that fungi will grow in TSB.

There are multiple, human manipulations in this old version of the sterility test, and the possibility of inadvertent microbial contamination of the drug samples by a human operator is unacceptably high. Retesting has been permitted. The present view is that a single retest is allowed.

## The MFST—Canister + Isolator System

Government regulatory bodies (e.g., FDA, UK MCA [Medicines Control Agency], etc.), related agencies (e.g., the USP and the British Pharmacopoeia [BP]), and the pharmaceutical industry are beginning to agree that this test must be done properly the first time, and that any contaminants detected must be assumed to originate in the product.

As a consequence, the more popular version of the sterility test employs two (U.S.) or three (European Community, EC) closed, sterile canisters, each with a membrane filter in the base, into and through which the liquid products are impelled by a peristaltic pump. Although contamination of the batch samples by sterility test personnel is dramatically lower when the canisters are used, false positives decline to virtually nil when sterility testing is done with canisters inside a vapor-sterilized isolator

(Akers et al. 1995; Wood 1996). Simple solutions and emulsions, as well as any drug formulation that can be solubilized with a nonmicrobicidal solvent system, can be tested for sterility in this way. An overview of particular sterility test methods, which is still current, can be found in Olson (1987), and background on the chemical sterilization of air and surfaces, including isolators and barriers, can be found in this text (see chapter 8).

There are two canister systems presently in use: (1) Steri-Test® (Millipore, Bedford, MA), a disposable; and (2) a Sartorius GmbH (Gottingen, Germany) device, which is cleaned, a new filter inserted, autoclaved, and reused. The Millipore patents have expired; hence, units from other manufacturers should be expected in the marketplace.

## Recent Developments in the MFST

At this writing (October 1997), the USP proposes to increase the number of vials or containers from a lot from 20 to 40 so as to improve the sensitivity of the test; the proposal also includes an increase in the incubation times from 7 to 14 days. It has been shown (Bathgate et al. 1993; Besajew 1992), and confirmed informally by a number of U.S. manufacturers, that some MFSTs are positive after 7 days. It is also proposed that where MFSTs are *not* performed in an isolator, that one retest will be permitted at this time. No retest will be permitted for sterility tests performed in an isolator. Any growth is considered a batch contaminant. This is applicable in both the United States and Europe.

## RECOVERY AND ENUMERATION OF ORGANISMS UPSTREAM OF FILLING

### Justification

The sterility test is qualitative, an all-or-nothing event. Control of the bioburden of a parenteral product as it is prepared, until it enters the final container, ultimately is more important than the sterility test. Control of that bioburden is contingent on quantitative information about the microbes in the product during the manufacturing process. Where this presently is done, reliance is on standard plate counts (SPCs). The great disadvantage of SPCs is that they provide data that are of historical interest. Manufacturing cannot be delayed until the SPCs can be read, evaluated, and recorded.

## Planktonic Organisms and Biofilms

This chapter addresses bioburden in a liquid product. Such organisms will be suspended in the solution as single cells or aggregates. Biofilms of organisms on manufacturing equipment and medical devices have been reviewed elsewhere (Olson 1997).

## Standard Plate Counts

In the various stages of drug production, control of the microbial population is necessary because microbes can metabolize the drug or excipients (additives used in drug formulation). Additionally, microbial by-products may contaminate the product and do harm to the recipient (e.g., gram-negative endotoxins, also called lipopolysaccharides or LPSs).

One of the ordinary ways to quantitate microbes in a liquid is to place 1.0 or 0.1 mL of the liquid onto a solid growth medium, such as nutrient agar in a petri dish. A series of 1-in-10 dilutions is usually made in sterile water or saline. From each of the dilutions, 0.1 mL is placed on nutrient agar; each of the droplets is spread on the surface of its petri dish with a bent glass rod (hockey stick) sterilized with flamed alcohol. The petri dishes are covered and incubated for 24 hours and, in many instances, a week or more. Visible colonies form where each single microbe had been and are noted as colony forming units (CFUs). These quantitative tests are referred to as the SPCs (Greenberg et al. 1981).

SPCs are useful only for counting large numbers of microbes. If the contamination level is 1 microbe in 10 mL, such low counts will not be seen or quantitated by SPCs. Furthermore, if the solution in some way inhibits microbial growth (e.g., an antibiotic in the solution), no counts whatsoever may be observed, and one might conclude, quite wrongly, that the solution is sterile. This often happens even in an innocuous liquid such as distilled or demineralized water (Byrd et al. 1991).

If the drug inhibits microbial growth (e.g., the solution contains a high concentration of salts that dewater but do not kill the organisms, rendering them temporarily unable to reproduce), low counts may not be seen. Therefore, one of the more effective ways to count small numbers of organisms in a solution is to filter a large volume of the solution through a 0.45 μm pore size filter (most often a 47 or 50 mm diameter disc), wash the filter with a sterile, nutrient-type solution, and then place the filter atop a nutrient agar. Water and nutrients move upward into the filter by capillarity, bathing and nurturing the organism(s) that subsequently form CFUs. One counts the CFUs. If 39 colonies are counted from the 1/100 dilution, then the original suspension contained 39,000 CFU/mL when 0.1 mL of each dilution was plated.

## Rapid Quantitation of Uninjured and Injured Microbes

The minimal incubation time required for the SPC estimation of CFUs is overnight. Therefore, in most instances, SPCs provide bioburden information too late to be of use in manufacturing. Several days or a week or more may be required for slow growers, if they grow on solid media. Many organisms, especially damaged organisms, do not grow on solid media (Olson 1996; Roszak et al. 1984; Stewart et al. 1994). Such organisms tend to be ignored because they seem not to grow. But such organisms may metabolize product and resuscitate to reproduce in enriched liquid media or in a patient. Whether they are a true concern remains to be shown. For example, if a slow grower is indistinguishable in standard tests from normal strains of *Pseudomonas pickettii*, it is reasonable to consider that the organism has been damaged, and the long growth interval relates to the regeneration of reproductive capability.

There are five basic approaches to the rapid quantitation of injured and uninjured organisms:

1.  *Count viable and nonviable bacteria and fungi directly from filters by staining the organisms with acridine orange on 0.2 or 0.1 μm rated black Nuclepore® filters (Jaster 1977):* This is more a measure of cleaning efficiency in removing microbes than of the effectiveness of disinfectants and their application. However, the recovery of such organisms from air, surfaces, and liquid product can be done fairly rapidly while filling is underway (there is no point in monitoring the air until there is activity in the room that results in the resuspension of organisms). Direct counts of this type are used widely in the dairy industry in preference to SPCs. The staining method is rapid.

2.  *Direct viable count (DVC) is a labor-intensive method for determining viable counts only and is an extension of the acridine orange method (Kogure et al. 1978):* Cells are recovered and incubated for 6 hours at 20°C in the following medium:

    | | |
    |---|---|
    | TSB base | |
    | Yeast extract to | 0.025 percent |
    | Nalidixic acid | 0.002 percent |

    Nalidixic acid inhibits deoxyribonucleic acid (DNA) synthesis. Cells lengthen rather than divide. The actively growing cells fluoresce red-orange because the acridine orange complexes primarily with ribonucleic acid (RNA). (Acridine orange that binds to DNA in cells fluoresces green-white [Yamabe 1973]). Viability is verified by measuring the increase in cell length, which is why the method is particularly labor intensive. To the best of this author's knowledge, this method has not been attempted with injured organisms.

3. *Stain viable bacteria with 4′,6-diamidino-2-phenylin-dole (DAPI) following microbial incubation on filters (Bianchi and Giuliano 1996):* Organisms are collected on 0.2 μm black Nuclepore® filters and cultured on the following medium:

> Agar base
>
> | | |
> |---|---|
> | Polypeptone | 1.0 mg/L |
> | Proteose peptone | 0.5 mg/L |
> | Bacto soytone | 0.5 mg/L |
> | Bacto yeast extract | 0.5 mg/L |
> | Sodium glycolate | 0.5 mg/L |
> | Sodium malate | 0.5 mg/L |
> | D-mannitol | 0.5 mg/L |
> | Sucrose | 0.5 mg/L |
> | Ferric citrate | 0.05 mg/L |

The authors incubated the recovered marine organisms for 48 hours at a low temperature (14°C) suitable for marine species and then stained with DAPI. Their data indicated approximately a 3-log improvement in counts. These authors interpreted these counts as viables in the original sample that were dormant and would not grow on conventional solid media. Tsuji et al. (1995) also have a viable stain method.

4. *CFUs observed on a resuscitation medium containing chicken fat:* The proposed method provides nutrients for the recovery of damaged cells (Martin and Katz 1993) and is a variation on the SPC (Greenberg et al. 1981). Cells recovered on a conventional sterility test membrane filter are plated on solid media containing the following:

| | |
|---|---|
| TSB base | |
| Yeast extract | to 0.5 percent (w/v) |
| Sodium pyruvate | to 0.25 percent (w/v) |
| Sodium thioglycolate | to 0.01 percent (w/v) |
| Chicken fat | to 0.1 percent (w/v) |

The present author would then place several drops of the sterile nutrient medium, agar free, on the filter (containing trapped microbes) atop the same nutrient agar, with a view to simulating the liquid environment that appears to resuscitate damaged organisms. Angela Strantz (Baxter Healthcare) suggests pouring the modified nutrient agar on the filter atop the same medium. This approach remains unverified at this writing.

The basic method was developed for the recovery of heat- and freeze-injured *Listeria monocytogenes*. The present author suggests the Martin and Katz chicken fat formula as a starting point for the resuscitation of other organisms damaged by other means. The recovery time of note was 4 hours.

McFeters et al. (1982) reported that resuscitation of injured cells required the avoidance of surfactants. In studies on the quantitation of injured organisms where metabolic acitivity or replication is required, the recommendation for using milk solids or peptone media and low temperatures during resuscitation (McFeters et al. 1982) bears repetition.

5.   *ATP quantitation (Klegerman 1996; Olson 1987):* Adenosine triphosphate (ATP) is extracted from bacterial and/or fungal cells concentrated on a membrane filter. This author prefers to extract into a small volume of n-butanol/n-octanol and to quantitate ATP by photometry during the luciferin-luciferase reaction. However, the ATP assay is not relaibly sensitive to fewer than

100 organisms (Olson 1987) and requires a greater investment in time and effort to validate, as compared with the other tests that require greater microbiological skill.

## A NOTE ON MONITORING AEROSOLIZED MICROBES

The basic approaches to organism **recovery** from air are (a) impaction into a solid medium (e.g., slit-to-agar [STA] sampler) and (b) impingement into liquid medium (e.g., the Millipore all-glass impinger). All impaction samplers are not equivalent. For example, the STA impaction sampler has been shown in comparative tests to be more efficient than FH2, and either of these yield higher CFUs in comparative tests than the RCS Plus™ (Ljungqvist and Reinmüller 1996). It follows from the environmental data on disinfectant-damaged organisms that the CFUs detected per air volume should be far lower when recovered by impaction than with a liquid impinger. However, this has not yet been shown.

The liquid impinger solution that this author recommends is 0.1 M phosphate-buffered saline. For the recovery of organisms from surfaces, one would think that a sterile swab is best. In fact, RODAC (Replicate Organism Direct Agar Contact) plates are more efficient for undamaged organisms on surfaces (Ljungqvist and Reinmüller 1996). The reason for this may be that a large proportion of the organisms taken up by a swab remain adsorbed to the large surface area of the fibers of the swab.

One possible, rapid quantitation of aerosolized microbes is the staining of organisms in suspension with the lipophilic cationic dye, rhodamine 123, and subsequently counting the apparently viable, rhodamine-staining organisms in a flow cytometer (Kaprelyants and Kell 1993). If this approach can be made to work, it would be far more rapid and less labor intensive than the membrane-related methods, and there would be no need to concentrate the samples. One of the problems in developing the method would be the reduction of nonspecific counts.

## IDENTIFICATION OF ORGANISMS

In large facilities requiring extensive microbial work, microbes are usually identified from gas chromatographs of methylated fatty acids (Smith and Siegal 1996) or by means of a series of stains and growth in various media (Clontz 1996). Almost invariably, identification is required if the source of microbial contamination in a parenteral product is to be discovered and corrected.

The identification of *Staphylococcus aureus* or *Staphylococcus epidermidis* as a contaminant in the final container usually indicates a human source, most likely flakes from the skin. Which human is the source of contamination can be determined only if the organism is identified to a specific strain, which is possible only if a substantial number of isolates of the same species can be done in-house and examined together with strains from a culture collection. The only simple, reliable, and cost-effective method for doing this is gas chromatography. Identification to a specific strain is possible only if a microbial library of isolates is established in-house.

## ECONOMIC JUSTIFICATION FOR MICROBIAL ANALYTICAL METHODS DEVELOPMENT

For example, a given aseptically filled SVP lot consists of 100,000 items that is sold for $750,000. Two lots of the 40 lots produced annually of this particular product are lost due to microbial contamination. The loss is the value of the product plus overhead and the business opportunity. This is greater than the $1,500,000 that can be accounted for immediately. But on the basis of 2 lots lost, let us assume that 3 people (the cost of whom, collectively, including benefits, totals less than $200,000 per year) are trained in the differentiation of viable versus nonviable organisms based on staining characteristics. The final assumption is that, with a realistic view of the microbial flora of the clean room, contamination levels can be consistently reduced to a point below the level of detection in final containers by the USP sterility test.

## A NOTE ON MEDICAL APPLICATIONS OF ANALYTICAL METHODS DEVELOPMENT

The emphasis in this chapter has been on the actual sterility of aseptically produced (filter-sterilized) product. However, medical applications of revised analytical tests for dormant, injured, or dysgenic microbes must also be considered.

A dormant organism that passes through 0.2 μm rated membrane filters appears to be associated with interstitial cystitis; the organisms resemble cell wall–deficient forms (Domingue et al. 1995).

About 8 percent of the bacteria recovered from middle-ear effusions in acute otitis media in children (5–17 years old) were dormant (Stenfors and Raisanen 1992a). In another study, 19 percent of the bacterial samples recovered from patients contained dormant *Haemophilus influenzae* or *Streptococcus pneumoniae*, which did not grow on standard agar plates (Stenfors and Raisanen 1992b). Dormant organisms may act as persisting antigens in inflamed joints or in the eye (Fox 1990).

Dormancy has been postulated as one of four mechanisms by which organisms manifest resistance to antibiotics (McGowan 1991). It may give rise to misleading results in antimicrobial screening (Zak et al. 1990) and may also be the reason why mycobacteria persist in patients under antimicrobial therapy (Rook 1990). However, until appropriate tests for dormant organisms are available in the marketplace, much of this is speculative.

## ACKNOWLEDGMENTS

The views expressed herein are those of the author, who benefited from technical discussions with Dr. Roger Dabbah (U.S. Pharmaceopeial Convention), Molly Pickett (Millipore Corp., Bedford, MA), and Bill Young (Vice President for Sterility Assurance, Baxter Healthcare, Round Lake, IL) together with Dr. Angela Strantz of his group.

# REFERENCES

Akers, J. E., C. M. Kennedy, and J. P. Agalloco. 1995. Experience in the design and use of isolator systems for sterility testing. In *Isolator technology: Applications in the pharmaceutical and biotechnology industries*, edited by C. M. Wagner and J. E. Akers. Buffalo Grove, IL: Interpharm Press, Inc.

Bathgate, H., D. Lazzari, H. Cameron, and D. McKay. 1993. The incubation period in sterility testing. *J. Parent. Sci. Technol.* 47:254–257.

Beloian, A. 1983. Methods of testing for sterility: Efficacy of sterilizers, sporicides, and sterilizing processes. In *Disinfection, Sterilization and Preservation*, 3rd ed., edited by S. S. Block. Philadelphia: Lea & Febiger.

Besajew, C. 1992. Importance of incubation time in the test for sterility. *Pharm. Ind.* 54:539–542.

Bianchi, A., and L. Giuliano. 1996. Enumeration of viable bacteria in the marine pelagic environment. *Appl. Environ. Microbiol.* 62:174–177.

Byrd, J. J., H.-S. Xu, and R. R. Colwell. 1991. Viable but nonculturable bacteria in drinking water. *Appl. Environ. Microbiol.* 57:875–878.

Clontz, L. 1996. Substrate utilization and the automated identification of microbes. In *Automated microbial identification and quantitation: Technologies for the 2000s*, edited by W. P. Olson. Buffalo Grove, IL: Interpharm Press, Inc.

Coles, T. 1998. *Isolation technology: A practical guide*. Buffalo Grove, IL: Interpharm Press, Inc.

Domingue, G. J., G. M. Ghoniem, K. L. Bost, C. Fermin, and L. G. Human. 1995. Dormant microbes in interstitial cystitis. *J. Urol.* 153: 1321–1328.

Fox, A. 1990. Role of bacterial debris in inflammatory diseases of the joint and eye. *APMIS* 98:957–968.

Greenberg, A. E., J. J. Connors, D. Jenkins, and M. A. H. Franson, eds. 1981. *Standard methods for the examination of water and wastewater*, 15th ed. Washington, DC: American Public Health Association.

Jaster, S. 1977. Use of Nuclepore filters for counting bacteria by fluorescence microscopy. *Appl. Environ. Microbiol.* 33:1225–1228.

Kaprelyants, A. S., and D. B. Kell. 1993. Dormancy in stationary phase cultures of *Micrococcus luteus*: Flow cytometry analysis of starvation and resuscitation. *Appl. Environ. Microbiol.* 59:3187–3196.

Klegerman, M. E. 1996. Quantitative ATP analysis. In *Automated microbial identification and quantitation*, edited by W. P. Olson. Buffalo Grove, IL: Interpharm Press, Inc.

Knudsen, L. F. 1949. Sample size of parenteral solutions for sterility testing. *J. Am. Pharm. Assoc.* 38:332–337.

Kogure, K., U. Simidu, and N. Taga. 1978. A tentative direct microscopic method for counting living marine bacteria. *Can. J. Microbiol.* 25:415–420.

Ljungqvist, B., and B. Reinmüller. 1996. Some observations on environmental monitoring of cleanrooms. *Eur. J. Parenteral Sci.* 1:9–13.

Martin, A., and S. E. Katz. 1993. Rapid determination of *Listeria monocytogenes* in foods using a resuscitation/selection/kit system detection. *J. AOAC Int.* 76:632–636.

McFeters, G. A., S. C. Cameron, and M. W. LeChevallier. 1982. Influence of diluents, media, and membrane filters on detection of injured waterborne coliform bacteria. *Appl. Environ. Microbiol.* 43:97–103.

McGowan, J. E., Jr. 1991. Abrupt changes in antibiotic resistance. *J. Hosp. Infect.* 18 (Suppl. A):202–210.

Olson, W. P. 1987. Sterility testing. In *Aseptic pharmaceutical manufacturing: Technology for the 1990s*, edited by W. P. Olson and M. J. Groves. Buffalo Grove, IL: Interpharm Press, Inc.

Olson, W. P. 1996. The fourteen-day sterility test: How we got there (letter to the editor). *PDA J. Pharm. Sci. Technol.* 50:198.

Olson, W. P. 1997. Biofilms in the pipeline and in the patient. *PDA J. Pharm. Sci. Technol.* 51 (Nov–Dec).

Rook, G. A. 1990. Mycobacteria, cytokines and antibiotics. *Pathol. Biol. (Paris)* 38:276–280.

Roszak, D. B., D. J. Grimes, and R. R. Colwell. 1984. Viable but nonrecoverable state of *Salmonella enteritidis* in aquatic systems. *Can. J. Microbiol.* 30:334–338.

Smith, A. R., and J. P. Siegal. 1996. Cellular fatty acid analysis for classification and identification of bacteria. In *Automated microbial identification and quantitation: Technologies for the 2000s*, edited by W. P. Olson. Buffalo Grove, IL: Interpharm Press, Inc.

Stenfors, L. E., and S. Raisanen. 1992a. Immunoglobulin- and complement-coated bacteria in middle ear effusions during the early course of acute otitis media. *Scand. J. Infect. Dis.* 24:759–763.

Stenfors, L. E., and S. Raisanen. 1992b. Occurrence of *Streptococcus pneumoniae* and *Haemophilus influenzae* in otitis media with effusion. *Clin. Otolaryngol.* 17:195–199.

Stewart, P. S., T. Griebe, R. Srinivasan, C.-I. Chen, E. P. Yu, D. DeBeer, and G. A. McFeters. 1994. Comparison of respiratory activity and culturability during monochloramine disinfection of binary population biofilms. *Appl. Environ. Microbiol.* 60:1690–1692.

Tsuji, T., Y. Kawasaki, S. Takeshima, T. Sekiya, and S. Tanaka. 1995. A new fluorescence staining assay for visualizing living microorganisms in soil. *Appl. Environ. Microbiol.* 61:3415–3421.

USP/NF. 1995. *The United States Pharmacopeia, 23rd ed./The National Formulary, 18th ed.* Rockville, MD: The United States Pharmacopeial Convention, Inc.

Wood, R. T. 1996. Points to consider in the use of sterility testing isolators. In *Isolator technology: Applications in the pharmaceutical and biotechnology industries,* edited by C. M. Wagner and J. E. Akers. Buffalo Grove, IL: Interpharm Press, Inc.

Yamabe, S. 1973. Binding of acridine orange with DNA. *Arch. Biochem. Biophys.* 154:19–27.

Zak, O., H. Mett, and T. O'Reilly. 1990. Remarks on the screening of antibiotics for antibacterial activity. *Eur. J. Clin. Microbiol. Infect. Dis.* 9:462–465.

# 14

# Parametric Release

*James Wilson*

Pharmaceutical Systems Inc.
Mundelein, IL

The U.S. Food and Drug Administration (FDA) has been approving submissions for parametric release for approximately 20 years, and the FDA has established a policy that a 510(k) with premarket notification for medical devices is required before a firm may change from product sterility testing to parametric release for terminally sterilized products. Implementation of parametric release, however, demands carefully controlled process parameters derived from a sound validation system.

The term *dosimetric release* is often applied synonymously with process control release or parametric release. Dosimetric release is a finished product release procedure based on an accurate measurement of the delivered radiation dose and reliable process controls to ensure product sterility in lieu of sterility testing of final product, or an inoculated unit presumed to be a reasonable facsimile (i.e., inoculated carrier) of the product. Process control release or parametric release, on the other hand, is a finished product release procedure based on a comprehensive process validation study, as well as effective control, monitoring, and documentation of the sterilization process, which may allow for product release without performing finished product sterility tests or chemical and/or biological indicator (BI) testing.

The European Pharmacopoeia (EP) Sterility Test, step 5 draft (3/22/96) states in the nonmandatory section:

> *In the case of products sterilized in their final sealed containers, physical proofs, biologically based and automatically documented, showing correct treatment throughout the batch during sterilization are of greater assurance than the sterility test. Hence, a validated parametric release of these products is acceptable.*

The United States Pharmacopeia (USP) also acknowledges that the product sterility test is not required as a release test in the General Notices section. The requirement for a product sterility test, as a release test, therefore, is based on Good Manufacturing Practices (GMPs), 21 CFR 211.167, not on the USP.

## ADVANTAGES OF PARAMETRIC RELEASE

When product is released as sterile as a result of approved processing (i.e., parametric release), the manufacturer benefits primarily in three ways:

1.  Sterility testing for lot release is no longer required. The product is declared sterile as a result of approved sterilization processing. This approach eliminates the need for the placement, retrieval, testing, and incubation of sterility test samples.

2.  Sterility assurance measured in terms of the probability of nonsterility on an appropriately validated process provides a greater margin of consumer protection than can be achieved with the USP referee test for sterility.

3.  Product can be released as soon as the process has been acceptably reviewed and approved. This approach eliminates the quarantine period for sterility testing

(i.e., on test inventory). Normally, this amounts to approximately two weeks. A one-time inventory savings is realized, and the responsiveness of the manufacturer to its customers is increased by the same amount of time.

*Note:* Ethylene oxide (EtO) residuals and EtO degassing (i.e., of residuals inside product containers) must be taken into account.

## DISADVANTAGES OF PARAMETRIC RELEASE

- Failure to meet a critical process parameter for a given batch ensures rejection; there is no allowance for additional testing.

- Residuals and degassing of product must be taken into account for product sterilized by EtO processing.

- Costs for comprehensive validation/evaluation elements may be high.

## FACTORS INFLUENCING STERILIZATION METHODS

From a technical standpoint, the reliability of a sterilization process can be stated as the inverse of the number of factors that must be controlled. Therefore, on this basis, radiation would be the easiest process to control and the EtO gas process the most difficult to control (see Table 14.1).

## GENERAL EXPECTATIONS OF THE FDA

To obtain FDA approval on a parenteral submission for parametric release, the following categories must be satisfied:

- Terminally sterilized product

- $10^{-6}$ probability of nonsterility demonstrated

**Table 14.1. Factors to Be Controlled in Various Sterilization Processes**

| Factor | Autoclaving | Gamma Radiation | EtO Gas |
|---|---|---|---|
| Time | Yes | Yes | Yes |
| Temperature | Yes | No | Yes |
| Pressure | Yes | No | Yes |
| Vacuum | Yes | No | Yes |
| Concentration (diffusion) | (Yes) | No | Yes |
| Wrapping | Yes | No | Yes |
| Humidity | No | No | Yes |
| Product Density | Yes | Yes | Yes |
| Approach to Sterilizer | No | Yes | No |

- Cycle description

- Cycle qualification: heat distribution and heat penetration

- Minimum parameters defined

- Heat resistance of BIs

- Comparison of BIs to bioburden

- Focus on container-closure system

- Written procedure on action level conditions

- Description of environmental monitoring program

- Cycle monitoring: quantitative levels

- Will not allow management review board release outside of specifications

## REGULATORY SCOPE

The aborted, proposed large-volume parenteral (LVP) GMPs issued in 1976 did not allow for parametric release [21 CFR 212.167(c) required end product sterility testing]. Both device and drug GMPs contain requirements that apply to parametric release. 21 CFR 211.165(b) states:

> There shall be appropriate laboratory testing, as necessary, of each batch of drug product required to be free of objectionable microorganisms.

For devices, 21 CFR 820.809(d) reads:

> Final acceptance activities. Each manufacturer shall establish and maintain procedures for finished device acceptance to ensure that each production run, lot, or batch of finished devices meets acceptance criteria. Finished devices shall be held in quarantine or otherwise adequately controlled until released. Finished devices shall not be released for distribution until: (1) the activities required in the Device Master File (DMF) are completed; (2) the associated data and documentation is reviewed; (3) the release is authorized by the signature of a designated individual(s); and (4) the authorization is dated.

The FDA position means that, even for a validated process, there must be some method whereby each lot is tested or checked to ensure that sterilization processing requirements are met.

An industry initiative by the Pharmaceutical Manufacturer's Association (i.e., now PhRMA, Pharmaceutical Research and Manufacturers of America) resulted in a 1980 position paper supporting an alternative to end product sterility testing for LVPs.

The FDA issued a compliance guide on parametric release for terminally heat-sterilized products dated 10/21/87. To date, there has been no official guide issued specifically for EtO and irradiation processing.

The FDA guideline for moist heat-sterilized products was issued by compliance mainly to focus on pre-1938 drugs (i.e., drugs not currently covered by a New Drug Application [NDA]), so as to prevent firms from eliminating sterility testing of final product without any safeguards at all. For NDA drugs, a prior approval NDA supplement would be required so that the FDA would have prior knowledge of and input into the implementation of parametric release for NDA drugs.

The FDA guidelines differed from the PhRMA position—in that the PhRMA specified periodic testing of bioburden and BIs or chemical indicators (CIs), whereas the FDA prescribed these tests for each lot.

As defined in the FDA compliance guide, the parametric release of steam-sterilized products, in lieu of end product sterility testing, may be acceptable when the following four parameters are met and documented.

1.  The sterilization process cycle has been validated to achieve microbial bioburden reduction to $10^0$, with a minimum safety factor of an additional six-logarithm reduction. Cycle validation includes heat distribution studies in the sterilizers, heat distribution studies of the product, bioburden studies, and lethality studies referencing a test organism of known resistance to the sterilization process. All cycle parameters must be identified by the manufacturer as critical (e.g., time, temperature, pressure) or noncritical (e.g., cooling time, heat-up time). Under parametric release, the failure of more than one critical parameter would result in the automatic rejection of the sterilizer load (see number 4 concerning BIs) [21 CFR 211.113(b)].

2.  The integrity for each container-closure system has been validated to prevent in-process and postprocess contamination over the product's intended shelf life. Validation should include chemical or microbial ingress tests, utilizing units from typical products (21 CFR 211.94).

3. Bioburden testing (covering total aerobic and total spore counts) is conducted on each batch of presterilized drug product. Resistance of any spore-forming organism found must be compared to that of the organism used to validate the sterilization cycle. The batch is deemed nonsterile if the bioburden organism is more resistant than the one used in validation (21 CFR 211.110).

4. CIs or BIs are included in each truck, tray, or pallet of each sterilizer load. For CIs, time/temperature response characteristics and stability are documented, and minimum degradation values are established for each sterilization cycle. CIs cannot be used to evaluate cycle lethality. It is relevant to note that CIs can be employed as an alternative to BIs for parametric release but not for validation.

Documentation is required for BIs. Documentation for each BI lot must include the organism's name, source, D value, spore concentration per carrier, expiration date, and storage conditions. BIs can be used to evaluate cycle lethality where equipment malfunction prevents the measurement of one critical cycle parameter. If more than one critical parameter is not met, the batch is considered nonsterile, despite BI sterility [21 CFR 211.165(e) and 211.167].

## PDA RESPONSE TO FDA COMPLIANCE GUIDELINE ON PARAMETRIC RELEASE

The PDA (Parenteral Drug Association) Task Force and Research Committee suggested that parametric release means the lot release of terminally sterilized product based on measured and documented information related to the operation and control of the sterilization process during routine manufacturing and sterilization of the product. The accuracy and reliability of such recorded information may be verified by comparing process information collected daily to data and information that was obtained during initial vessel, process, and/or product validation programs. Structured periodic performance qualification studies should be conducted to assure process performance consistent with the operational ranges in validation.

The PDA concluded that the FDA Compliance Policy Guide is a qualified statement that does not fully recognize parametric release (i.e., reliance on measured, predefined operational parameters). The FDA guide additionally requires the use of process monitors in each load and autoclave tray. The FDA Compliance Policy Guide, therefore, represents a substitution of alternative methods for sterility testing, not acceptance or recognition of a true parametric release program. Specific PDA technical concerns relative to the FDA Compliance Policy Guide are given below.

The PDA pointed out that paragraph D of FDA Compliance Policy Guide 7132a.13 describes a substitute or alternative method that may be used in lieu of sterility testing. Mention is also made of CIs; however, unlike BIs for moist heat sterilization, no defined, proven, universally accepted CI system is available. Further, large variabilities could occur with a chemical assay system when using a wide variety of temperature and time conditions. Extensive data to establish the variability associated with a chemical system using many conditions of use would have to be generated. Number 4 (see above) also does not consider the use of thermoprobes or thermocouples that can be very accurately calibrated to measure the temperature of products and/or the sterilizing environment. The data obtained from thermoprobes (converted to $F_o$ computations if desired) can be used to calculate microbial inactivation accurately when correlated to a preexisting microbial inactivation database.

In effect, the PDA recommended that the FDA Compliance Policy Guide be amended to provide different levels of parametric release criteria based on the type and magnitude of initial validation, revalidation, and microbial inactivation data that each firm may possess.

## CALIBRATION AND MAINTENANCE

Since parametric release will depend on information obtained from measurements and the majority of the data generated critical to the success of the process, a comprehensive calibration program is essential for parametric release. Documented calibration and preventive maintenance programs, consistent with FDA regulations and/or another national standards body, should be utilized on instruments that control, monitor, and record the sterilization process. Equipment should be designed and installed with provisions for periodic maintenance, calibration verification, and correction. Equipment to be maintained and/or calibrated on a routine basis may include, but is not limited to, the following equipment:

- Monitoring equipment (i.e., environmental and/or chamber)
- Monitoring gauges
- Recorders
- Timers
- Pressure transducers
- Gaskets and seals
- Door safety interlocks
- Safety pressure relief valves and rupture disks
- Filters (for periodic replacement)
- Volatilizers/vaporizers
- Chamber jacket system (for corrosion and insulation)
- Audible and visual alarms
- Boiler system for steam and heat supply (i.e., steam quality and quantity)
- Evacuation equipment
- Weigh scales
- Valves

## OPERATIONAL REQUIREMENTS FOR PARAMETRIC RELEASE

### Steam Sterilization Process

The major elements of proper steam sterilizer design for parametric release are summarized as follows:

### *Design Considerations*

- *A reference temperature measurement device, separate from the controller/recorder, for use as a cross-check or in the event the principal temperature controller fails.* Accurate devices such as thermocouples or resistance thermal detectors (RTDs) for temperature range, legibility, response time, and location are needed. Continuous steam bleed past the bulb (or probe) is a positive feature.

- *A permanent record of chamber temperature through the entire process.* A temperature recording device is required, and it must agree with the reference temperature standard; a means of preventing unauthorized changes in the adjustment is required; for the chart, the size of graduations and the maximum temperature range should be specified; the means of sensor installation (with constant bleed) should be indicated.

- *Measurement of sterilizer pressure.* A pressure gauge equipped with a watertight pressure-transmitting diaphragm (i.e., a self-draining pressure gauge) is required; criteria for calibrations and installations must be specified. (Note: The gauge is often used in conjunction with the temperature monitoring device to confirm that saturated steam conditions existed in the sterilizer.)

- *Control of steam to the sterilizer.* The required steam controller is to be activated by temperature, not pressure, and a provision to prevent unauthorized changes to the control setting is required.

- *Adequacy of steam distribution.* The steam inlet must be large enough to provide sufficient steam for the proper operation of the sterilizer(s) under conditions where the maximum number of autoclaves were venting simultaneously).

- *Effectiveness of steam distribution.* Steam entry is opposite the vent; the design should provide uniform steam distribution. Steam spreaders are tightly controlled with regard to size (vs. inlet line) and perforations (size, location, number); alarms for blowers, if used, should be provided.

- *Assurance of complete removal of air from chamber.* The required vents must achieve removal of air before they are closed; valves must not restrict the flow of vented air (i.e., gate or ball valve preferable to globe valve); atmospheric breaks required; the vent line and manifold are sized to prevent back pressure (i.e., resistance to venting).

- *Uniformity of temperature, during exposure, removal of traces of air left in sterilizer or product load after venting.* A requirement for bleeders of specified size and location, with a stipulation that at least one be located within sight of the operator.

- *Assurance that product containers are exposed to the sterilizing medium (steam or steam/air mixture).*

Equipment and microbiological validation studies required to accomplish parametric release have not been thoroughly described in publications. They include, but are not limited to, the following:

- Bioburden data
    - How measured
    - Frequency of testing
    - Historical data summary profile
    - How data are dispositioned

- Environmental data

  - Tests performed

  - Frequency

  - Action levels

  - Investigative/corrective action system

- BI data

  - Source

  - Resistance

  - Population

  - Stability

  - Storage and handling

- Qualification of equipment

  - Penetration and heat distribution studies to evaluate heating characteristics of the vessel, the carrier system, and heating medium used

  - Evaluate ability of sterilizer to hold medium used

  - Evaluate ability of sterilizer to hold required cycle parameters

  - Ability of control mechanisms to operate as intended

- All equipment is suitable for the intended function, correctly adjusted and maintained, and, if it is measuring equipment, calibrated.

- Some of the essential elements of process control for steam sterilization are (1) complete removal of air from the chamber before starting the exposure period; (2) use of saturated steam at near 100 percent quality (i.e., water vapor with neither superheat nor entrained condensate); and (3) a steam delivery and control system that can assure uniform temperatures on the order of ±1°C throughout the empty chamber.

- All appropriate process operations, such as packaging, sealing, sterilization, and so on, are validated. The steam sterilization process is established and validated such that it will destroy all bioburden on the product and package, or destroy all spores on a standardized BI and includes, for either case, a specified safety factor (e.g., $10^6$). All of the selected equipment and cycle parameters are documented and become part of the specifications for routine production.

- Subprocess lethality verification studies on product and BIs.

- Load composition.

- Periodic revalidation of process and equipment.

- The product and package are designed to be compatible with steam sterilization.

- In case process parameters are not met for a production lot, there should be procedures for reprocessing or rejecting the lot.

## Reliability of Process Control

Reliability in process control may be enhanced by adoption of the following features:

- Microprocessor-based programmable controllers.

- In situ, sterilizable, vacuum-breaking, air inlet filter with stainless steel sanitary housing and sterilizing replaceable filter element.

- Pipes connected to the vessel with at least 3/4 inch diameter opening and equipped with at least a 1/16 inch (or larger diameter) bleeder opening.

- Cycle event sequence recorder.

- Sterilizing medium equipped with visual and audio alarms to signal improper operation.

- Suitable valve on air line to prevent leakage into sterilizer.

- Water level indicator (where applicable).

- Recirculating pump water temperature (where applicable).

- Time/temperature integration control.

- Digital printouts.

- Temperature recorder to have a response time of not more than 2 seconds for 90 percent of scale response.

- Temperature recorder chart working scale of not more than 30°C of 50°C per inch with a range of 10°C or 20°C of the processing temperature; graduation shall not exceed 1°C or 2°F.

- The steam spreader pipe, if present, must individually be no longer than the steam inlet line and extend the length of sterilizer. Perforation shall be along a 90° sector of the spreader and shall direct steam uniformly toward the longitudinal axis of the vessel.

- Cooling water system with means to prevent backflow.

- Computer recording of data.

## Ethylene Oxide Processing

Gaseous sterilization is considered by some practitioners to be inappropriate for parametric release; nevertheless, there seems to be a consensus that the number of variables to be controlled for EtO processing do indeed present a challenge.

It is important to bear in mind that the Association for the Advancement of Medical Instrumentation/American National Standards Institute/International Organization for Standardization (AAMI/ANSI/ISO) document 11135 defines parametric release as processing without BIs. Some practitioners suggest that parametric release for EtO would require only monitoring/control of water vapor, temperature, gas concentration, and load density.

## Design Considerations

Bioburden, environmental data, and BI submission requirements would be the same as for steam processing.

### Sterilizer Chamber

• General Description

The actual sterilization process occurs within the sterilization chamber.

Successful EtO sterilization requires that the following four conditions be present within the chamber in order for the chemical reaction between the EtO gas and the contaminating organisms to occur:

1. *Moisture:* Almost all chemical reactions take place more efficiently in the presence of water (i.e., moisture).

2. *Temperature:* The rate of a chemical reaction is directly proportional to the temperature at which it occurs. Generally, the reaction rate doubles with every 10°C rise in temperature.

3. *EtO concentration:* There must be sufficient EtO gas present to react with all of the microbial cells present.

4. *Time:* The product must be left in contact with the sterilizing conditions long enough to achieve the desired effect when using 100 percent EtO. The

chamber and all ancillary equipment must be explosion proof.

• Chamber (Vessel)

Chambers are constructed of stainless steel to prevent rust and corrosion that are associated with catalytic formation of polymerized EtO. This also increases the risk of explosion when using 100 percent EtO gas.

Vessels are heated by means of a series of channels that fully jacket the vessel with softened, hot water. Channels are constructed of noncorrosive materials.

Doors must be jacketed to prevent them from becoming a large heat sink.

The exterior (jacket) must be insulated to prevent conductive heat loss and minimize temperature variation. Insulation with a value of R-19 or better is essential and must include the doors, sides, floor, and top of the chamber.

Heating and cooling of the chamber is accomplished through heat exchangers linked to the jacket circulation system.

The temperature range within an empty chamber during gas exposure should be ±3°C.

The points of temperature control and monitoring must represent the status of the process vessel. Redundant systems are preferred.

- *Manifold recirculation system:* One of the most critical components and must contain manifolds throughout the vessel length. Recirculations establish an even temperature distribution and eliminate stratification of the sterilant gas and/or the presence of air pockets. The system must be designed to prevent pooling of condensed moisture that may accumulate between cycles. The manifold system must also be insulated and free of deadlegs.

- *Automated door opening/closing:* Doors must interlock in operation.

- *Automatic pallet handling:* Allows for rapid load transfer into the sterilizer chamber. Transfer time should be as little as a few minutes.

- *Chamber dimensions:* Chamber size must allow maximum product utilization on standard size pallets. High strength plastic or metal pallets should be used.

- *Chamber leakage:* Chamber seals and joints must withstand extremes in pressure and vacuum in order to prevent leakage. The chamber shall be capable of low vacuum (< 1 inch of mercury) and pressures of 30 psig. Leakage of chamber atmosphere is dangerous to exposed employees and results in numerous gas makeups to maintaining specified pressure (e.g., EtO concentration). Leakage of air into the chamber can result in explosive air/gas mixtures.

- *Jacket temperature and circulation flow:* The temperature of jacket water and its flow rate must be continuously monitored to maintain an even temperature within the vessel. An alarm system must be a part of the design to detect changes in these two parameters.

- *Humidity injection ports* shall be located in such a manner that the steam does not enter directly and cause product "hot-spots." It is preferable to inject steam via the circulation manifold.

- *Chamber monitoring probes are needed* (e.g., 1 each per 100 cubic feet of usable chamber volume for temperature monitoring).

- Gas Supply System

  100 percent EtO is supplied as a liquid under pressure with nitrogen and is stored in secure, explosion proof, well-ventilated areas, with provisions for temperature control.

  The gas room must be equipped with scales for the verification of gas quantities. Scales must be capable of continuous monitoring throughout all phases of the sterilization process.

  – *Noncorrosive, insulated feedlines:* Sterilant gas is fed from the supply force via stainless steel, Teflon™-lined feedlines. This connection must contain a bleed valve that directly exhausts to the EtO disposal system so that pressure buildup can be relieved when connections are made or broken.

  – *Gas volatilizer:* Cold, liquid sterilant must be vaporized and heated prior to entering the chamber. The volatilizer is a tank filled with water or glycol through which the supply line is coiled; the volatilizer has a blade-type recirculation system to circulate the heating medium around the feedline coils.

  The temperature of the volatilizer-heating medium is kept above the chamber temperature.

  The effectiveness of the volatilizer is monitored by placing a thermocouple at the injection site into the chamber. Sterilant injection rate into the vessel is controlled by the gas supply pressure and the inlet valves.

  The sterilant injection port should be located and baffled so that incoming gas does not impact the product.

- *Gas supply system alarm:* This alarm is activated when the gas temperature drops below a specified temperature.

- Capability of *variable rate control* and equipped with automatic switching when cylinders are empty.

- *Nitrogen* must be added via an independent part(s) and have in-line filters rated to 0.2 micrometer (μm). A bulk supply system may be considered as an alternative.

- Evacuation System

  Prior to product removal from the chamber, the sterilant atmosphere must be removed from the chamber and as much as possible from the product. This prevents exposure of employees to toxic EtO; reestablished atmospheric pressure within the chamber eliminates the explosive mixtures of air and EtO.

  - The vacuum pump must have a recirculating design that is kept cool with a heat exchanger. Mechanical seals are required because they eliminate the accumulation of EtO in the service liquid.

  - The piping system must be impermeable to EtO and noncorrosive to prevent polymerization of EtO.

  - The inbleed system must contain 0.22 μm filters in addition to prefilters.

  - Capability of variable rate control.

  - Guarded to prevent accidental blockage by product(s) or vessel content.

- Residues of EtO and its reaction products (ethylene chlorohydrin and ethylene glycol) can be toxic to the patient on whom the product is used as well as employees assigned to the sterilizer area.

Temperature, dwell time, forced air circulation, chamber loading, and product and packaging materials affect the efficiency of aeration.

- *Chamber airwashing/hot cells:* Following evacuation, aeration is performed in both chambers by airwashing during evacuation and in rooms called "hot cells." The temperature in the hot cells is maintained at the same level as during sterilant exposure, with ventilation in the range of 10–20 air exchanges per hour.

- *Fugitive emissions* from the hot cells are connected directly to the EtO disposal system. Fugitive emissions of EtO from aeration rooms and gas storage rooms are scrubbed using catalytic converters.

- *Automated pallet handling* into the aeration (hot cell) areas.

- Gas monitoring systems for all accessible areas must be capable of detecting emissions in the breathing zone, with warnings (audible and visual) per the U.S. Occupational Safety and Health Administration (OSHA). Adequate ventilation and other engineering controls must be installed to assure compliance.

- Computer Hardware/Software

  The chamber and its ancillary equipment must be computer controlled. The process must be capable of monitoring all phases of the cycle and controlling parameters. Continuous printouts must be available. Real-time data in a digital format are needed for cycle parameters.

  - Temperature

  - Time

- Humidity

- Pressure

- Gas weights

- Gas inlet temperature

- Gas volatilizer temperature

- Flow monitoring (circulation)

- Rate control of pressure changes

• Other Design Features

- *Relative humidity/temperature monitoring:* Required for both the preconditioning control room (PCR) and product loads.

- *Relative humidity/temperature by steam injection:* Steam must be nontoxic and free from boiler water additive chemical carryover. Steam is to have a dryness value of not less than 0.95, containing not more than 3.5 percent (V/V) of noncondensable gases and not superheated more than 5°C.

- *Insulation:* Sufficient to allow control of room parameters within established specifications and avoid areas of condensation that cause water pooling and/or surface deterioration.

- *Room surfaces:* Must be smooth and have a durable surface (e.g., epoxy paint) to allow frequent cleaning and disinfection, especially for mold growth. Accessibility of covered drains to expedite cleaning.

- *Recirculation:* Circulation blowers are required to provide air circulation within the room and in and around product sufficient to maintain specified temperature and humidity ranges.

- *Traffic in and out of the PCR can interrupt the temperature/humidity balance of the PCR.* Controlled entrances are a necessity. Airlock door designs will allow traffic in and out of the room without degrading the PCR environment. Doors must be self-closing with auditory/visual preset alarms in case a door does not close.

– *Automated pallet handling* allows for rapid load transfer into the sterilizer chamber. Transfer time can be as little as a few minutes.

## Specific Monitoring Requirements

Sterilization cycle validation is based on a carefully chosen microbial challenge and correlation to selected key process parameters.

- Temperature of load during preconditioning

- Temperature of load during sterilization

- Humidity during conditioning (by direct measurement)

- Humidity during sterilization

- EtO concentration in chamber (by direct analysis)

- Defined and validated load configurations

- Defined and validated time limit for product transfer between load conditioning and sterilization

- Actual volatilization temperature

- Temperature of gas as it enters the chamber

## Irradiation Processing

Implementation of dosimetric release for terminally sterilized medical devices and drugs has become more popular with the radiation process as compared to other processes. It is important to remember that dosimeter values reflect dose delivered, not microbial lethality. Bioburden, environmental, and BI submission requirements are the same as for steam processing. Data required to change from end-product release testing procedures to dosimetric release procedures include the following elements:

- Bioburden and its resistance to process

- Environmental monitoring data

- BI tests/data

- Process controls during processing

- Sublethal dose studies on product and BIs

- Calculation of appropriate sterilization dose

- Dosimetry studies

- Overall control of process

- Correlation of BI sterility test validation data and dosimetric results

- Description of procedures used to monitor and document data

- Audit procedures for both process and product

## Design Considerations

- *Accountability:* The receiving report must agree with the shipping order. Packers are checked to assure satisfactory condition (no punctures, tears, etc.).

- *Material handling:* Nonirradiated and irradiated storage areas are identified and separated by a physical barrier. Severely damaged packers are taped, and all packers are irradiated regardless of condition. The lot number on each packer label should be verified. Only those lots scheduled for a run are staged for loading. Carriers/totes are loaded sequentially, starting with the carrier/tote in the first position. Similar density components are loaded in consecutive carriers/totes. Units of different item codes should not be mixed in the same carrier/tote. The quantity of component should be sufficient to complete at least one full layer in the carrier/tote. Empty carriers/totes are not separating loaded carriers/totes. Packers are typically centered between side walls and pushed to the back of the carriers/totes. All end labels should face the front.

- *Dosimetry:* Dosimeters should be placed in every carrier/ tote of each different type component. Dosimeters are placed in the proper positions of the component loaded in the carrier/tote per specifications. For partial carriers/totes, a dosimeter should be in the top position. Dosimeter batches should not be mixed in a run.

- *Sterilizing process:* Carriers/totes are placed in the sterilizer in their loading sequence. The master timer and overdose timer are set to the correct cycle times. The irradiation cycle continues until the minimum dose has been delivered. The cycle may be interrupted. All mechanical failures or run irregularities are recorded in the batch record.

- *After each run:* Packers are properly palletized, with labels facing in the same direction. Only one lot should be placed per pallet. (Labels should face toward the fork entry point of the pallet.) The pallets of sterilized product are protected by on corner boards and banded for shipment. "Processed" labels contain the run number and date of irradiation and are attached to at least one side of each pallet.

- *Dosimeter evaluation:* The program for the dosimeter readout system agrees with the batch of dosimeters being used. The dosimeter batch and spectrophotometer are verified daily. The thickness gauge must be within its calibration date. Dosimeters are clean and handled by the edges when being read. Spectrophotometer readings and thickness readings should be taken at the same location on the dosimeter. The dosimeter readings are to indicate, by design, that the minimum dose was delivered, and that the maximum specified dose was not exceeded.

## DOSIMETRY SYSTEMS AND DOSE MEASUREMENT

Accurate measurement of the radiation dose is essential for irradiation processing. The accuracy of measuring absorbed dose in processed product is directly related to the quality of the dosimeter and the accuracy of its measurement system.

The most common form of dosimeter readout is the determination of optical absorbance changes in the ultraviolet or visible region. Absorbance measurements are performed using spectrophotometers or instruments designed for a particular type of dosimeter.

Other forms of dosimeter readout other than optical absorbance include temperature measurement (in the case of calorimeters) and electron paramagnetic resistance (in the case of alanine dosimeters). These dosimeters can be purchased commercially.

Many dosimeters require a measurement of thickness. In such cases, the thickness-measuring device becomes part of the dosimetric system. Some dosimeters require the control of environmental conditions, such as humidity. In such cases, the equipment used to control these environmental conditions becomes part of the dosimeter system.

Primary dosimeters are usually established, calibrated, and maintained by national standards laboratories. These dosimeters are either sanitation chambers or calorimeters and will measure absorbed dose to carbon or metal with an uncertainty of about $\pm 1$ percent. A theoretical step is required to convert the measured dose to the dose normally used in radiation sterilization, namely, the absorbed dose to water. This will typically lead to an overall uncertainty of $\pm 2$ percent.

Calibration conditions can, at best, only approximate those conditions that occur in the irradiator. Significant errors, due to differences in conditions when calibrating or using dosimeters, can still occur. One way of checking for this is to irradiate a number of reference dosimeters (e.g., dichromate or alanine) in close proximity to routine dosimeters in the irradiator. Often, this is done as part of qualification dose mapping; however, it should always be considered whenever a new batch of dosimeters is introduced. To ensure traceability to a national standards laboratory, reference dosimeters should be calibrated by a national standards laboratory or a formally accredited secondary standards laboratory. The use of a facility not having formal accreditation for calibrating dosimeters must be regarded as a break in the traceability chain to national or primary standards.

In irradiation processing, it is vital to understand the following terms.

- *Random Error:* A variation on measured value of unpredictable sign and magnitude occurring when measurements of the same quantity are made under effectively identical conditions.

- *Systematic Error:* A contribution to the total uncertainty comprised of the combined effects of all nonrandom error sources, known or unknown, that tend to offset uniformly or predict all results or repeated applications of the same measurement process at the time of the measurement.

- *Uncertainty:* An estimate of the range of values about the measured value in which the accepted value is believed to lie. The total uncertainty represents the sum of a measure of the random error and estimates bound to the systematic error. The term is inversely related to the term *accuracy*, in that an instrument or measurement with the greater uncertainty is said to have the lesser accuracy and vice versa.

A good dosimetry system has the ability to provide a reasonable and accurate measure of absorbed dose in a given material; however, there are several significant sources of errors that contribute to the total measurement uncertainty of the dosimetry system used to quantitate the absorbed dose of healthcare products being sterilized by radiation. These error sources are as follows:

- Dosimeter characteristics

- Dosimeter calibration uncertainty

- Optical density measurement uncertainty

- Dosimeter thickness measurement uncertainty

- Measurement procedures

- Dosing error (i.e., in calibration)

Unfortunately, industry working groups on dosimetry discuss theory routinely, but very little data are exchanged or utilized to set target standards for acceptable variability. There is, evidently, a widespread fear among dosimeter users to release such "confidential data" to the public. It is apparent among sterilization experts that sources of error/uncertainty can result in a combined measurement uncertainty of absorbed dose of up to 25 percent. It is largely left up to individual companies to adopt their own path since no regulatory standard exists for dosimeter variability. Even the National Institute of Standards and Technology (NIST) guideline on the subject offers little light from a technical standpoint.

Systematic difference in the mean response of dosimeters, for any given dose level, between different batches of dosimeters is compensated for by calibrating each batch of dosimeters; however, variability in the response of each dosimeter with the same batch can result in a significant error in the absorbed dose measurement. The random variability of individual dosimeter response for any single dose value may result in a significant dose measurement uncertainty.

This measurement variability may be reduced by using multiple dosimeters at each dose measuring point, and using the average of the dose measurements as the predicted dose value. Measurement variability can be reduced by

$$R = 100\left(1 - \sqrt{\frac{1}{N}}\right)$$

where *R* is the percent reduction in measurement uncertainty and *N* is the number of dosimeters used at each measuring point.

Regardless of what declared variability value is used, each dosimeter result should be viewed as X (kGys) ± Y (kGys) (declared variability in percent at a given confidence level). For example, this means that variability may be declared as ±5 percent at the 68 percent confidence level (i.e., one standard deviation) and ±10 percent at the 95 percent confidence level (two standard deviations). When one measures dose for lot release using a single dosimeter, variability needs to be considered near the edge of specification limits. Setting the specified dose range should take single dosimeter variability into account. When dealing with dosimeter readings to determine delivered product dose, it is prudent to have a high confidence level. If the confidence level is unknown, or is statistically insignificant, interpretation of dosimeter results can be more treacherous.

The performance of routine dosimeters with respect to environmental influencing factors is not as good as that of reference dosimeters; however, their low cost and ease of use make them suitable for use when large numbers of dosimeters are required (i.e., for dose mapping exercises and routine monitoring of radiation sterilization processing).

The distribution of dosimeters within a load is important for the identification of maximum and minimum dose points. Multiple dosimeters should be read, and the dose values graphed to yield penetration profile characteristics. These data should in turn be used to determine dosimeter placement. Two different types of dosimeters should be used for dose validation, but only one system need be used for sterilization subsequent to validation.

As required by Section 510(k) of the medical device amendments of 1976 and in conformance with 21 CFR 807, a change to dosimetric release requires prior approval by the FDA. The 510(k) submission for changing from finished product and/or BI testing to dosimetric release must contain the basis for selecting and controlling the dose and process controls. Firms may use one of the AAMI/ANSI/ISO dose-setting protocols—or another approach—which demonstrates for a particular product and microbial load that the sterility assurance level will be equal to or better than that selected as appropriate for the device. Dose mapping of the load is required to ensure that the routine dosimeters are at the low dose zones. If the dose is based on bioburden, the firm must describe the environmental control and monitoring used to ensure that bioburden limits are not exceeded.

The validation selected process variables must be monitored for routine processing to assure adherence to process validation boundaries. The following parameters would typically be controlled for routine processing on a validated process:

- Source position

- Source strength at time of irradiation

- Product position

- Conveyor motion (continuous) or dwell period (batch)

- Product density

- Low- and high-dose zones

- Approach to source

## DATA MANAGEMENT

Regardless of the sterilization mode employed, the presentation of data is important. On one hand, there is potential to be confronted with a number of technical decisions concerning the handling of data. For example, questions on how parameters such as pressure, $F_o$ readings, and so on would be rounded for reporting purposes must be defined. On the other hand, data on the repeatability of the sterilization process, from a functional standpoint, is important for a submission to the FDA. Repeatability can be viewed overall by using histograms. A normal distribution of data is a must if the traditional approach to setting process limits is to be followed (i.e., mean ±3 sigma). An assessment of the "mean time between failure" may be required on critical process variables to assure eligibility of the process for parametric release from a financial viewpoint. It is common industry practice for practitioners to complete statistical surveys on process data, and it may be important to evaluate the impact of process variable interactions specifically.

## REFERENCES

PDA. 1987. Parametric release of parenteral solutions sterilized by moist heat sterilization. Technical Report No. 8. *J. Parent. Sci. Technol.* 41 (2S).

PDA. 1988. Parenteral Drug Association response to the FDA Compliance Policy guide, #7132a.13, entitled, "Parametric Release—Terminally Heat Sterilized Drug Products." *J. Parent. Sci. Technol.* 42 (4):105.

# Index

Porter, M.E., 275, 307, 310
potassium dichromate, 413, 414
potassium octatitanate, 199
potassium permanganate, 254
potentiometric techniques, 331
povidone, 374, 375
poxvirus, 440
PRANDTL Number, 82
Prasil, Z., 371, 391
precipitation, 21, 22, 453, 459
preconditioning control room,
   507–508
premarket notification, 487
preservatives, 375, 378
pressure
   in barrier isolator, 270–271,
      273, 276–277, 284–288,
      290, 307, 308
   bubble point and, 221, 225
   for chamber test, 279
   constant, for filtration, 211–
      216, 218
   diffusion and, 222–224
   effect on filtration, 209, 211,
      215, 217, 454
   effect on microorganisms,
      30, 473
   of gas and plasmas, 162–163,
      187, 189
   head loss and, 80
   leakage of, 503
   piston pump and, 141
   piston pump operation and,
      125, 126, 133
   profile of, 277, 278
   temperature and, 57
   water activity and, 115
pressure gauge, 104, 496
pressure-hold test, 211–212, 213,
   454, 459
pressure relief valve, 495
pressure testing, 141
pressure transducer, 279, 287, 495
preventive maintenance, 94, 95
preventive maintenance
   program, 495
primary cooling, 101
primary standard, 330, 331, 511
Prince, H.N., 434, 435, 468
prions, 7, 28, 43, 462–465

process and instrumentation diagram. *See* P&ID
process control, 3, 498, 499–500, 509
process control release. *See* parametric release
process filtration, 2
process flow diagram. *See* PFD
processing pump, 87
processing strategy, xvi
process instrumentation
   diagram, 85
process qualification, 354, 356,
   364–369
process validation, 369, 380, 487, 515
product–contact surface sterilization, xxi
product processing system, 87
product qualification, 354, 356,
   357–362, 380, 381
product recall, xix
programmable controller. *See* PLC
programmable logic controller.
   *See* PLC
progressive cavity pump, 126, 128
prokaryotic cells
   chemical action on, 20
   functions of, 9
   protein cloning in, 434
   structure of, 6
pronase, 464
propylene glycol, 255
propylene oxide, 20, 236, 255
propyl paraben, 378. *See also*
   parabens
prospective validation, 355
protease, 463, 464
proteinase K, 464
proteins, 10, 260, 339
   acid-soluble, 243–244
   adsorption of, 213–214
   alkylation of, 20
   binding of, 204
   chemical reactions with, 240
   cloning of, 434
   cold sterilization of, 197
   conformation changes in, 19
   cross-linking of (*see* cross-
      linking)
   denaturation of (*see* denatura-
      tion, of proteins)